A C. S. Rafinesque Anthology

Jeremiah Rea engraving from the Falopi portrait of C. S. Rafinesque.

A C. S. Rafinesque Anthology

C. S. Rafinesque

Edited by Charles Boewe

McFarland & Company, Inc., Publishers
Jefferson, North Carolina, and London

Also edited by Charles Boewe

The World or Instability, by Constantine Samuel Rafinesque
(Scholars' Facsimilies & Reprints, 1956)

Fitzpatrick's Rafinesque: A Sketch of His Life with Bibliography
(M & S Press, 1982)

With Georges Reynaud and Beverly Seaton,
Précis ou Abrégé des Voyages, Travaux, et Recherches de C. S. Rafinesque
(North-Holland Publishing Co., 1987)

John D. Clifford's Indian Antiquities; Related Material by C. S. Rafinesque
(University of Tennessee Press, 2000)

Mantissa: A Supplement to Fitzpatrick's Rafinesque
(M & S Press, 2001)

Profiles of Rafinesque
(University of Tennessee Press, 2003)

Library of Congress Cataloguing-in-Publication Data

Rafinesque, C. S. (Constantine Samuel), 1783–1840.
[Selections. 2005]
A C. S. Rafinesque anthology : Constantine Samuel Rafinesque / edited by Charles Boewe.
p. cm.
Includes bibliographical references and index.

ISBN-13: 978-0-7864-2147-3
(softcover : 50# alkaline paper) ∞

1. Natural history. 2. Science. 3. Antiquities. 4. Linguistics.
5. Phytogeography. 6. Education. 7. Lectures and lecturing.
8. Metaphysics. I. Boewe, Charles E., 1924– II. Title.
QH45.2.R35 2005 508'.092—dc22 2005010697

British Library cataloguing data are available

On the cover: background ©2005 PhotoSpin; Cladrastis kentuckensis (Yellowwood) from Morton Arboretum Quarterly (1980); engraving from the Falopi portrait of C.S. Rafinesque by Jeremiah Rea

Manufactured in the United States of America

McFarland & Company, Inc., Publishers
Box 611, Jefferson, North Carolina 28640
www.mcfarlandpub.com

Whatever be my future fate and field of exertions, I shall not have lived in vain .... My works, researches, travels, collections, &c., will remain as a proof of uncommon zeal....

—C. S. Rafinesque, 1836

Contents

INTRODUCTION: Reprinting Rafinesque 1

I. *Antiquities*

1. Archaeological Site 17
2. Origin of Native Americans 24
3. Big Bone Lick 32

II. *Linguistics*

4. Chinese 39
5. Ruter Reviewed 54
6. Hebrew Studies 59

III. *Society*

7. Life in Lexington 67
8. Utopian Society 81

IV. *Education*

9. Colleges 97
10. Natural History Surveys 108

V. *Public Lectures*

11. Inaugural Lecture 117
12. Greek Independence 129

VI. *Popular Science*

13. The Cosmonist 135
14. Lightning 158
15. The Milky Way 162
16. Phenology 171

VII. *Phytogeography*

17. Kentucky Botany 179
18. North American Botany 185

VIII. *Natural Science*

19. Natural Science in America 193
20. Review of Elliott's *Botany* 209
21. Review of Maclure's *Geology* 218
22. Discovery of Aerosols 225
23. Sea Serpents 232
24. Evolution. 241

IX. *Metaphysics*

25. The Psyche Papers 251

Index of Genera 263

General Index 266

Illustrations

Frontispiece Jeremiah Rea Engraving of Rafinesque ii
Fig. 1 Map of Trigg County Site 19
Fig. 2 Perspective View of Trigg County Site 20
Fig. 3 Map of Adena Site 21
Fig. 4 Map of Mississippian Site 22
Fig. 5 Map of Big Bone Lick 34
Fig. 6 Chinese Woman, Afong Moy 41
Fig. 7 Chinese Ideograms 49
Fig. 8 Bory de Saint-Vincent 68
Fig. 9 Kentucky Belle 76
Fig. 10 Transylvania University in 1820 102
Fig. 11 Lecture Manuscript 118
Fig. 12 Trilobite *Isoctomesa* 137
Fig. 13 *Cladrastis* Tree 156
Fig. 14 The Milky Way 164
Fig. 15 Stephen Elliott 210
Fig. 16 William Maclure 220
Fig. 17 Gloucester Sea Serpent 237

Introduction: Reprinting Rafinesque

T. J. Fitzpatrick's *Rafinesque: A Sketch of His Life with Bibliography* (1911) indicates that the first reprint of a Rafinesque article occurred in 1814, when the author was 31. Since then, bibliographers have logged more than a hundred reprints as well as numerous translations (from French to English and from both to German). When these titles are closely examined it is seen that most of them—both reprints and translations—were made to facilitate access to technical papers; taxonomists needed them to trace the history of Latin binomials denoting either the plants or animals for which Rafinesque had published scientific names. This need explains why, in the 1940s, Elmer D. Merrill reprinted in facsimile such rare books from Rafinesque's pen as his *Autikon Botanikon, Flora Telluriana, New Flora of North America,* and *Sylva Telluriana.* At that time Merrill was preparing his *Index Rafinesquianus* (1949), a book intended to list with reference citations every one of the 6,700 Latin plant binomials Rafinesque had published. Merrill observed that "there are about 14,400 references in Rafinesque's own publications to genera, subgenera or sections, species and varieties [of plants] where the names used originated with him" (p. 17). Merrill knew that the publication of his book would cause a storm of protest from fellow botanists, because according to the current International Code of Botanical Nomenclature many of these hitherto unknown names ought to be adopted. If they had been validly published in the first place they should supplant names made familiar by long use. Merrill's Rafinesque facsimiles, which since have become rare books themselves, were documentation for the publication status of many of the technical names listed in *Index Rafinesquianus.*

No similar broad scale effort has ever been made by zoologists, though to be sure Rafinesque's onomastic contributions in the animal kingdom resulted in far fewer new names to be dealt with. Still, from time to time, specialists have tried to sum up his contributions in whatever branch of zoology they practice. In 1864, William G. Binney and George W. Tryon, Jr., collected in about a hundred pages *The Complete Writings of Constantine Smaltz [sic] Rafinesque on Recent & Fossil Conchology.* Richard Ellsworth Call, Rafinesque's first biographer, reprinted in 1899 his *Ichthyologia Ohiensis,* a slim book of large importance to ichthyologists concerned with the fresh-water fishes of North America. In the period 1917–20, David Starr Jordan, in his *Genera of Fishes,* tried to list with bibliographical citations all the known Rafinesque names for fish genera from whatever source, including those published in French and Italian. Charles W. Richmond issued "A Reprint of the

Ornithological Writings of C. S. Rafinesque" in two numbers of *The Auk* journal in 1909. L. B. Holthuis brought together in 1954 "An Annotated Compilation of the Information on Crustacea" described and named by Rafinesque, and published it in Leiden in an issue of *Zoologische Verhandelingen*. Joseph Ewan edited a reprint of Rafinesque's *Florula Ludoviciana* in 1967.

Produced in small numbers, none of these reprints and compilations have remained available very long. *Ichthyological Ohiensis* was issued again in facsimile in 1970; a facsimile of the book *Caratteri di Alcuni Nuovi Generi e Nuove Specie di Animali e Piante della Sicilia* was published in the Netherlands in 1967, and again in Sicily in 1995. However, the needs of taxonomists in both botany and zoology were pretty well met during the period 1965–76 when the Inter-Documentation Company in Switzerland put all these Rafinesque titles, and others as well, on collections of microfiche, which it has continued to keep available—albeit at a steep price.

It was his scientific name-giving that has kept Rafinesque's own name memorable among naturalists—some would say, notorious—but his taxonomic writings made up only a part of his oeuvre, though unquestionably they are a major part. On the last page of his autobiography, *A Life of Travels* (a book that also has been reprinted twice), he wrote: "I have been a Botanist, Naturalist, Geologist, Geographer, Historian, Poet, Philosopher, Philologist, Economist, Philanthropist." Then after the ellipsis sign, inserted as if to catch his breath, he went on: "By profession a Traveller, Merchant, Manufacturer, Collector, Improver, Professor, Teacher, Surveyor, Draftsman, Architect, Engineer, Pulmist, Author, Editor, Bookseller, Librarian, Secretary." After another gasp for air he concluded: "and I hardly know myself what I may not become as yet" (p. 148).

All this is more or less true. For those unfamiliar with the details of his life, it might be explained that having compounded a botanical concoction alleged to cure tuberculosis, he treated lung diseases as a "Pulmist" with the medicament called Pulmel that he produced as its manufacturer. He was less successful and less well known as a manufacturer of such commodities as "syrup of cream" and "incombustible paint." As an engineer he devised a "new system of reciprocal wheels reducing friction & enhancing movement, which may be adapted to any present machine," according to a letter to his family in France, but that is all we ever hear of it. Likely it was his incombustible paint, as well as other unnamed devices, that justified his advertising in 1833 his availability as "Architect" to build fireproof buildings. As far as we know, he never obtained a client, and similarly his contributions as a "Philanthropist" were hampered by his lack of money, but he wrote about what he would do for the world if he *had* been rich. And sanguine to the end, he specified in his will that any surplus left over when his estate was settled should go to the founding of a college for young women. But as events transpired, the estate turned out indebted to its executor. Hence, some of these professions are those he would have practiced had he not been hampered by circumstances beyond his control.

Surely Rafinesque exaggerated what his "versatility of talents and of professions" accomplished, but there is no denying that he did dabble in many activities, the majority of which found some expression in his writing. Yet, readers who know only those fragments of his prose that naturalists have chosen to keep in print would hardly be aware of them. It is the purpose of this book to make available the less well-known—and even, to some extent, the unknown—writings of C. S. Rafinesque.

Certainly it was not Rafinesque's fault that he is known today mainly as an author of descriptive scientific prose, some of it in a kind of Latin shorthand of his own devising. As early as 1821, he printed up a four-page pamphlet in Lexington, Kentucky, titled *Proposals to Publish by Subscription a Selection of the Miscellaneous Works and Essays of C. S. Rafinesque.* There he proposed to sell at a dollar a volume five volumes "of about 240 pages each" containing, in the first instance, eighteen "Selected Tracts &c., already published." Many of these would have been translations of material he had published in Sicily, in his magazine *Specchio delle Scienze.* A few cannot be identified among earlier Rafinesque publications, but at least two are now included in this book, achieving, after 184 years, part of Rafinesque's promise.

More thought-provoking are the titles listed as "Selected unpublished Tracts, &c." in section two of the pamphlet. If only enough readers had been ready to pledge five dollars we might now possess Rafinesque's "Theory of the invisible world," his "Outlines of a general plan of public instruction," his "Dissertation on the modern Languages and dialects of Europe, and America; the ancient dialects of Greece and Italy, &c.," his "Tracts on American agriculture and manufactures," his "Memoirs on the antiquities of America," and even his "Essays and thoughts on moral and philosophical subjects." Alas, the copy of the pamphlet examined by bibliographer Fitzpatrick contained the names of only eight subscribers, and so the edition never was printed. In 1971, the *Proposals* pamphlet itself was reprinted as a kind of joke and distributed as a "Keepsake" issued to commemorate a conference on the history of science at Edinburgh University.

In 1821, when he made his proposal to publish by subscription, much of the writing for which Rafinesque has been remembered still lay in the future. However, already he had published three of his major books (two in Sicily and one after he returned to New York). Earlier that year, the magazine he intended to edit and publish in Lexington, the *Western Minerva,* had been suppressed by "secret foes," enemies of science he believed, among whom he numbered Horace Holley, president of Transylvania University. The pamphlet proposing to publish five volumes by subscription was the author's attempt to salvage some of the material already set in type for the *Western Minerva.* In addition, several other essays were mentioned whose loss biographers regret. Their titles are: "A visit to the Island St. Michaels, in 1815," "Journey to Mount Etna, in 1809," "Sketches of Italy, France, Spain, Greece, &c.," and "Journals of several excursions in the United States."

Readers of Rafinesque's scientific contributions might not be aware that he also aspired to be a poet. Among the poems he would have included in the five volumes "The Banks of the Ohio" appears to be his own translation of "Les Rives de l'Ohio" intended for the *Western Minerva.* Most regrettable is the loss of what sounds like an attempt to write epics based on the studies of "Indian Antiquities" that he and his friend John D. Clifford had been carrying out. These are: "The Adelphiad or Tecumseh and Dolloweh," and "The Alleghawees." Seventeen years later he was still pondering the subject when he urged John Howard Payne to write the "American Iliad" with material on Amerindian history Rafinesque himself had collected. His own book-length poem *The World or Instability* (1836) was reprinted after 120 years, when the author of the only review it garnered observed that "this trundling would-be epic in blank verse" is "a curiosity more reminiscent of Erasmus Darwin than of Milton (to whom Rafinesque likened himself)."

Augmenting the poems in the proposed five volumes would have been Rafinesque's critical essay on "The laws of universal Prosody and Poetical Harmony," and perhaps most surprising of all a number of "American Tales"—stories with such titles as "Charles Cleers, a tale of truth," "Peter Swallow," "The three Sisters," "Feliciana"; the list—as customary with him—ends with "&c." No fiction by Rafinesque, either published or in manuscript, has ever been found.

As far as we know, Rafinesque's earliest published writing was produced during his first short stay of two and a half years (1802–04) in the United States, when he sent to Paris brief notes on fauna he had observed in Peale's Museum in Philadelphia. Exotic fauna at that—he described stuffed birds brought back from Java. Once relocated in Sicily in 1805, he began writing for the *Medical Repository* in New York, edited by Samuel Latham Mitchill, whose friendship he had obtained during his American residence. Some of these longer articles also were cosmopolitan. Now he described plants he himself had discovered in the United States, printed notes on medical properties of North American plants, and published a lengthy article on "Exotic Plants, Mostly European, Which Have been Naturalized, and Now Grow Spontaneously in the Middle States of North America." Nobody, on either side of the Atlantic, was more qualified by experience than he was to write about the naturalization in America of European plants.

During the decade he spent in Palermo, Rafinesque began branching out beyond descriptive natural science. With a friend, Giuseppe Emmanuele Ortolani, he issued the first part of a booklet titled *Statistica Generali di Sicilia* (1810), giving physical details of the island illustrated by two maps, but the authors were not permitted by the press censors to issue the second part, called the "civil geography." It was in Palermo, too, that Rafinesque made his first attempt to edit a journal. The contents of his *Specchio delle Scienze*, which lasted through two volumes (1814–15), was the kind of olla-podrida that characterized the three other magazines he launched later after his return to the United States. With most of the copy written by Rafinesque himself, the journal had departments devoted to "Agricoltura," "Economia Politica," "Legislazione," "Meteorologia," "Geografia," "Mineralogia," "Medicina," "Chimica," "Archeologia," "Storia," and, of course, "Storia Naturale," "Botanica," and "Zoologia." Contributions by others usually took the form of letters to the editor—a type of scientific communication that persisted in the English-speaking world for several decades as well. The contents of *Specchio delle Scienze* illustrate the breadth of its editor's interests; from time to time the magazine even contained his comments on current "Letteratura" and "Belle Arti."

After his return to the United States in 1815, Rafinesque settled for a time in New York City, where friends like Mitchill introduced him to the city's small coterie of resident naturalists. He supported himself through international trade carried out with his younger brother, still located in Palermo, thus helping to repair his fortunes lost by shipwreck. As one of the founders of the Lyceum of Natural History (ancestor of today's New York Academy of Science) he arranged to report on the Lyceum's meetings in the pages of the *American Monthly Magazine and Critical*

Review. This practice may have been copied from the procedure in London of the *Annals of Philosophy*, which carried similar reports of learned societies, including the Linnean Society. The publication of these brief reports (in which Rafinesque's various contributions to the Lyceum of specimens and lectures never failed to get prominent mention) soon led to his having a monthly "department" called "Museum of Natural Sciences"—later changed to "Museum of Natural History." The Museums dealt exclusively with Rafinesque's own contributions in articles ranging through both botany and zoology. One of them—on sea serpents—appears in this book; most of them are taxonomic in nature. In addition, he also published a number of articles in the magazine without putting them under the "Museum" label. One is his lengthy account of the development in the United States of natural sciences that also is included in this book.

It may have been Rafinesque's proposal that the magazine begin carrying reviews of natural science books, which he was prepared to write. His first review, of the *Flora Philadelphica Prodromus* of William P. C. Barton, appeared in the September 1817 issue, and was strikingly negative. This was followed in subsequent months by nine more reviews, mostly about books on botany, also mostly negative. Frederick Pursh's *Flora Americae Septentrionalis* threw its reviewer into such a sputtering rage that he needed two numbers of the magazine to list all the author's shortcomings (chiefly publications by Rafinesque that Pursh had overlooked). Only two of Rafinesque's reviews were temperate, and according to the reviewer these books were not without flaws. These are his reviews of publications by Stephen Elliott and William Maclure; they, too, are included here.

Both Rafinesque and the editors of the magazine were conferring a favor on the nascent scientific profession, because in this period, 1817–18, there really was no other review medium for books making a struggling attempt to identify and catalog the flora and fauna of the nation. Older, staid quarterlies like the *North American Review* paid scant attention to such books and, when they touched on science at all, were likely to reprint a review that had appeared already in Britain. On the other hand, it is hard not to believe that reviews of books having so little interest to the general public hastened the early demise of the *American Monthly Magazine and Critical Review*. Nor were the reviews any more beneficial for the reviewer himself. Justified as some of Rafinesque's harsher criticisms may have been of books he reviewed, his colleagues saw only overweening egotism in his constant harping about his own publications—some of these being very obscure indeed. When his own first book in America came out in the midst of this period—his *Florula Ludoviciana* (1817)—the only "review" it received, anywhere, was an anonymous, noncommittal, factual paragraph in the *American Monthly Magazine and Critical Review*, written, according to Fitzpatrick, by Rafinesque's long time friend, Samuel Latham Mitchill.

Book reviewing was still so uncommon that the standing of most scientific books was established largely through personal correspondence among scientists. A representative observation about *Florula Ludoviciana* is in a letter by botanist William Baldwin to botanist William Darlington. The book, Baldwin said, "is a shocking production, to come from one who has placed himself at the head of the botanical profession in our country,—and who finds fault with, and criticises all his predecessors and contemporaries."

During a visit to Philadelphia, in 1816, Rafinesque had become a member of its Academy of Natural Sciences. In his published review of the Academy's maiden effort to issue a Journal, he chided the editors for rejecting "several valuable papers," presumably because the editors were unable to vouch for the accuracy of their contents. It is not hard to guess whose papers had been rejected. During another visit to Philadelphia, Rafinesque met his friend John D. Clifford whose business firm had been responsible for bringing him to the United States in the first place. Clifford, now established in Lexington, Kentucky, urged Rafinesque to visit the West, an invitation he took up in the spring of 1818. Letters that he sent to New York and Philadelphia about his discoveries in natural history during his trip down the Ohio River and back were among the last of his contributions to the *American Monthly Magazine and Critical Review*, and they too were printed under the Museum of Natural History rubric. His farewell to the magazine, published just before he departed to live in Kentucky, consisted of a series of letters concerning the horticulture of the date palm and the tea plant.

Although Clifford, a trustee of Transylvania University, arranged for his friend to teach there, Rafinesque returned East for the winter of 1818–19 to settled his affairs and, during his residence in Philadelphia, read several scientific papers to the Academy. At least two were rejected by the publication committee designated to review them. One that was accepted and published in the Academy's *Journal* in November 1818 was a description of three new genera of fishes. But ever impatient to see his words in print, the author had also sent a description of one of these fishes to Benjamin Silliman's new *American Journal of Science*, where it was published about the same time in that journal's second number. Such duplication was not to be tolerated by the Philadelphia academicians; never again was a Rafinesque contribution permitted to appear in their journal.

Rafinesque continued publishing short, descriptive articles in the *American Journal of Science* through 1819; two of his longer contributions published there are included in this book. Once settled in Lexington, with the university's well-stocked library available for reference and facilities for field research supplied by Clifford, he fired off a stream of articles to the *American Journal of Science*. By June the next year he was complaining in a letter to Zaccheus Collins that "Prof. Silliman has not published any of my essays in his late Journal[;] he has had 12 Memoirs of mine, some for 2 or 3 years!" Nevertheless, within a month he was sending John Torrey another article describing and naming three new genera of plants he had discovered in the West, to be "published in Silliman's journal if you can induce him to do it." But on the first of September in another letter to Torrey he muttered that "Prof. Silliman has behaved as a *Sillyman* with me (that is between us). I complained that he did not insert a *single one* of my tracts in his N 5, having 18 on hand, and he has called my just complaint a recrimination & sent me back all my tracts." Not until after Rafinesque's death do we get Silliman's side of the dispute. Then, in a footnote he appended to the Rafinesque obituary his journal carried, he remarked that "I became alarmed by a flood of communications, announcing new discoveries by C. S. Rafinesque, and being warned, both at home and abroad, against his claims, I returned him a large bundle of memoirs.... The step was painful, but necessary; for, if there had been no other difficulty, he alone would have filled the Journal, had he been permitted to proceed."

Even before he left Philadelphia Rafinesque had begun to send articles, in French, for publication in Europe. The first of these, published in June 1819, in the Paris *Journal de Physique, de Chemie et d'Histoire Naturelle, et des Arts,* described seventy new genera of animals he had discovered during his trip down the Ohio River and back. The article acknowledged that the author had already described these same animals in the *American Monthly Magazine and Critical Review,* and he was grandly identified now as "Professeur de Botanique et d'Histoire naturelle dans l'Université de Lexington en Kentucky." Still smarting from his mistreatment by Silliman, Le Professeur remarked to Torrey that he was prepared to send more of his manuscripts abroad, but "It is hard to be compelled to send my discoveries 5000 miles off for publication." Nevertheless, he continued to do so. Four more of his articles appeared in the Paris journal that year, followed by twenty-three, also in French, published in subsequent years in the *Annales Générales des Sciences Physiques,* in Brussels. In 1820 his *Annales* articles were gathered and reprinted by the Weissenbruch firm in Brussels in pamphlets of about a hundred pages each. The *Annales Générales des Sciences Physiques* was edited by Baron Bory de Saint-Vincent, a rolling stone no less colorful than Rafinesque himself. There is no reason to believe they ever met in person, but they became so well acquainted that Rafinesque chose to write to Bory about skullduggery in Lexington and the Ohio Valley in general. Rafinesque's own translation of that indiscretion also is included in this book. It is understandable that Rafinesque had long hoped to attain a favorable reputation in Europe, especially in France, going so far as having the last of his books that were issued in Sicily printed not in the Italian language but in French, which, he expected, would replace Latin as the language of science. His articles published in Paris and Brussels went some distance toward achieving for him the European reputation he sought. Furthermore, some of them were translated into German, giving rise to the mistaken notion that he knew that language, too. He did not. In addition, he managed to place a few articles in Britain, in the *Journal of the Royal Institution* in 1820 and much later in the *Journal of the Royal Geographic Society,* while Alexander Tilloch reprinted an abridged version of his report on sea serpents in his *Philosophical Magazine,* and on one occasion the *Gardner's Magazine* gave Rafinesque a few pages of space. Long after Fitzpatrick's 1911 bibliography appeared, a few other European publications of Rafinesque were identified, one being in the *Actes* of the Linnean Society of Bordeaux, and two in the *Bulletin Universel des Sciences* published by Baron André Férussac in Paris. During his Lexington residence, Rafinesque also began sending contributions to the Société de Géographie in Paris for its *Bulletin.* One of these, in translation, is included in this book. Despite American publication media being closed to him one by one or ceasing publication altogether, Rafinesque did continue to publish, and he did receive more international recognition during his lifetime than most other American naturalists of the same period.

Once he was in residence in Lexington, Rafinesque took advantage of every publication avenue available in that university town. He had been hired by Transylvania's trustees, who gave him only room and board but no salary. About the same time they also brought in a new president, Horace Holley, a truly great educator but one steeped in the classical tradition, who had no wish to disturb the traditional curriculum by the introduction of newfangled fads like natural history.

Rafinesque had been given the privilege of charging admission to his lectures—like professors of Transylvania's medical faculty and of medical professors at most other universities at that time. Accordingly, before he could teach natural history he had to create an audience willing to pay for it.

William G. Hunt gave him the first opportunity to make himself known. Hunt had come to Lexington to edit a Federalist newspaper, the *Western Monitor*. Harvard graduate and unsuccessful lawyer, he aspired to publish writing of greater substance than the usual fare found in weekly newspapers. With the Transylvania faculty as contributors, including the university's new president and his bluestocking wife, assisted from time to time by students and literary-minded clergymen and lawyers of the town, Hunt commenced publication of the *Western Review and Miscellaneous Magazine* in August 1819. Every month during the magazine's two years of existence its pages were embellished by one or more essays by C. S. Rafinesque. Other contributors—including Rafinesque's friend, trustee Clifford—were modestly identified by initials only, often not their own. But Rafinesque, needing to make known his area of expertise, signed his name boldly to such articles as the nine he contributed on "Fishes of the River Ohio," for which the publisher retained tear-sheets that he bound up and issued in 1820 as *Ichthyological Ohiensis*, a book that has been commended since as one of Hunt's most important publishing ventures.

The first number of the magazine contained Rafinesque's observations on Midwest thunderstorms (now reprinted in this book), and his other contributions ranged through both botanical and zoological subjects, meteorological records, Indian antiquities, book reviews, poems, and word puzzles. Once again he threatened to swamp the vessel by overloading it. After a year of this, Transylvania's president, Horace Holley, wrote to his brother that "Rafinesque writes so much nonsense that I have told Hunt he ought not to publish any more of his communications if the work is intended to be kept alive." Surely the brother, Orville L. Holley, appreciated this warning, for, as editor in New York of the *American Monthly Magazine and Critical Review*, it was he who had published Rafinesque until that magazine went broke.

Nor did Rafinesque overlook the opportunity of advertising his talents in the Lexington newspapers. On his initial trip to Lexington in 1818, the *Kentucky Reporter* had carried two different paragraphs about his discoveries en route there, anonymously "Communicated." Now Hunt's newspaper, the *Western Monitor*, made a similar announcement on the occasion of his return in 1819, also anonymously "Communicated." In subsequent issues, the *Western Monitor* carried many Rafinesque contributions, some of them signed, such as his horticultural notes on tea plants and ginseng, others over the name "Constantine" alone, such as the five "Psyche" essays now reprinted in this book. Other newspaper contributions include notices of his university lecture series, of his gratuitous public lectures (that on Greek independence being included here), reports on the botanical garden he tried to found, copies of his "open" letters on Indian antiquities, and the prospectus for the *Western Minerva* magazine and shortly thereafter the notice that it had been suspended. Most important of these short newspaper articles is the series of fifteen "Cosmonist" papers published during 1822 in the *Kentucky Gazette*. All thirteen of the extant ones have been reprinted here.

Articles intended for the suppressed *Western Minerva* did not see light of day until 1949, when a unique set of proof sheets that had escaped destruction were at last published. Two of Rafinesque's most illuminating essays on social life in Lexington are from that source and are now included in this book. After the *Western Minerva* debacle he found an outlet for some of his writing in the *Cincinnati Literary Gazette*. Among those essays were several on North American prehistory and the location of promising archaeological sites, and, despite the magazine's title, several others giving technical descriptions of plants he had discovered, one being a more methodical description of the tree first described in Cosmonist XV. It was in the *Cincinnati Literary Gazette* also that he printed a review of a Hebrew grammar published in that city. Because of a renewed interest in recent years in Rafinesque's philological studies, that review also is included in this book. Like his attempt to clarify the unrecorded history of the Native Americans through a study of their languages, Rafinesque's curiosity about Hebrew began in his effort to understand what he believed to be the esoteric meaning of the Pentateuch.

After his return to Philadelphia in 1826, Rafinesque established a business relationship with Samuel C. Atkinson who, in partnership with Charles Alexander, published the weekly *Saturday Evening Post* and the monthly *Casket*. To the weekly periodical Rafinesque contributed dozens of short illustrated articles under the title "School of Flora," many of which were reprinted in the monthly. Another of Rafinesque's attempts to popularize botany, many of these articles were collected and, with the addition of extra material, were published from the same office in 1828 and 1830 as the two-volume *Medical Flora*.

Unhappy about the cost of the 200 woodcuts he had commissioned to illustrate the "School of Flora," which he valued at $12 each, Rafinesque resolved to recover some of the expense not only by using many of them again in the *Medical Flora* but also by reprinting a selection of a few of the more attractive of them twice, first as a broadside, later as a pamphlet, both under the title *American Florist*. Then, from the second volume of *Medical Flora* he extracted the 70 pages devoted to the genus *Vitis*—they had little to do with medicine anyway—and reprinted them as a pamphlet titled *American Manual of the Grape Vines*. Priced at twenty-five cents, the pamphlet became a proven money-maker when retailed at Bartram's Garden.

The success of such modest publishing ventures encouraged Rafinesque to consider larger projects. His extant account book for this period—the early 1830s—shows that he was in no sense well off, but by means of a variety of commercial activities including the sale of Pulmel, his TB nostrum, he now had funds to launch his third magazine, the *Atlantic Journal and Friend of Knowledge* (1832–33). Again, its varied contents reflected his own fertile mind and again he was almost the sole contributor. This journal differed from his earlier periodicals in having an occasional crude woodcut to enliven its pages.

The financial failure of the *Atlantic Journal and Friend of Knowledge* after two years has caused many to wonder how, beginning in 1836, he could issue a whole string of books and pamphlets, totaling 1,048 pages that year alone, 550 pages in 1837, 1,032 pages in 1838, 96 pages in 1839, and even 316 pages during the ten months remaining to him in 1840, the year of his death. Francis W. Pennell hazarded the guess that the publication of these 3,042 pages—in books and pamphlets mostly manufactured by job printers and marketed by the author himself—might have been made possible by profits from the small workingmen's bank Rafinesque founded in 1835 and which his old friend Samuel C. Atkinson assisted him in seeking to incorporate. The bank did pay for printing Rafinesque's 138-page *Safe Banking* (1837), according to that pamphlet's title page. No doubt the bank did make some money—Rafinesque claimed that it paid its shareholders a dividend of 9% the first year of its existence—but the real explanation for his sudden affluence appears at the conclusion of another pamphlet, titled *Plan* of the *Philadelphia Land Company*, and reprinted here.

There we find Rafinesque listed as secretary of a development company whose president is Charles Wetherill, a company that just happened to share office space with the bank, in a house that also happened to be Rafinesque's own residence. Wetherill, whose job had been supervising the grinding of white lead for paint, was forced to retire in 1836 because of ill health. After a settlement with his brothers who were partners in the business, he possessed no less than $60,000 to spend on projects of interest to him in the two years he had left to live, a fortune equivalent in purchasing power to well over a million of today's dollars. His illness, probably due to lead poisoning, likely impaired his judgment. If, as the *Plan* states, "$16,000 were subscribed at the outset by the promoters" of the land company, we can be fairly sure that most of it came from Wetherill. Nowhere have records been found accounting for the disposition of any assets remaining after the company failed, but it is a plausible assumption that some of these funds went to pay Rafinesque's printing bills during Wetherill's lifetime and, probably, after his death.

For several years Rafinesque had mentioned, both in print and in letters to friends, his intention to publish a multi-volume collection of his research on New World prehistory. He expected this great work to occupy ten or twelve volumes. When two small volumes of it titled *The American Nations* actually did appear in 1836 but with a dedication page dated 1833 it seems likely that Wetherill's wealth made them possible. That certainly was true of the 1837 reprint of *An Original Theory or New Hypothesis of the Universe* (1750) by Thomas Wright of Durham, for which Rafinesque supplied eleven pages of notes and a slightly different title. Manufactured by Rafinesque's job printer, the book was labeled "Printed for Charles Wetherill," because Wetherill wanted Wright's book to prepare the world for his own revolutionary "electric theory of the solar system"—which, sadly, he never lived to publish.

The centerpiece of the colony Rafinesque and Wetherill hoped to found in the vicinity of Bloomington, Illinois, would have been a university with a free extension service called the "Eleutherium of Knowledge." Such was Rafinesque's enthusiasm for having his own university that during the course of a week he wrote out the manuscript for its first textbook, *Celestial Wonders and Philosophy* (1838), and had it printed "For the Central University of Illinois," surely with funds the land company had collected. A reason for believing these funds continued to subsidize some of his publishing ventures even after the death of Charles Wetherill is that "Printed for the Eleutherium of Knowledge and the Central University of Illinois" was his 264-page *Genius and Spirit of the Hebrew Bible* in 1838. Printed on behalf of the Eleutherium alone was the 96-page *American Manual of the Mulberry Trees* in 1839, the 18-page *Improvements of Universities* also in 1839, and two publications the next year, 1840, the year of his death: the 32-page *Pleasures and Duties of Wealth* and—Rafinesque's final attempt to publish a periodical—*The Good Book, and Amenities of Nature*, whose first and only issue numbered 84 pages.

On the other hand, Rafinesque himself probably was able to finance the publication of his thin autobiography, *A Life of Travels* in 1836, and that year also saw the publication of his book-length poem, *The World or Instability*, by Judah Dobson, a respected commercial publisher whose best-known product may be the American edition of Audubon's *Ornithological Biography*. Dobson also had a well-deserved reputation for losing money on books by naturalists.

What appears to be a great flurry of Rafinesque reprints began in 1833 and continued for at least five years—all in the pages of the numerous successive "editions" of a single book: *American Antiquities and Discoveries in the West*, an omnium gatherum cobbled together in Albany by Josiah Priest, a saddler turned author. This time it was not Rafinesque's descriptive natural history reports that were being reprinted, but rather his observations on "American antiquities." Everything Priest used from Rafinesque was taken from the *Atlantic Journal and Friend of Knowledge*. Fitzpatrick listed thirty-four of these snippets and four more have been identified since. There may be others, because Priest, who often quoted Rafinesque only to controvert him, usually failed to identify whom he was quoting; and the various printings of his book were issued under such a jumble of spurious "edition" numbers that it is a laborious task to wade through them. In casual observation of all these reprints it appears that Rafinesque was a devoted contributor to Priest's publications; however, just the opposite was true. Rafinesque bitterly resented Priest's unauthorized use of his prose, and his position on prehistoric antiquity was entirely contrary to Priest's. The dispute led to Rafinesque's throwing down the gauntlet for a duel of wits with Priest in a public debate in Albany. The challenge, published in a newspaper, is the most succinct statement we have of Rafinesque's hypotheses about remote prehistory. It is reprinted here for the first time. While in the vicinity of Albany on an earlier occasion he had the opportunity to publish his ideas on the need for higher education in agriculture. That essay also is included here along with his writing on improvements of universities.

The first attempt to identify and list Rafinesque's publications was that of Richard Ellsworth Call, in 1895 in connection with his Rafinesque biography. Call's bias was evident when he stated that he hoped his list of titles "will prove useful to men of science who are interested in the historical phase of their subjects." He was cocksure about what would *not* prove useful to the men of science. Without having seen a copy of the suppressed *Western Minerva* he had no hesitation declaring: "It is really fortunate that the journal failed to secure subscribers!" Since he was unable to find an article on "The Chinese Nations" that Rafinesque said he had published in the *Knickerbocker* magazine, Call categorically declared that "the article never appeared."

Sixteen years later, T. J. Fitzpatrick was less negative about Rafinesque's nonscientific writing, but he personally had little interest in it. He did, however, try to transcribe faithfully the record as he found it. Nevertheless, of the articles that now make up this book, two were unavailable until 1949, six were listed only in 1982, and four remained unknown until 2001. The recovery and now the reprinting of these twelve contributions help to broaden our understanding of Rafinesque's lifetime achievement. Their unavailability until so long after his death helps to explain why such a collection as this was not undertaken sooner.

With a varied collection of his prose now assembled in one place it is possible for anyone interested to make an assessment of the quality of Rafinesque's writing. French was his native language; during his impressionable youth he learned Italian by immersion in it for nearly a decade in Tuscany; and it was there, too, that he began the study of English under the direction of a tutor. Planning a commercial career like his father, he looked on English as a valuable tool language. He began to perfect his knowledge of English during his first trip to America when he was 19. Writing to his younger brother (in French, of course) he gloated that he was learning the language the best possible way, by "spending all my evenings in the company of various young and charming American women, who are teaching me English delightfully." Although essays he sent back to New York for publication by Samuel Latham Mitchill during the decade he lived in Sicily were in English, he continued thinking in French for some while after he returned to live in the United States in 1815. This is attested by his extant travel and field journals of the period. But on June 1, 1817, the journals dramatically switch from French to English and continue in that language thereafter. References by others throughout his life to his foreign demeanor—usually that of a Frenchman, but sometimes a Sicilian—suggest that his spoken English retained his French accent. Only two of the essays in this book were written originally in French. One was translated by the author; the other by the editor. An excerpt of three short paragraphs from a pamphlet printed in Italian is given in Arthur Cain's translation.

Rafinesque's written English was influenced by some of the lexical practices of French, and, to a lesser extent, of Italian. For instance, all his life he wrote "correspond*a*nce" in his English-language correspondence, apparently never noticing a vowel differs between the French and English cognates. His knowledge of French led him to create such English nonce words as "proeminent" from *proéminent*, "circonference" from *circonférence*, and "foundator" from *fondateur*. His knowledge of Italian caused him to use "venture" to mean future, as does the Italian *venturo*; from Italian *differie* (to postpone) he was misled to write "differed" when he meant "deferred." All his life he used "actual" to mean present or current, influenced by *actuel* in French and its cognate *attuale* in Italian. French *s'illustrer* (to become illustrious) misled him to write that he had a "desire of illustrating myself"; he carried over into English the Italian *infaticabile* when he meant to say "indefatigable." Words spelled the same in French and English could also cause him trouble with gender; thus he might use the pronoun "his" to refer to a *journal* or construct the plausible "montaneous" out of the French *montagneuse*, mindful that "region" is spelled the same in both languages but has feminine gender in French.

In those of his essays included here that first passed through the hands of a contemporary editor, most—but not all—of these lexical quirks were eliminated at the source. Indeed, when the texts of his earliest English-language essays are compared with his English-language letters of the same period, it is evident that the essayist profited from heavy-handed copyediting. It is hard to say whether or not he appreciated it—or even noticed it, for, in those few examples we have, he is shown to be remarkably cavalier about proofreading.

The organization of his longer essays usually was slapdash, but effective enough for his objectives. Whenever possible, he liked to arrange material alphabetically; when not, he was likely to write short, numbered paragraphs. Whatever happened to be neglected, overlooked, or turned up out of natural order he would throw into a supplement at the end. He was so fond of supplements, and of the word "mantissa" to mean supplement, that his whole book *Sylva Telluriana* is called a supplement to his book *Flora Telluriana*, which is itself a "mantissa synoptica" to plants published between 1796 and 1836, while his *New Flora of North America* is "a supplemental flora" to everything that had ever been published before on North American flowering plants.

Sometimes criticized by his successors for not leaving visual records of his discoveries, Rafinesque tried to do so whenever he could commandeer the necessary facilities. His mentioning that in Palermo he cooked and ate the rare fish he studied in the market has caused amusement since, but it should be remembered that before a fish went to the pan he drew it in outline form. These drawings were then engraved by a capable artist and printed in the naturalist's first professional book, *Animali e Piante della Sicilia* (Palermo, 1810).

None of his articles published in Lexington were illustrated, for at that time no one there was making woodcuts. The only illustrations appearing in the town's newspapers were tiny standardized cuts of houses for sale or rent, rewards for the apprehension of runaway slaves, or offers by horse owners to "stand" a stallion. But when Rafinesque shipped off to Paris his article on a prehistoric town in western Kentucky, he included four sketches that were engraved by the publishers to illustrate the article. Once back in Philadelphia he drew on the talents of its cutters of woodblocks to make plant illustrations from his sketches, and even for a few crude pictures to enliven his *Atlantic Journal*. It appears that his use of pictures mostly was restrained by the cost of their preparation.

Here, the four illustrations originally published in Paris are included with the essay they pertain to. Elsewhere, other illustrations have been added from contemporaneous sources that Rafinesque might have used had it been possible for him to do so. He intended to print the Chinese symbols now inserted in his essay on that language (Fig. 7). They have been taken from a Rafinesque manuscript prepared to illustrate his article on "The Graphic Systems of the Ancient American and Chinese Nations." The article was published in *The Good Book, and Amenities of Nature*, 1 (January 1840), 76–81, but the illustrations were not included. The entire seven-page manuscript of these symbols and others is at the American Philosophical Society. Rafinesque indicated at the head of the page reproduced here that these Chinese characters derive from *Mémoires sur les Chinois*, by which he meant Joseph Marie Amiot, *et al.*, *Mémoires concernant l'histoire, les sciences, les arts, les mœurs, les usages, &c. des Chinois* (17 vols., Paris, 1776–1814). The drawing of Transylvania's principal building (Fig. 10) was done by the Kentucky artist Matthew Jouett to illustrate Charles Caldwell's *Discourse on the Genius and Character of the Rev. Horace Holley* (Boston, 1828). The "Kentucky Belle" in the big hat (Fig. 9) is Rafinesque's own sketch of a young girl named Charlotte Chapman. One of a number of pen portraits he made in Lexington that are extant at Transylvania University, this, like all his drawings of people, is of the left profile. Since he drew his fishes from the same point of view, they and the people sometimes bear an unfortunate resemblance to each other.

The frontispiece portrait used here was adapted from the frontispiece of Rafinesque's *Analyse de la Nature* (Palermo, 1815) and is a line engraving done by

Jeremiah Rea for its publication in *Potter's American Monthly* magazine in 1876. The Sicilian picture it is based on was signed by an artist called "Falopi" (otherwise unknown to fame) and rendered as a stipple engraving by one "P. Vaincher," who also engraved in Sicily Rafinesque's own drawings of fishes and plants. The likeness of Stephen Elliott that now illustrates Rafinesque's review of his book derives from an image on the banknotes issued by the bank Elliott founded. William Maclure's profile comes from a physionotrace portrait he had made while living in Paris that is preserved today at the Working Men's Institute he founded in 1838 in New Harmony, Indiana. The physionotrace machine (physiognotrace, in English) was invented in 1783–84 by Gilles-Louis Chrétien, and enabled its operator to trace the precise outline of a subject's profile by means of a pantograph; details of the face and dress were filled in by hand. Sources of other illustrations are given in their captions.

Since original copies of most of the texts reprinted here are difficult to access, efforts have been made to insure the intelligibility and accuracy of their transcriptions. Whenever possible, first names have been inserted—between square brackets—of persons otherwise mentioned by last name only. All other additions have been similarly treated, including numerous insertions required on such pages as 165 and 166. In the original printing of this essay, gaps had been left where the sense required the repetition of words, probably because the Kentucky printer had a limited supply of type. We know, for instance, that when a Rafinesque poem in French was printed in Lexington, a note advised the reader "to supply the accents" because the printer's fonts had none. The same editorial principles have applied in the few instances where supplementary text has been transcribed direct from manuscript (e.g., p. 77). In addition, canceled words have been shown as cancellations in those transcriptions and interlined expressions have been lowered to the line and enclosed between angle brackets. The wording of all titles has been retained as originally printed, even when excessively long, but the typography of titles has been conformed to present-day standards. A bibliographical notation at the end of each passage identifies its source.

This collection does not exhaust Rafinesque's known uncollected prose, but it includes some of the best of it. The essays included here also have been chosen to illustrate various aspects of his career that continue to be of interest. They are intended to aid in clarifying some of the confusions about his activities, and to make public a few of his accomplishments hitherto unknown or otherwise neglected. Generously aiding this effort have been Marc Cazalets, John Frederick, Sam Gon III, Sherri Graves, Rudolf B. Husar, Gary W. Kronk, Rolf Ludvigsen, Daniel McKinley, Frank Smith, and Mary Stiffler—to all of whom the editor is grateful.

I.

Antiquities

1

Archaeological Site

Among several reports on different aspects of American prehistory that Rafinesque sent to the American Antiquarian Society, one was his description of this site in Trigg County, Kentucky, written about six months after his visit there. Neither this written description (dated 10 January 1824) nor the map accompanying it was published by the Society, although the manuscript map finally was reproduced in the year 2000 to illustrate John D. Clifford's brief reference to the place in his *Indian Antiquities*. Rafinesque probably was made aware of this Trigg County site by Clifford, and he visited it only after Clifford's death. Nearly a decade later, when he was hoping to interest the Société de Géographie de Paris in publishing an account of his adventures, Rafinesque wrote to its president, Edme François Jomard, 15 January 1833, offering, among several other contributions, to submit "Some extracts of my travels since 1800 which I continue every year (for I always travel every Summer) in the Mediterranean, Atlantic Ocean, the Azores, and North America—Starting with the description with a drawing and view of an ancient city on the Cumberland river & my trip to the falls of that river with maps & sketches."

The offer of the article about the ancient city was accepted, for receipt of the manuscript was noted in the June 1833 issue of the Society's *Bulletin* (p. 362). The article, with professionally redrawn maps, was printed in the *Bulletin de la Société de Géographie*, 20 (1833), 236–41, where it remained unknown to Rafinesque scholars until listed in the 1982 revision of the Fitzpatrick bibliography. Thirteen years later, the article was resurrected by Charles Stout and R. Barry Lewis in their "Constantine Rafinesque and the Canton Site, a Mississippian Town in Trigg County, Kentucky," *Southeastern Archaeology*, 14 (Summer 1995), 83–90. They gave an English translation of the text, reproduced the figures, and provided a modern context for the earthen mounds found there. They identified the two bracketing sites as Adena (Fig. 3) and Mississippian (Fig. 4). Their study led them to conclude that Rafinesque's "dating of the Adena circle complexes ... is in remarkable accord with current dating" (p. 89), and that he "was indeed capable of producing work of lasting consequence" (p. 85). It may be that his work has even greater consequence when buttressed by additional information from related unpublished documents (now included here) and when viewed through the lens of a more accurate translation (which clears up, for instance, the puzzlement Stout and Lewis expressed over Rafinesque's reference to the elevation of the site).

Sixty more pages of Rafinesque's descriptive articles that contribute to New World archaeology appear as an adjunct to the speculations of his friend John D. Clifford in the latter's *Indian Antiquities*, first collected in book form in 2000.—*Editor.*

* * *

Description of an Ancient Town in Western Kentucky on the Cumberland River (Extract from the Travels of Professor Rafinesque, of 1815–1833)

The 23 June 1823, at night, I arrived at the village of Canton on the Cumberland river, about 50 miles east of the mouth of the Ohio. Before reaching there, one must descend the limestone plateau 400 feet above the river, which leads to the valley of Ramsay Creek and to the floodplains. Canton is situated in Trigg county, on a hill that borders the Cumberland and rises still more 170 feet above the village on the north side, but very little on the south. The floodplain is 50 feet above the river bed, when the waters are low, but it is flooded in the high waters; there is forming on the bank and in the Ramsay valley a belt of compost and sand without stones.

The soil of the hill has 2 feet of rich compost, then 12 feet of clay, 8 feet of limestone, then comes a bed of black felsite, beyond which has not been penetrated.

Canton is not yet put on the maps, but *Boyd's landing* can be seen in its place, because it has been for only five years an entrepôt for shipping out produce. It is yet a hamlet of only eight houses. The location is very choice, on the site of an ancient town of the indigenous primitives which occupies 35 acres of land; the present village covers only the northwest corner; but the gnats swarm there, a gloomy sign of the places liable to the autumn fevers.[1]

The primitive village is still quite visible in its entirety, they have cut down all the trees that were encumbering, and cultivated all the covering fertile soil; however the large monuments have been respected. The next day, 24 June, I drew the map and took the sketch; the outline was still visible everywhere, but the plow will make it disappear one day here as elsewhere.

The shape is nearly square, but a bit irregular and elongated.[2] It stands against the hill to the north; having a circumference of about 3,800 English feet and surrounded by a parapet, except to the west where the hill is perpendicular. The parapet is a remnant of an ancient wall of earth and wood which forms a protective barrier 3 to 5 feet high, and 15 to 20 wide, with an exterior ditch 10 to 20 feet wide half filled in by rain, time and the plow. On the hill, to the north, there are four small circular mounds, on the parapet, many small elevations.[3]

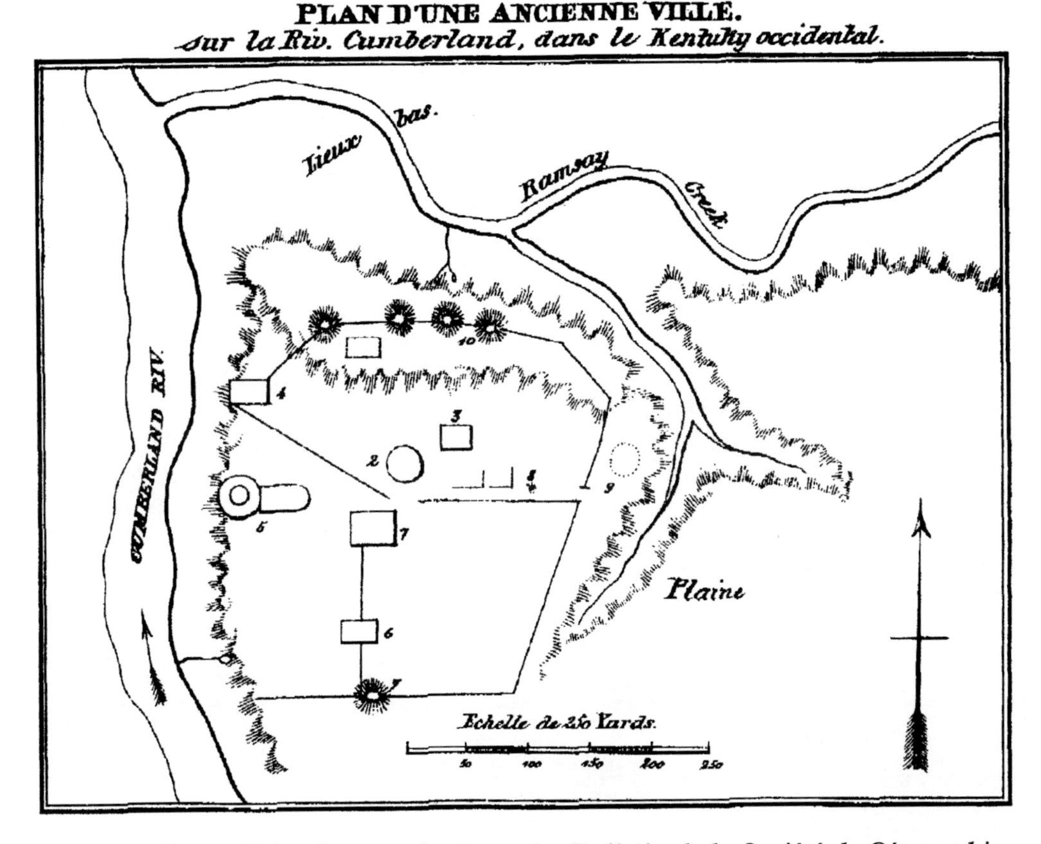

Figure 1. Map of Trigg County site from the *Bulletin de la Société de Géographie*.

This is therefore a fortified town, because the ditch is not inside as in the temples. The monuments of this town are numerous, but all less lofty than those of older towns without walls or ditches. Here they are:

No. 1. Is a square altar (or platform), 22 feet high, of which the base is 600 feet around and the top 360, each side being 150 and 90 feet.[4]

No. 2. Circular altar, 12 feet high, 360 feet around, flat top, 180 feet high [in circumference?].[5]

No. 3. Square altar, 8 feet high, 300 feet around at the base, 200 feet at the top.

No. 4. Square platform, 3 feet high, 70 feet long, 40 feet wide and 220 feet around, made out of a pile of pebbles, of stones and crushed shells, while all of the others are of compacted clay.[6]

No. 5. Two altars, that one to the west is circular and 200 feet around; it rises 20 feet, but slopes down 10 feet on the west; it bears on the east another elliptical platform, 8 feet high.[7]

No. 6. Square altar to the south, 10 feet high, 400 feet around at the base and 240 feet around at the top.[8]

No. 7. Is a gateway to the south with a mound nearby.

No. 8. Is an altar nearly obliterated, 4 feet high, on two sides 60 to 70 feet.[9]

No. 9. A gateway to the east with a large circular basin close by half filled in, once an amphitheater.[10]

No. 10. The high hill to the north with small mounds is a square site, ancient sepulcher.[11]

This small town, strengthened with ancient altars on a single level, provides us an example of towns and monuments of the middle age of the primitive peoples of America. To better understand this idea, for comparison I give below plans of two other monuments, one older and the other more recent. The figure 2 is a view of this same town taken from the side of No. 1.

The first (figure 3) is a large amphitheater or solar temple (No. 1), nearly obliterated by the centuries and the plow, near Mud Creek in central Kentucky. It is 180 feet in diameter and about 600 feet around, the parapet or exterior wall is no more than 5 feet high and 20 feet wide; the ditch, nearly filled in, has 30 feet of width. It has a circular altar of 150 feet at the center, a large gateway on the west side towards the river, and two pathways at the east, separated by a small mound. Very near, to the east, there is an elliptical altar (No. 2), 220 feet around, 8 feet high, with a flat top; and farther still, to the east, a small circular altar (No. 3), 100

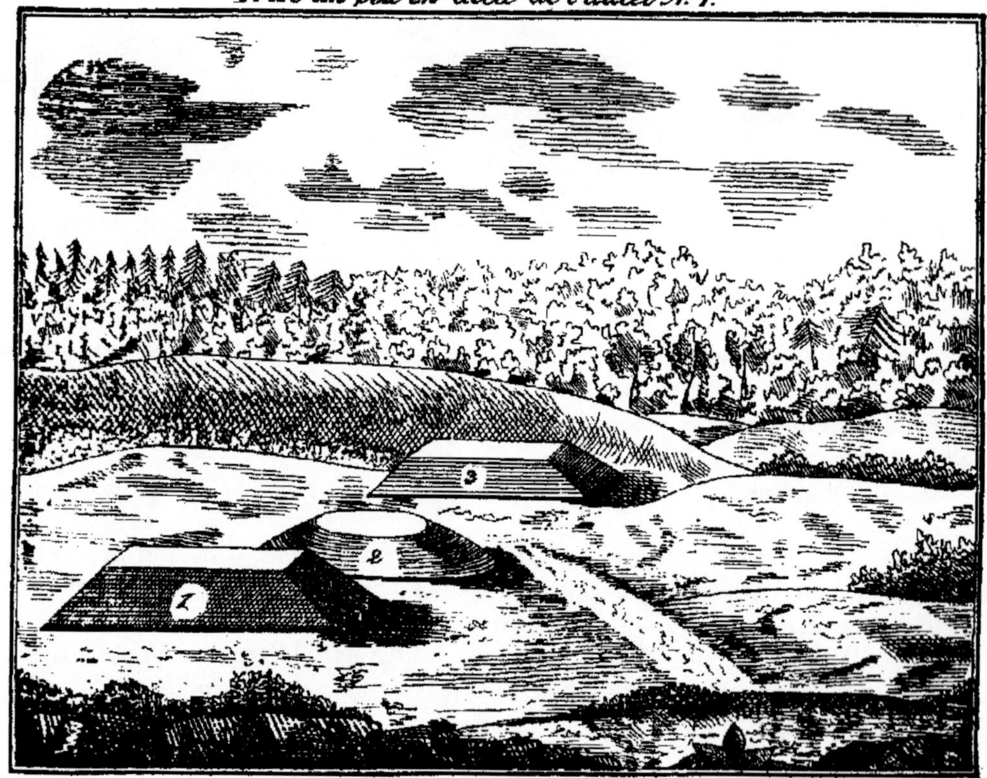

Figure 2. Perspective view of Trigg County site from the *Bulletin de la Société de Géographie*.

PLAN D'UN CIRQUE OU TEMPLE SOLAIRE.
dans le Kentuky.

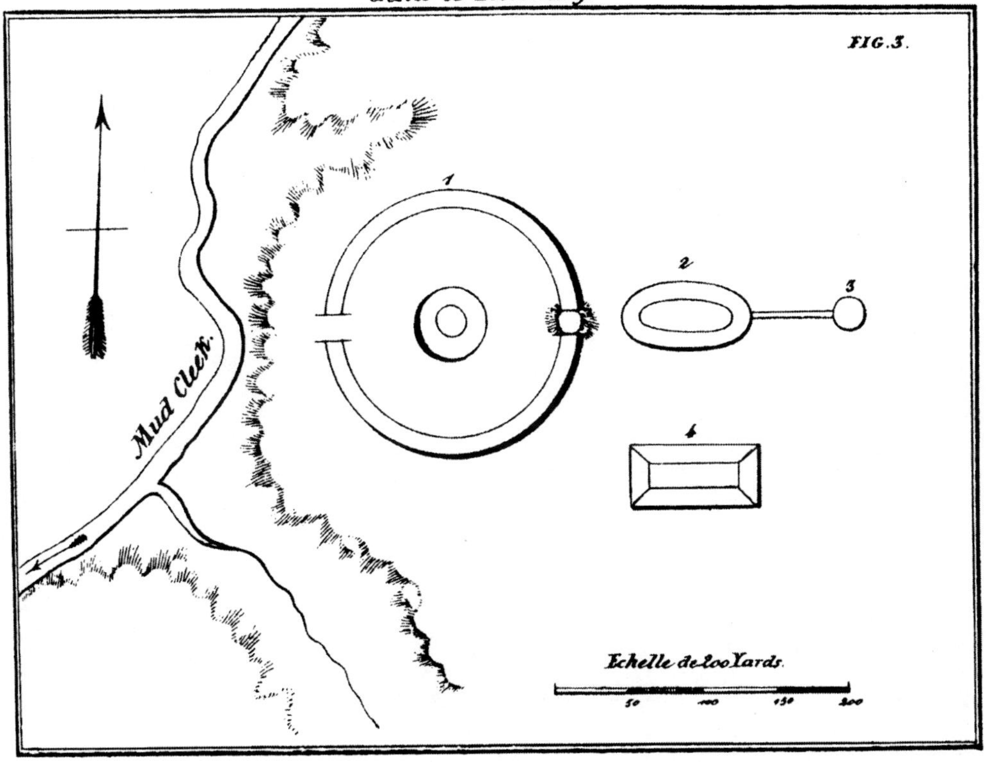

Figure 3. Map of Adena site from the *Bulletin de la Société de Géographie.*

feet around and 4 feet high, joined by a causeway or a raised road. Still farther, at the south, there is a large square platform (No. 4), 120 feet long, 60 feet wide, 360 feet around, and 8 feet high; the flat top occupies only half of this space. All this is a mass of clay covered by 3 feet of fertile earth. One cannot see walls, nor vestiges of houses.

The figure 4 represents the plan of a fortified village on the Bigbarren river (which flows into the Green river), a mile from the town of Bowling-Green, in western Kentucky. It is only an irregular octagon area, 1,385 feet around, enclosed by a nearly obliterated parapet, 2 to 3 feet high and 6 to 8 feet wide, without a ditch, and with small mounds at the angles. Here there are neither altars nor monuments, but all of it is wooded and has never been cultivated, one can still see, just as in the other villages of this sort, the plan of the houses it contained and forming small raised furrows. There were 7 of these clearly visible in 1823 (Nos. 1 to 7), at the time of my visit. The largest[,] or the hall of the chief[,] No. 1, is 100 feet long and 50 feet wide, No. 2 is nearly 60 feet long and 40 feet wide, No. 3 is 40 by 25 feet, the others are smaller. All that is covered by a deep bed of very fertile earth.

I presume that the amphitheaters, the temples, and other similar monuments that one finds by hundreds in Kentucky and Ohio, are the most ancient monuments of the civilized people that lived here in times past, and that, according to their remains covered by a deep compost and the 500 year old third or fourth growth

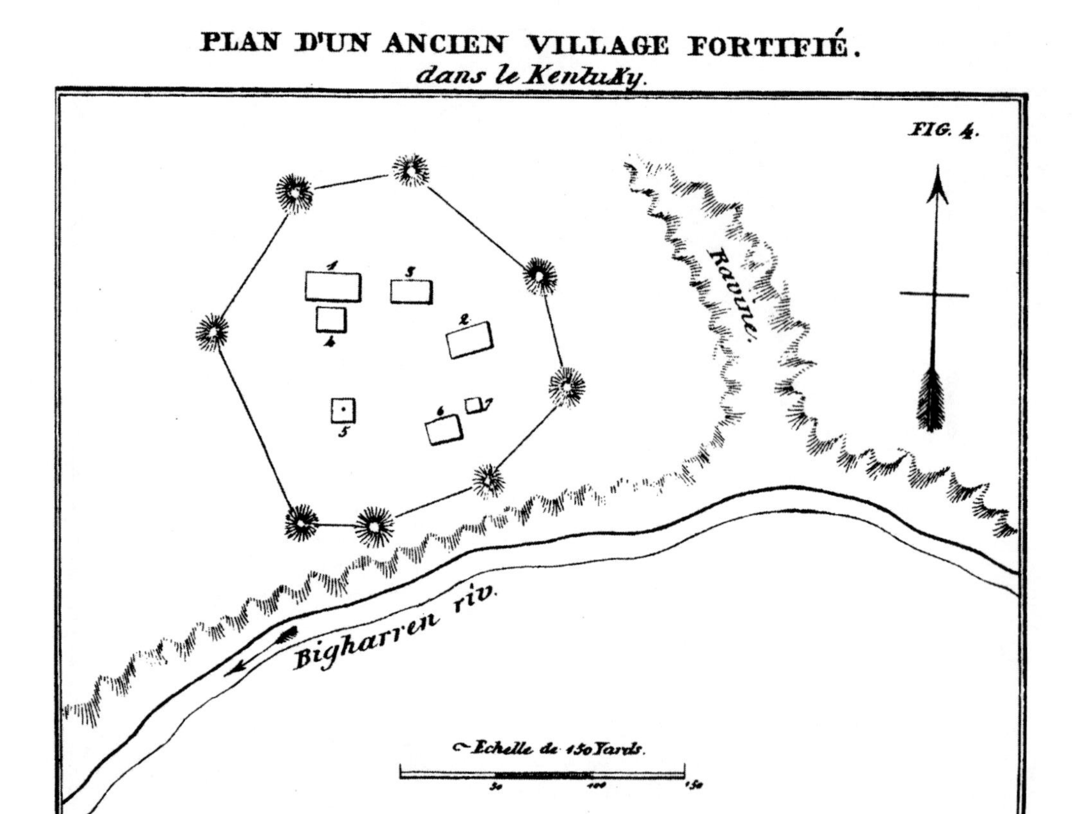

Figure 4. Map of Mississippian site from the *Bulletin de la Société de Géographie.*

of forest, they date to at least 2,000 years, or were abandoned then, because they may have been built 1,000 or 2,000 years previously. This nation had no enemies, because the monuments are without defence, one does not see walls, nor outside ditches (the interior ditches were in fact only for religious purposes), and often they are built on the mounds which they dominate completely.

The ruins of the second age, abandoned for the last 1,000 years or thereabouts, are of a people less civilized, who had enemies, who fortified their towns with walls and ditches, had a different religion of sun worship, of the altars often square or hexagonal or octagonal; but not the large circular temple. The town of Cumberland gives a sufficiently exact idea. One can not see the remains of the houses, which only begin to appear in the third series of the more modern monuments.

The era of their abandonment dates back 500 years (or 400 to 600 years), although they may be more ancient. They were built by the present race of our Indians, people not civilized, without knowledge of geometry, of raising neither temples nor altars, but with the necessity of defending themselves against their enemies, because these villages are all enclosed by walls, fragments of their palisades without ditches. I have seen some of them which contain many houses, small mounds, and a sunken elliptical amphitheater in the ground for their games no doubt.

The first French maps of Canada place in western Kentucky the *Chaouanons*, which are our *Shawanis*; but this name designates in general the small southern tribes of the *Lenni* nation. One map of 1680 gives them 40 villages, and the Cumberland river is named the river of the Chaouanons and the Shawanis in the first French and English maps. Meanwhile the rest of Kentucky had been abandoned for a long time and remained deserted; this was the battlefield of the Iroquois, Miamis and Hurons of the north with the Shawanis, Cherokis and Catabas of the south. The villages of the Shawanis were without monuments or defense, with wooden huts. They had in Canton, without doubt, a place that was a good location; but they occupied this ancient town as the Americans occupy it today. They did not leave even the least trace of their existence; they were driven out by the Chicasas, their enemies, who must have burned their huts; their remains have vanished.

---

1. The overall aspect of this site is further clarified by the observation of Rafinesque's friend, John D. Clifford, who alludes to the place as an example of the "circumvallations" he writes about in his *Indian Antiquities*, ed. by Charles Boewe (Knoxville, 2000), p. 3. There he says that some "circumvallations are seen on the summit of the precipices which border some of our rivers, and are from three to five hundred feet above the water courses. I visited one of these on the Cumberland river. The earthen rampart occupies the highest ground, and is bounded on the north by the precipice. The ground gradually slopes from its southern walls about three hundred yards to a spring of water."

2. In his unpublished 1824 English-language description, Rafinesque described the shape of the ancient town as "oblong, irregular."

3. The 1824 account at the American Antiquarian Society adds to this paragraph that "The wall and ditch were very plain towards 1796: they are now partly destroyed by the plough...." In Rafinesque's *Ancient Annals of Kentucky* (1824), where this site is briefly mentioned (p. 34) though he had not yet visited it, the circumference is said to be 7500 feet.

4. In the 1824 manuscript, Rafinesque called this artifact "a square mound or Teocalli" and remarked further that it is the largest of the monuments, that the ascent to it is from the south, and that "This fine monument has been laid out for a garden, and will not be destroyed like the others." In *Ancient Annals of Kentucky* he called it "a square Teocalli."

5. The French text concludes with an uncorrected misprint. In English this "is a fine circular flat mound 12 feet high; the circumference of the base is 360 feet and of the level top 180 feet or exactly one half of the base."

6. In English, this monument was called "a flat oblong mound of a square form, formed of limestone, pebbles and Shells, while all the others are of earth. It is 3 feet high, 70 long and 30 broad, a warehouse stands on it."

7. Perhaps this description is clarified by its English equivalent, which remarks on "a conical mound 10 feet high on the East side and 20 on the west side, circumference about 200 feet; it has easterly an appendage or apron, 8 feet high north and east, nearly level to the south."

8. In English, this monument is said to be "similar" to Nos. 1 and 3 but "smaller" than No. 1. It also is said to be "fast sinking under the plough."

9. The 1824 English-language text calls this "another square monument of which two sides are now visible[,] the East and South, each 65 feet long and 4 feet high."

10. In 1824, Rafinesque merely said that the depression "appears to have been an ancient well, now reduced to a hollow."

11. The 1824 text calls this "the grave yard of the town, where many bones &c are found."

2

Origin of Native Americans

Josiah Priest, who had been an Albany coach trimmer, saddler, and harness maker before he turned to stitching together books, was described by his printer, Joel Munsell, as the "greatest inventor of ancient history and biography of his time." Priest began reprinting unauthorized snippets from Rafinesque's *Atlantic Journal* in his enormously popular *American Antiquities and Discoveries in the West*, starting with its "second revised edition" of 1833. There were six different "fifth" editions of Priest's book (one each in 1835, 1837, 1841, and three in 1838) all of which carried the legend "22,000 volumes of this work have been published within thirty months, for subscribers only." The jumble of so-called editions is reflected in the jumble of the contents of the books. Rafinesque actively opposed publication by Priest—probably the only time in his life he fought to keep his words from being printed.

Rafinesque first read Priest's book during a trip to Albany in 1833, and he may have accosted the harness maker in person, for in his 1834 edition Priest dropped four of the Rafinesque pieces he had carried before. This did not satisfy the naturalist; back in Philadelphia he challenged Priest to a debate in the published letter reprinted here. Priest refused to accept the challenge, but he did drop two more Rafinesque pieces from his next (1835) edition, while stubbornly keeping the remaining five.

Rafinesque must have resented the fact that Priest was making money on words of his that he could not sell for enough to break even; however, the ideological crux of their dispute was Priest's steadfast support of the ten lost tribes of Israel theory for the origin of the American Indians. Rafinesque had long argued that "the American Nations and Tribes Are Not Jews," as he titled his August 1829 letter to the Rev. Ethan Smith, first published in the *Saturday Evening Post*, then reprinted in his *Atlantic Journal*, where Priest picked it up and printed it for a third time, only to dispute it. There was no common ground between them on this issue, for Priest viewed Rafinesque's accounts of Indian ethnology in terms of "primitive words, primitive usages, and primitive religion" as a diabolic attempt "to overturn the Scriptures."—*Editor.*

* * *

Letter of Professor Rafinesque, of Philadelphia, to Mr. Josiah Priest, of Albany, on American Antiquities.

I have lately read the Second Edition of your Work on American Antiquities.[1] I ought to be very grateful to you for the handsome manner in which you have mentioned and made use of some of my labors on American history; although I perceive that you have distorted a few of my remarks, to suit your own views. I have analyzed carefully your work, so as to obtain the facts and spirit of your labors; the result is that you have brought together about sixty important facts relating to the ancient state of America; but omitted one hundred and fifty quite as valuable and needful to support any theory! Your own theory has also fifteen truths or correct results; but it has also fifty errors or incorrect views. This of course must render your work quite imperfect; but still you deserve commendation for having done so much, and begun to make this subject popular. I readily assent to your conclusive remarks, that mighty nations and empires have flourished in succession in America, and many cataclysms or convulsions of nature have here taken place. This being an evident truth, it behooves us to become acquainted with these revolutions of the soil and mankind in this our hemisphere: they belong to Geology and History as well as Arch[a]eology and Ethnology. Therefore let us hope that more attention will be bestowed thereon in future. Meantime it is the duty of the learned or enlightened to open the way as you have done; but they ought in all cases to give evident facts, and not to mislead by adding new hypothetical theories to the numberless dreams of the learned dreamers of yore!

If I was to write a review of your work and your learned dreams, I might have a fine scope for ridicule and blame; but I write these remarks in the best feelings, and with the wish to render the subject popular in its true colors, instead of false dresses. Therefore I offer you a fair chance by wishing to debate with you the matter either privately or in print, or in public, as you may deem best. If you are not afraid to sustain in public the popular theories which you adopt or modify, I tender you my assistance by offering to debate in public with you, any matter relating to American History, Antiquities, &c, before any public audience in Albany, Troy or Lansingburgh, at any place you may select or procure, and at any time during next summer.—You may arrange the whole matter, time and place, with any of my friends in those places or the members of the Lyceums. The mode of debate you may also select yourself, either by alternate lectures or discourses, or by questions and answers, or any other way.

That you may prepare yourself for this debate, I herewith state what are the main errors of your book that I will controvert; and what are the main truths that I will sustain in public, against you or any body else that you may bring to help you.

1. I will sustain and prove that the central regions of Asia were the real cradle of mankind, and that Eden was there and not in Syria nor Arabia.

2. That America was peopled partly before the flood of Noah, and that it is doubtful whether that flood covered all the mountains of America.

3. That this flood was not universal, and the bible does not say so.

4. That the English Bible is not the Hebrew Bible, being badly translated in 10,000 instances, and none but the Hebrew text is to be quoted.

5. That the days of Creation were divine days, and long periods of time, that the Bible says so, and they were not days of 24 hours, nor polar days.

6. That Noah did not build the ark in America, as you say, but in Central Asia, and there also came out of the Ark and not in Arminia [*sic*].

7. That the three sons of Noah were not of three colors in your sense as individuals since we know not their wives['] colors; but might be if implying nations saved in the ark, although of different hues.

8. That Moses is not the only author who has given an account of the flood of Noah, and that he has given none of Peleg's flood.

9. That Moses's account is perfectly rational when well translated and explained, and the Ark was a refuge for men and beasts, but no mammoth was there, nor huge and carniverous beasts.

10. That many geological floods have happened before that of Noah, and even since the second historical flood of Peleg.

11. That the origin of Stone Coal is not from forests sunk in the pulpy earth at the flood, as you state.

12. That the Chronological System of [Archbishop James] Us[s]her is false, preposterous and absurd, although adopted by many: that there are twenty better systems based on the Bible, and that the Seventy and Josephus are by far older and better.

13. That the flood of Noah happened over 3000 years before Christ, and over 2000 years after Adam: that all the Eastern christians from Russia to Egypt reckon so, and whoever reckons otherwise is a[n] Us[s]herian.

14. That the date of the flood by the absurd short chronology, refers by a blunder to that of Peleg, or Rabbinic Jew, instead of Noah, and the two events have been blended by the oversight.

15. That the flood was not caused by the recoil of the earth being stopped in its course, but a change of vibration, or by an irruption from the Caspian Sea—and the year of 360 days became 365¼.

16. That many nations came to America before and after the floods; but no Jews ever came there before Columbus.

17. That the language of Noah was not the pure Hebrew or Hebrew of the Bible, as no language can stand unchanged so long.

18. That hogs, cows, dogs, sheep, &c., are born of various colors, but men are not, and their complexions change by means of climate and food, although they are also liable to other variations.

19. That the confusion of Babel was not a total change of speech, but a confusion of families, tribes, synonymous of tongues.

20. That Melki Sedeh was not Shem but a high priest or holy king.

21. That the Asiatic nations took nothing from the Jewish rites and books, but Moses and the Jews many things from the Asiatic nations.

22. That the spirit of the Bible teaches better than the literal script.

23. That the black nations do not all come from Ham.

24. That the American nations came all from Japhet and Ham, but none from Shem, before Columbus at least.

25. That the Chinese, Hindoo, Egyptian and Persian traditions are older than Moses, and corroborate better than if borrowed from him.

26. That our American Indians have no Jewish custom, but what is common to ancient usages before Moses.

27. That all the dates in your book are false.

28. That the Romans never came to America, and never built the ruins at Marietta[, Ohio], which are older than the Romans.

29. That the Osages are not Wyandots, being as different as Jews and Greeks.

30. That there are mounds in the West over 100 feet high.

31. That the Guanches, Peruvians, &c., had mummies as well as the ancient people of America, and the Egyptians never came to America.

32. That Maleys, and Scandinavians, and Welch, did never spread in America, nor from [form?] peculiar nations.

33. That no American nation came early by the north-east, but by the east from Spain and Mauntania [*Mauretania*].

34. That even the Esquimaux, the last nation come in America before the late Europeans[,] came from west to east.

35. That fire worship was known 1000 years before Moses and the burning bush, but that Confucius was not a fire worshiper, as you state.

36. That the ridiculous stories of Lord Mondobo [*Monboddo*], are of no account: that the Americans are not Celts, nor Barks [Basques?], although ancestor[l]y related.

37. That the Compass was known in China 2800 years before Christ, but *after the flood*, and not before, as you make me say to suit your Chronology.

38. That the Eries who gave their name to Lake Erie, were not of Malay stock but Senekas, rebelled against parents.

39. That your omission of Russel, Heckemodder, M'Culloh, and one hundred travelers, evince you lack much extant knowledge.[2]

40. Lastly, that all the modern Sciences unite and combine to prove the Mosaic traditions, provided we translate them word for word correctly, instead of depending on erroneous additions and unlearned glossaries of ignorant translators.

Thus you perceive that there is an ample field open to you to exert your talents and memory. By coming with me before the public you will assert conspicuously no doubt, whatever you can prove, for you must know that I will ask your proofs for every thing. You must have a Hebrew bible by you, because Moses did not speak English to the Jews. If you can get one in both languages, word for word, that will be better still, since we are to speak now in English. This becomes needful because you have dared to condemn some of my sentiments based upon the original bible, by quoting the erroneous English translation. I will bring for my authorities two hundred millions of christians who believe in better translations, and one hundred English christian writers who have shewn the absurdity of the literal passages badly translated.

Although some devout christians and the worthy William Penn have believed in the Jewish origin of the Americans, ten times as many have not, and Roger Williams who knew so well the Nar[r]aganset[t] language, is at their head. Also Dr. [Edwin] James of Albany, who knows so well the Chip[p]eway Language, and translated the Bible in it. In fact all those who know well the Americans and their languages deny it.

Hoping we may make these topics interesting to a large public audience,[3] I await your pleasure and remain respectfully your well wisher.

C. S. RAFINESQUE
Prof. of Historical and Natural Sciences.

Philadelphia, 5th Jan., 1835.

First published in the Troy (N.Y.) *Daily Whig*, Tuesday, January 13, 1835, and reprinted a week later in the weekly *Troy Whig*, on January 20, 1835.

1. Josiah Priest, *American Antiquities, and Discoveries in the West* ... (Albany, 1833). In the second edition and onward through many subsequent editions, Priest's book drew heavily from Rafinesque's *Atlantic Journal*, sometimes with attribution, other times without.

2. Michael Russell, *A Connection of Sacred and Profane History, from the Death of Joshua to the Decline of the Kingdoms of Israel and Judah* (London, 1827); John Heckewelder, "An Account of the History, Manners, and Customs of the Indian Nations Who Once Inhabited Pennsylvania and the Neighbouring States," *Transactions of the Historical & Literary Committee of the American Philosophical Society*, 1 (1819), 1–348; James Haines McCulloh, *Researches, Philosophical and Antiquarian, Concerning the Aboriginal History of America* (Baltimore, 1829).

3. Rafinesque's friend Amos Eaton tried to encourage the debate by addressing the following letter to the editor of the newspaper, James M. Stevenson, who published it in the *Troy Daily Whig*, January 14, 1835.

> Mr. Stevenson:
>
> I am delighted with the article in your paper of the 13th inst. on this subject [of American Antiquities]. Not that I approve or disapprove of any of Professor Rafinesque's forty articles, proposed for debate. But because it is time to commence a liberal discussion on this subject. The facts already collected are very numerous; and the proposed debate will elicit inquiry among those who reside where these ancient monuments are perpetually in view.
>
> My only fears are, that either Mr. Priest, or his friends, or even some of Mr. Rafinesque's friends, will deem some of his 40 articles too visionary for discussion[.] But suppose his positions are half, or even three-fourths, untenable—I affirm that the discussion will still be very profitable to auditors. Even those who are disposed to pronounce Mr. R. an extravagant enthusiast, all agree, that he is a scholar of the first order; of vast reading and great classic learning. His nice discriminating talents have never been questioned. Morose Naturalists of the old school, like myself, have accused him (face to face, not in the insinuating manner) of being disposed to assume *shades of variety* in the productions of Nature, as *evidence of specific difference.* This, I believe, comprises all his supposed scientific heresies.
>
> I venture to give my name to this assertion. Those Trojans who expressed their high gratification with the originality and amusing eccentricity of the lecture he gave them, at my call, a year and a half since, will be delighted with the proposed debate.
>
> Amos Eaton.

* * *

[Letter to C. S. Rafinesque from Josiah Priest]

To Mr. C. S. Rafinesque:

Sir—The pointed and lengthy address of your honor, came this day, Jan. 24, to hand, through the press of the Troy Whig, in which it appears you are anxious of debating the difference there may be in our opinions, on the subject of American Antiquities. But, however patriotic your intentions may be, in you[r] challenge to argument, I deem it proper, at once to announce to you, that such a measure is not agreeable, because not compatible with my engagements. That you were pleased to say I have treated your labors on American History, in a handsome manner, is grateful to me; whereof I will now remark, that the Atlantic Journal of which you are the author, has aided me much in my work: and for another reason I am grateful to you, which is, because in your letter upon my book, you have said that I have produced sixty important facts, and fifteen true results, relative to ancient America.

This is a compliment unlooked for, and especially from authority so high; as the benefit of those two ideas is certainly of no small amount to my production, and will aid in its favorable reception. As for compliments on my work, I have never dreamed of receiving any; and it is with wonder I contemplate the amount of its sale, which already counts seventeen thousand volumes. As it respects the errors which you have also alluded to, as well as to its truths, I have it to say in excuse, that I took the hard and unpromising subject in hand much as a man may be supposed to do, who proposes to himself to make a road through a dense and wide wilderness; and as a pioneer to his road, passes through by merely marking the trees, keeping his course, as well as he can, by having an eye to the supposed position of the cardinal points: thus I have done in pursuit of the secrets of the history of ancient America. My conclusions have arisen from the examin[a]tion of such evidence as came under my investigation; but if I have omitted to speak of a multitude of authors, of western repute, that is no evidence that I lack, as you say, much extant knowledge, for how do you know what I *have* or what I have *not* read—were you a little less abrupt, it would be somewhat more agreeable, yet dear sir, I know very well that every man has a right to his own way.

Were you again to look at the preface of my book, you will find that I have aspired to no higher object than to arouse the public attention to the fascinating subject of *American* Antiquities, which object, if it may be judged from its sale, I have even attained too: and further in the said preface, it is distinctly stated that on account of the subject, being every where surrounded with its mysteries, I have been compelled to *wander* widely in the field of *conjecture*. If I have made some blunders therefore, is it to be wondered at? yet I am far from being *conscious* of such errors, and have my doubts whether you or any other man can *dream*, as you say, any better than I have on the subject and prove it. Should I consent to meet you as a Gladiator on the philosophic arena, agreeable to your challenge, I could not add at present more light on the subject than I have already laid before the world; and as you say you can refute me or any one that I can get to help me, I do not therefore see the benefit of the proposed debates, for we should have to part exactly as we should meet, as the advantage is all on one side. Before I close it is certainly due to your feelings, that I should state so far as I have animadverted on your opinions, as found in the Atlantic Journal, about the Ark of Noah, where you have said that the Ark was a *country* instead of a vessel, in which Noah and his family with other people and animals were saved—and also that the *whole* earth was not overflowed by that deluge, as is otherwise stated by Moses, that I have indeed founded such animadversions, upon the Bible history as translated into English, and for no other reason. If therefore, I am deceived, so also is all christendom, with thousands of learned men whose fame we should blush to find fault with, who have risen up in all ages, commencing with David the father of King Solomon, till the present time. I have not yet learned that the subject of the creation, the Ark and the universality of the flood, have been differently told in the great variety of translations of the Bible into other languages from the original Hebrew, or that learned men in the oriental languages have found *so many* mistakes in *our* translation, as you have named, amounting to the enormous sum of *ten thousand*; more subjects, by far, than are in the whole old Testament, with the New superadded thereto.

But if you or any other man can plainly show that the scriptures are egregiously erroneous, such a man will acquire to himself the title of benefactor of mankind, as truth is better than error, and thus prevent further bad translations, which are now being made in various languages, all telling the same story. And because we have *relied* on *our* translations, I have, as you have expressed yourself in the printed letter directed to me through the Troy Whig even *dared* to condemn *some* of your sentiments, which I suppose to be about the Ark and about the universality of the flood of Noah, as I have controverted no other that I recollect, and shall still *dare* to do so till better informed.

As to all the other problems of my work, with which you are prepared to find fault, and offer to correct in debate, I consider each and every one of them, infinitely inferior, to this *one* subject, the *truth* or *falsity* of the Mosaic account, as translated into our language, about the creation, the flood, the ark, &c. Because, if we are misled by our translation, the foundations of religious society are overturned; for, if indeed it is true—as you say it is—that the six days' work of creation does not mean so many days of twenty-four hours each, as commonly understood, or rather of so many *periods*, or *divine* days, of vast and indefinite length, how can the Jewish Seventh day Sabbath be founded on that circumstance, as there is no affinity or likeness between your *divine* days, in their quantity of time; unless indeed you would also contend that the twenty-four hour day Sabbath, as kept by the Jews, should embrace a period of ages of indefinite length. As to the appointment of the Jewish Sabbath, it is found in Exodus 20, 11, as translated into English, that "in *six days* the Lord made heaven and earth, the sea and all that in them is, and rested on the seventh day," on which *very* account, Moses makes the observance of every seventh day in the order of time, binding on the Jews, as a remembrancer of the great works of creation, and for other purposes. That circumstance is also the true and prime origin of the Christian Sabbath. But if those six days, there spoken of, were so many indefinite *periods* of time, then is the Sabbath of no moral or divine authority; because it is nonsense to suppose Moses intended to tell the Hebrews that as God was six divine days or ages in making the heavens and the earth, that therefore they were to keep holy, abstaining from labor, &c., every seventh age, *drid* or divine day, which according to your discovery, he must have done, or he taught no Sabbath at all. If the six days of creation did not consist of twenty-four hours each, then did not the Sabbath which he ordained; and what is a consequence of this discovery? why, that one branch of the venerated, and supposed *divine* Decalogue, whose moral influence has always been acknowledged, is at a buffet driven out of the number of the sacred *ten*, in less time than was occupied by the finger of God to engrave it on the table of stone on Mount Sinai.

Now, sir, if this is one of the ten thousand mistakes which you speak of, which has been made in our translation by ignorant men, then if there are wise ones enough to be found who can remedy this difficulty, it may not be long ere we shall have a bible wholly on a different plan, imposing perhaps, new laws for morality, and new every thing else for the edification of mankind.

As to the Chronological dates of my book, they are as good as those of the Bible, and all profane history; and when you say *all the dates* of my book are wrong, you mean, I suppose, that the whole Chronology of Christendom is wrong also, and my work among the rest; no great wonder this, as I have drifted in the same channel.

But if you can show that a thousand years or more, as you aver in your letter, have been dropped out of the age of the earth, ever since the time of Adam's creation, you will succeed to establish *another* trait of confusion on the face of the Bible account of time, as deduced from the lives of the Patriarchs, and given in the book of Genesis, and rendered into English; commencing at Adam and running down the course of ages to the time of Isaac, the son of the patriarch Abraham; a lapse of years, according to the common reckoning of 2158; but according to you a thousand at least should be added. But in my present state of feeling and condition of information, I cannot subscribe to opinions which go to unhinge, and overturn all the foundations to which the morality of the christian world is at present moored, and consent to float off with you into the interminable ocean of, I know not what; based on your better knowledge of the Hebrew, which to me appears quite incredible, unless we are to suppose that all ages have agreed to keep up the delusion of the Bible errors, as it now reads, and that you are the only man who is honest and courageous enough to face the immense billows of the popular belief, and also to pour contempt on the talent, the research and the honesty of all other men who have not made these discoveries in the Hebrew Bible with yourself.

As to the zeal you manifest in wishing to get matters straight about the history of ancient America, it is praiseworthy; but nevertheless I am by no means persuaded that the excited altercation of public disputants is the best way. But if to debate with you the subject, were agreeable to me, yet is your forty positions full work enough to occupy us a year at least in their investigation, allowing one evening in a week for that purpose; and which of us has the time to spare? Surely I have not. Suppose you were to examine a work which is very popular, recently published on geology, by *Fairholm[e]*,[1] in which I see the very subject of the six days work of creation is attended to; possibly you might think differently, should you do so, unless you are predetermined, which no true philosopher will submit to.

But to close: Believe me, sir, that I have received in good part your letter, and take no exceptions to any thing therein contained, except a little abruptness and considerable dogmatism, as also the bravo words "*dared* to condemn some of my sentiments," as if your sentiments are not to be molested. Surely, if you require so much courtesy you ought not to shew less toward others whose opinions do not correspond to yours; but such illiberality is not the manner of writers in free America, nor of its free press; notwithstanding, venerated and much honored sir, I wish you, success in the development of truth, and health, and happiness, equal with your fellows of mortal, yet intellectual stamp.

JOSIAH PRIEST.

Albany, January 24, 1835.

Published in the *Troy* (N.Y.) *Daily Whig*, Wednesday, January 28, 1835.

1. George Fairholme, *A General View of the Geology of Scripture* (Philadelphia, 1833).

3

Big Bone Lick

Herbivores' need for salt caused both prehistoric and modern ungulates to seek out salt springs. Several of these "licks"—which Rafinesque mentions here—gave to localities in Kentucky the place names they still have. Big Bone Lick was the most famous of them, because in ancient times it had been a bog where animals now extinct became trapped. As a result, bones found there began to interest travelers in the eighteenth century, and both Jefferson and Franklin wished to study them. At his request, Jefferson received some 300 specimens, which he displayed for a time in one room of the White House. Early in the nineteenth century, Dr. William Goforth of Cincinnati collected there about five tons of the bones of mammoths and other extinct animals, which he was swindled out of by Thomas Ashe, who took them to England, where he realized a handsome profit by exhibiting them. In all, Big Bone Lick has yielded osseous remains of the Pleistocene giant ground sloth (*Megalonyx jeffersonii*), of both the mastodon and mammoth, horse, tapir, and musk-ox of that geological period; there were also bones of such modern animals as deer, moose, elk, caribou, bison, and black bear.

Although the water at Big Bone Lick was not very salty—five or six hundred gallons were needed to make a bushel of salt—it early became a source of this precious commodity that otherwise had to be hauled over the mountains and shipped down the Ohio. Thought to have medicinal value because, as Rafinesque said, it had "an abominable taste," the water also caused the construction of an inn at Big Bone Lick to accommodate the "the idlers who come there to loiter, drink, bathe, and kill the game."

As a result of its celebrity, Big Bone Lick had been described by many visitors[1] before Rafinesque, but his account may be unique in the attention given to vestiges of ancient human activity at the site, a consequence perhaps of his earlier collaboration with John D. Clifford in the study of "Indian Antiquities." Based on what he saw there and his knowledge of other Kentucky licks, Rafinesque raised—but did not answer—the interesting question of what connection there might be between the bones of prehistoric animals and nearby earthworks built by ancestors of the Indians.—*Editor.*

* * *

Visit to Big-Bone Lick, in 1821. By C. S. Rafinesque, Professor of Historical and Natural Sciences, &c.

MR. COOPER, in his account of Big-bone Lick,[2] has craved further information from other explorers. I shall, perhaps, add some additional facts to his. He has omitted Mr. John D. Clifford and myself among the explorers. To my knowledge Mr. Clifford visited the place in 1816 or 1817, and dug for bones. He procured many, which I have seen in his museum, in Lexington, among which a fine tusk of mastodon, and some horns of the oxen found there. His collection of bones has been removed, by purchase, to the museum of Cincinnati, and latterly to the Academy of Natural Sciences, of Philadelphia.

We proposed to visit this lick together in 1820; but his death that year prevented us. In 1821, I went with Dr. [Charles Wilkins] Short, from Lexington to Northbend, [Ohio,] at the mouth of the great Miami [River]. I left him there at his brother's seat on the Ohio, and went on purpose to the Lick by myself to explore it, and wait for him on his return. A horse having been lent to me, I went by the road of Cincinnati, following the banks of the Ohio. I visited in the way a beautiful elliptical mound, near the banks of the river, and the house of major Pratt. It has been preserved intact, with the trees that grow on it. The base measures 550 feet in circumference; it is 25 feet high, and the top is level 100 feet long from N. E. to S. W., by 50 feet broad. This mound, or altar, is nearly half way between the stone fort, at the mouth of the Miami, and the ancient city, temples, circus, and mounds on which Cincinnati has been built, now mostly levelled and destroyed. All are on the second bank of the Ohio.

Without stopping long in Cincinnati, I crossed there the Ohio to Covington, in Kentucky, on the west side of the mouth of Licking river. I went to survey the singular ancient monument near Covington, at Mr. Jacob Fowle's; the main road passes between two circular mounds of unequal size; the eastern is 12 feet high; the western 25, and has a pavilion on the top; but the singularity consists in a long sickle-shaped esplanade, running out of it to the south, which is 350 feet long, about 80 broad, and 8 feet high.

From Covington to Big-bone Lick, the distance is only 18 miles, nearly S. W. over the limestone upland, gently undulating: near the Lick the ground is more broken into ravines which open into the Big-bone valley.

I remained several days at the Lick, which is a watering place, with ample accommodations; but I found the actual owner a very surly man, who would no longer allow any excavations, having imbibed the notion that digging would take away the water from the spring, around which a pavilion and seats had lately been erected. Seeking for bones was then out of the question, and I spent my time in taking an ample survey of the place, the valley, and the landing on the Ohio, with the surrounding hills and monuments, now only two miles from the lick, where steam boats land their passengers. I made some maps and drawings, and collected several plants and fossils.

Mr. Cooper's account of the place is tolerably correct, but his map does not show all the streams, ravines and springs around the place, and omits entirely the remarkable ancient mound, connected with the Indian traditions mentioned long

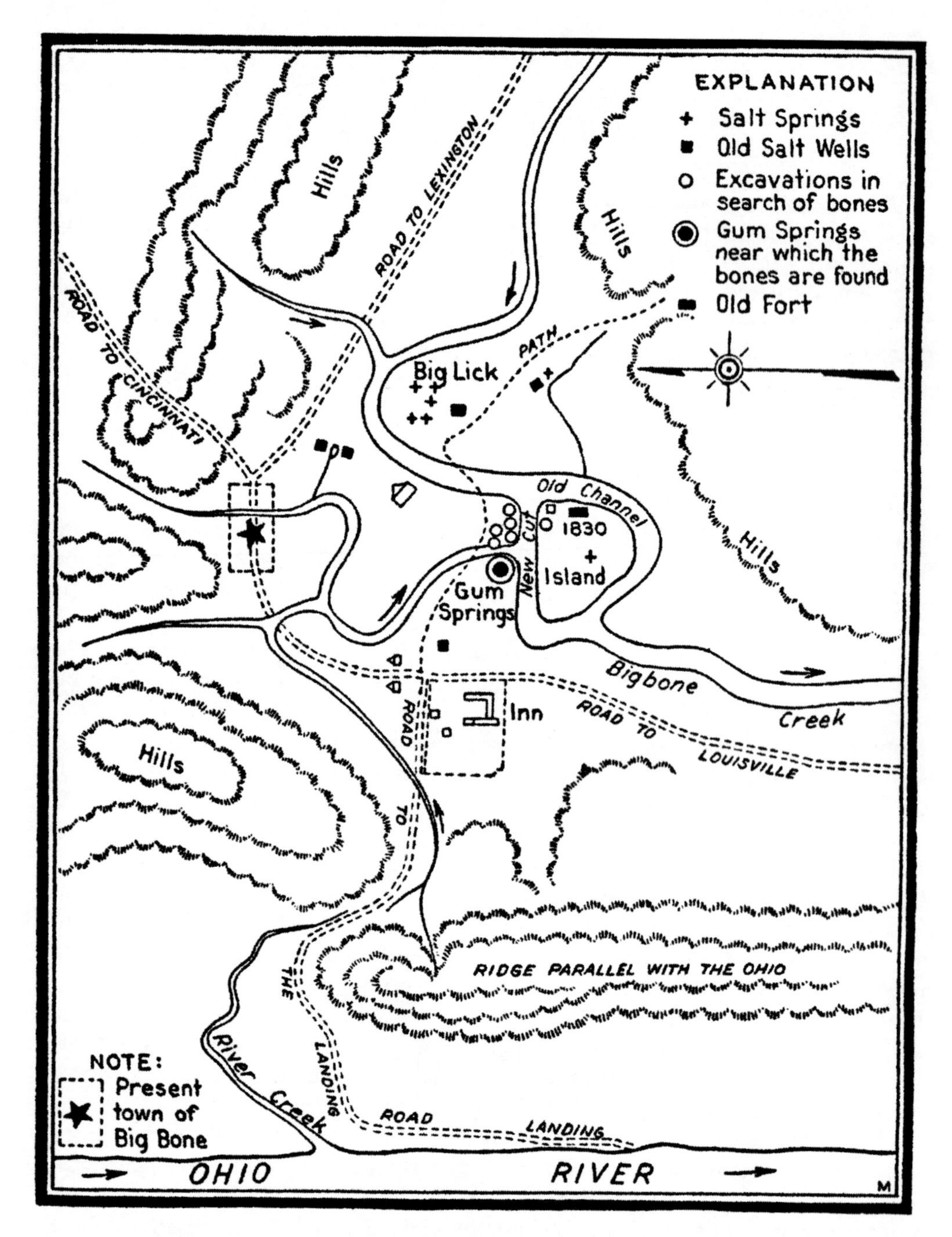

Figure 5. Map from Willard Rouse Jillson's *Big Bone Lick* (1936).

ago by Jefferson, in his notes on Virginia. Yet this mound is only 300 yards from the large boarding house, but in the woods on the steep hill behind it, towards the S. E. It is elliptical, 10 feet high, 430 feet in circuit at the base, 150 feet long, from N. to S. and is level on the top, with a hollow in the centre, which I ascribe to some late excavation, but am not positive, as no rubbish is seen.

This was the mound from which the Great Spirit destroyed the last mastodon, according to the tradition recorded by Jefferson.[3]

Behind this mound, and towards the landing, are three small sepulchral mounds near one of the springs of the western branch of Gunpowder creek, which empties itself into the Ohio at the landing; but the main branch comes from the north. The ridge separating the waters of Gunpowder and Big bone creeks, is not very high, and forms a kind of gap where the road crosses it: the lick may have once communicated with the Ohio by this gap.*

I walked to the landing, where there was a very inconvenient landing place; near it was a farm house only, the cliffs being there very near to the Ohio, quite steep, and subject to avalanches. I was told by the farmer, that not long ago, in a storm at night, he was frightened by a dreadful noise like an earthquake, which lasted a long while; and in the morning found a small ravine south of his house almost filled up by an avalanche of huge stones from the cliffs. I went to see the place, and found it so; the stones were of all sizes and shapes, but all angular; some must have weighed many thousand pounds, and yet had rolled 200 yards or more. These cliffs, as usual, are of limestone, in horizontal strata, and 200 feet at least above the river.

The water at the Lick springs contains salt and sulphur; it has a bluish cast, like that of the Blue licks, on Licking river; both are limpid, but of an abominable taste, although readily drank by the idlers who come there to loiter, drink, bathe, and kill the game—very plenty yet on the hills.

I should have wished to follow Big-bone creek to its mouth, but had not time. I have since regretted it, when I heard some years afterwards that a very singular ancient tomb had been found there. It was formed by two large slabs uniting into an angle above, and covered by the soil; some human bones were found in it, the fate of which I could not learn. I am inclined to believe it situated in the alluvion of the creek, which is ample in some places, and even contains many fossil shells, or unios, the same as those now inhabiting the creek and the Ohio. It would be interesting to know what connection may exist between this tomb, the mound on the hill, and the regular arrangement of the fossil bones at the lick, although I should myself be inclined to believe in the diluvial eddy which may have brought the bones there in a regular heap, in the bend of the valley.

At Blue licks, in a rocky valley, no bones and no monuments are found, but Drennon's lick has bones and mounds. Out of the limestone region, in the sandstone hills, many licks are found with fossils, but no bones and no monuments. Is it not strange that there should be an apparent connection between them, or rather their locality? as if some Indian tribe had collected these bones as relics.

The valley of Big-bone creek is nearly a mile wide at the lick and above it, but becomes much narrower below it, as if the lick had been formerly a basin, or small lake. All the hills are of horizontal blue limestone, with some shells, chiefly terebratulites, productus, &c. But the valley, with the sides of the hills, are of clay. This clay is of various hues and consistency, often mixed with sand and gravel, damp in the middle, dry and arid on the sides of the valley. It contains in the ravines several fossils, chiefly alcyonites and entrochites. The hills rise 120 to 180 feet above

**Which is badly laid out in the map, as well as Gunpowder creek, erroneously called River creek.*

the valley. They are wooded and full of game, but with a very thin soil. The soil in the valley, near the lick, is rather sterile, but higher up becomes fruitful, and is well cultivated.

Many pretty plants are found in the valley and hills, but no saline plants. The stream of Big-bone often changes its course, and washes away its banks when it overflows in the spring. The back-water of the Ohio, when very high, comes near to the licks and may have reached it formerly.

No bones were protruding or visible in the banks, in 1821; but some were visible as late as 1810, at least. The first European discoverer of this place was [Charles LeMoyne de] Longueuil, in 1739, who took away many bones to Louisiana and France. They were then quite out of the ground. He was led there by the Indians, who held the place as holy, and never took away the bones.

Having well explored the lick and valley, I returned to Lexington with Dr. Short, as soon as he called for me. This was in September, 1821.

Published in the *Monthly American Journal of Geology and Natural Sciences*, 1 (February 1832), 355–358.

1. See Willard Rouse Jillson, *Big Bone Lick* (Louisville, 1936), especially its annotated bibliography.

2. William Cooper, "Notices of Big-Bone Lick," pp. 158–174, 205–216, of the same volume of the *Monthly American Journal of Geology and Natural Sciences.*

3. As recounted by Jefferson in *Notes on the State of Virginia* (New York: Library of America edition, 1984):

> A delegation of warriors from the Delaware tribe having visited the governor of Virginia, during the present revolution, on matters of business, after these had been discussed and settled in council, the governor asked them some questions relative to their country, and, among others, what they knew or had heard of the animal whose bones were found at the Saltlicks, on the Ohio. Their chief speaker immediately put himself into an attitude of oratory, and with a pomp suited to what he conceived the elevation of his subject, informed him that it was a tradition handed down from their fathers, "That in ancient times a herd of these tremendous animals came to the Big-bone licks, and began an universal destruction of the bear, deer, elks, buffaloes, and other animals, which had been created for the use of the Indians: that the Great Man above, looking down and seeing this, was so enraged that he seized his lightning, descended on the earth, seated himself on a neighbouring mountain, on a rock, of which his seat and the print of his feet are still to be seen, and hurled his bolts among them till the whole were slaughtered, except the big bull, who presenting his forehead to the shafts, shook them off as they fell; but missing one at length, it wounded him in the side; whereon, springing round, he bounded over the Ohio, over the Wabash, the Illinois, and finally over the great lakes, where he is living at this day." It is well known that on the Ohio, and in many parts of America further north, tusks, grinders, and skeletons of unparalleled magnitude, are found in great numbers, some lying on the surface of the earth, and some a little below it. (p. 165)

II.

Linguistics

4

Chinese

This essay was unknown until 1982, when it was brought to light in the revised edition of T. J. Fitzpatrick's bibliography. Fitzpatrick had innocently accepted the mistaken notion of biographer R. E. Call about "a paper on 'The Chinese Nations,' said by Rafinesque to have been published in the *Knickerbocker Magazine* for 1834," which Call declared "never appeared" (p. 136). Others, such as E. D. Merrill, were all too ready to believe that Rafinesque padded his bibliography—after all, he had been repeatedly accused of worse intellectual sins, including that of having invented new plants so he could describe them.

Although signed merely with the initial "R."—which was the magazine's practice; not Rafinesque's preference—the article's authorship is not in doubt, because Rafinesque's correspondence shows that three years later he was still trying to collect the payment he had been promised.

One might suppose that the harsh comments here on the "paradoxical opinions" of Pierre Étienne Du Ponceau resulted from the two Philadelphians being rivals at this time for the Prix Volney in Paris—which, in fact, Du Ponceau won. However, the documents submitted for that contest had to be anonymous, and there is no reason to believe that either contestant was aware of the other's effort. Rafinesque *was* resentful that the American Philosophical Society, of which Du Ponceau was president, had never elected him to membership, despite his contributions of linguistic manuscripts and copies of his own publications. However, his quarrel with Du Ponceau seems to have been entirely on professional—not personal—grounds. Rafinesque wrote in his autobiography that he decided to "keep aloof" from Philadelphia's professional societies because they were "disgraced by their tenacious learned errors, and by admitting unworthy members for sake of mere fees" (p. 95). Instead, he said, he "cultivated chiefly the friendship of old friends or liberal Savan[t]s" (p. 96), among whom he named Du Ponceau, and elsewhere he remarked that Du Ponceau loaned him manuscript copies of vocabularies to study.—*Editor*

* * *

The Chinese Nations and Languages.

MANKIND have always been subject to errors, either popular or learned. They have prevailed in every age, and become by turns the current beliefs or opinions: many are at this day prevailing among the most enlightened nations.

Popular errors arise chiefly from superstition and prejudice, but learned errors from fanciful theories and systems. The human mind is but too prone to admit mistaken notions and paradoxical opinions. Every good modern work commonly explodes some of these learned fancies; but the erudite are as tenacious of their notions as the superstitious vulgar. Even in matters of fact, when there ought to be no uncertainty, ascertained truths are denied, or discredited, if they contradict the previous opinions. Some learned men cannot be convinced: they appear to wander on, and waver between egregious credulity and proud incredulity.

It is not meant here to enter the wide field of learned errors. Our libraries are filled with works of mistaken opinions, now discarded or disputed. A single subject shall receive our present attention: viz. the state of our knowledge and errors in relation to China, and the Chinese,—subjects, concerning which so many singular impressions are entertained.

The intercourse between China and Europe has now lasted three centuries, and half a century with the United States. We ought now to know China tolerably well; and indeed we have some excellent works on that country, chiefly old ones, written by the Chinese themselves, or by Europeans residing in various parts of the empire. Latterly, since the intercourse has been limited, all our accounts have become garbled and local. Hence we appear to entertains as many wrong opinions of the Chinese, as they do of us, when they call us Western barbarians, and deem us inferior beings, or foreign devils.

We have lately been wondering at a Chinese Lady,[1] and her small, cramped feet, who has been exhibiting herself for money! Grave and learned doctors went to measure her feet, and gave a certificate of their inches, as if they doubted the fact stated in an hundred words. This singular fashion is not, however, general in China. It is both a late and partial custom, introduced by a princess in the fifteenth century, as hoops and bonnets were by our ladies. It is not used by the country women, nor even by many ladies of various provinces. Yet it is probably believed by many, that all the Chinese ladies have stump feet, and also that all the Chinese gentlemen wear long claws to their hands: while it is well known that the fashion of never paring the nails is only an oddity, adopted by the sons of clowns, grown wealthy, to show that they labored no longer. The well-born, the nobles, and the learned, despise this custom, as absurd and inconvenient.

One of the greatest blunders upon China, is, that the whole empire has adopted a uniformity of dress, manners, customs, languages, and religions. This vast empire ought to be considered as if the whole of Europe was under one monarch, with the same laws and a common written language; but retaining the various [spoken] languages, customs, dresses, and religions; as is also the case in Russia and Turkey.

China Proper was formed by the amalgamation of many kingdoms and nations, of peculiar habits and languages. The actual dynasty has united to it many regions of Tartary and Thibet, that retain their languages and customs; nay even peculiar laws, and princes, some of whom lead a pastoral life. Other countries in the vicinity, like

Corea, Luchu, Anam, Siam, etc., are merely nominal tributaries; and Japan is quite independent, although of acknowledged Chinese origin, by a colony settled there in 1197 before Christ, which was dependent on China till 660, before Christ. There are even now independent tribes and nations in the mountains of China itself, such as the *Miaos* and *Lolos*, speaking dialects of the Thibet and Birman language.

Figure 6. Afong Moy, adapted from James S. Moy's *Marginal Sights* (1993).

The Chinese nations ought to be divided into six distinct classes, never to be blended in our ideas and works. 1. The old Chinese proper. 2. The conquered nations in China. 3. The conquered nations of Tartary and Thibet, including the Manchus of the North, who have conquered the whole, and are the ruling dynasty of China. 4. The tributary States in or out of China. 5. The Independent States in or out of China, but of Chinese origin. 6. Lastly, the modern Chinese colonies in Java, Siam, Borneo, Philip[p]ines, etc., which live under the rulers of those countries.

When we speak or write about the Chinese, we should state which we mean: and even in China proper, we ought to say whether we mean the Chinese of Canton or Peking, or Fokien or Sechuen, etc. since they are as distinct people as the Greeks, Italians, English, and Russians. As to the *Miaos*, *Lolos*, Manchus, Mongols, Coreans, Japanese, etc., they are as different from those Chinese as the Turks, Basks, Laplanders, etc. of Europe are from the English, or our American Indians from the white people of the United States.

The latest work upon China is that of Gutzlaff,[2] the Prussian Missionary, which has been very properly pronounced a catch-penny work by many reviews. In fact, it is a mere compilation, without much discrimination: yet coming from him, it will become popular among the religious community, and may propagate many errors, because he has misrepresented the antiquities, history, nations, religions and languages of China, in several instances. His Travels on the coast of China,[3] are by far better, being original and instructive.

Another work on China, by an American Missionary, Daniel [*sic*] Abeel, published in 1834, contains more valuable information on the islands near China, than China itself.[4] Like many of our missionaries, the writer appears to be profoundly ignorant of the different religions and sects of the country he was sent to convert. How can we hope to convert the Chinese, or other Pagan nations, if we do not know well their tenets, mythologies, and superstitions? It was not so with the Jesuits of old, who were learned,—inquired deeply into the Chinese dogmas,—visited the whole empire, and resided in many distant provinces. To this day, in their works, the best accounts of China are to be found; and they form the base of all the best geographical and historical compilations on the country. Although liable to a few errors, they are mere trifles compared to later blunders.

At present there are three opinions about the Chinese, as upon nearly all the other nations: 1. The impartial friends of China, who try to know the truth. 2. Those who praise the Chinese above all other nations. 3. The revilers, who desire to depreciate or neglect their history, learning, religions, languages, customs, etc. Whenever we read a work upon China, we should ascertain to what opinion the writer leans, in order to beware of those who overpraise or despise the Chinese nations.

Guignes, Remusat, Klaproth,[5] and others, have opened a new path, by consulting and translating the original Chinese works, for information. This appears to be the true path to follow in future inquiries, upon a nation that has books on all subjects, more numerous than all those of Europe put together. The mere classics of China comprise 180,000 volumes, of which we know but few through translations. These, as well as quotations from Chinese writers, are to be multiplied to make us better acquainted with their learning. The Chinese Encyclopedias have never been translated as yet, nor many ancient authors on geography, antiquities, languages, mythologies, etc.

Robert Waln published in 1823 a very good book on China,[6] and it is to be regretted, that he lived to publish the first volume only, in which is found a tolerable account of the religions and history of the country; notwithstanding he unfortunately adopted the erroneous theory of Sir Wm. Jones, that the real Chinese came from the Chinas of Hindostan. But his survey of the main religions of China is better than can be found in almost any other work, although he did not enter fully enough into the details of their sects, tenets, and mythologies.

In order that our Missionaries may inquire into this subject, it is needful that they should know at least that there are three great national religions in China, besides the Jews, Mahometans, and Christians; but it appears that Hinduism and Sabeism have never prevailed there.

Thus there are: 1. The Celestial religion or ancient primitive Deism, coëval with the Chinese people. It is stated to have been antediluvian. The first government of China was a theocracy. The Emperor was the Pontiff, the worship was on mountains, or *Tan*, (raised altars,) directed to the Spirit of Heaven, and latterly to Heaven and Earth, as this religion did not always exist pure, but became divided into several branches. Chiefly the *Ju-kiu*, begun by *Se-tu* 2207 years before Christ. 2d. The revived worship of God and Ancestors, by *Cong-fu-tze*, towards 550 years before

Christ. 3d. The modification by *Chao-Kang*, in 1070 of our era, which is the actual religion of the learned. This religion and its three branches, answer to our primitive religion of Adam and Noah, with the three branches of Judaism, Christianity, and Mahometanism. As these admit of many sects, so do the Chinese branches. Some sects admit of idols, or images of spirits. One of the sects worship chiefly *Pu-sa*, or the goddess of nature.

2d. Tao religion, begun as early as 2800 years before Christ by Po-y, was restored towards 700 before Christ, by *Lao-tze*. It admits of many gods and sects, saints and idols, being therefore comparable to the Polytheism of the Greeks, Romans, etc. But while we have a thousand works on their gods and mythology, we hardly know the theory and practice of the Tao sects, except by ridiculous, partial details. They appear, however, to be as yet the most prevailing sects in China and Japan.

3d. Fo religion was introduced into China from Thibet, sixty-five years after Christ. It is a branch of Bud[d]hism or Lam[a]ism, admitting of incarnations and metempsychosis, monks and idols. It is now the religion of the Manchu dynasty; but has several sects blended with those of the last creed.

Judaism was introduced into China as early as 206 years before Christ, but has never spread. Towards 720 the Mahometans appeared and spread their religion in various parts. In 636 Olopuen, a Nestorian priest, introduced Christianity; but it did not spread much till the missions of the Jesuits.

It is evident, that before missions can succeed well in China, we ought to acquire a better knowledge of these Chinese creeds and sects, than we now have, in order to contend with them; else we must go blindfold to work. They have adopted very few things from us, and whether they ever will adopt our creeds, may be deemed doubtful. But we know they have converts to many foreign religions. Ignorant missionaries will not succeed with such a highly civilized people. Missions in China must be on a different plan from those among barbarians.

We hardly know of any European invention borrowed from us by the Chinese, except perhaps glass, clocks, watches, vaccination, and astronomical calculations, which they however claim as revived inventions of their own. Whether they will adopt steam power, steam boats, rail roads, machinery, etc., remains to be seen. Meantime we have borrowed from them several inventions, arts and customs, such as the compass, block printing, silk, China ware, paper, gunpowder, the use of teas, rhubarb, fans, etc. And we might with great advantage introduce among us many others, such as dividing the hulls of ships into tight compartments, the Chinese school system, the examinations for every civil and military office, etc.

Some of these Chinese inventions we falsely claim as our own. Klaproth has lately proved that the Chinese knew the polarity and compass at least 1100 years before our era; they even claim it as known 1500 years before that time, as we have some doubtful traces of it among the ancient Pelagians. But our actual compass was borrowed from China by the Arabs and introduced to us by them.

Block printing was used in China as early as 1120 years before Christ, and was known to some Gothic nations, but only recently revived among us. Papers made of reeds, barks, hemp, silk, cotton, flax, and straw, were used in China at very early dates. The Jesuits have given their history out of a Chinese work.

Gunpowder was known to the ancient Pagan priests, but kept a secret like the telescope, ventriloquism, etc. The Chinese have used it from time immemorial for fire-works, but seldom for war.

It is generally thought that the Chinese are indebted to us for the vice of smoking tobacco: this is another mistake: they knew the vice of smoking hemp, tobacco, opium, etc., very early, and before the discovery of America, which perhaps owes this custom to them. Tobacco grows wild in China as in America.

Barrow's travels in China[7] is a popular work; and although he does not overpraise the Chinese, we find in it the unanswerable statement, that between the eleventh and sixteenth centuries, while the European nations were semi-barbarous and intolerant, without gardens, canals, chimneys, or comforts, their learning low, and calendar wrong, the Chinese were a learned, polite, rich, and tolerant people, having fine canals, gardens, luxuries and comforts, with a proper calendar! But since 1560, the Europeans have wonderfully improved in all respects, while the Chinese have remained stationary.

Their ships appear to us clumsy and unwieldy, yet they are better than the old galleys and Dutch ships, and have partitions to prevent shipwrecks. The Chinese sent colonies far away as early as 3000 and 4000 years ago. They went as far as Sofata in Africa, the Persian Gulf, New Guinea, Kamschatka and perhaps America, about 1000 years ago. Barrow says, that in the 7th century, they called Kamschatka *Tashan*, and the north-west coast of America *Fou-Shan*.

He says that there are no drunkards in China, among 300 millions of human beings, while many millions of drunkards are found in the Christian lands. There are less vices in China than in some cities of Europe. Slavery is allowed, but it is very different from our slavery. Some slaves are convicts, sold for crimes, as in the state of Delaware; others are willing slaves, who sell themselves or their children; they are treated mildly and as children of the family. The slaves can acquire property and keep it, but they are few in number. In Japan there are none, but vassals instead.

If they have no social worship and no Sabbath, they have instead many holydays and festivals. Their diversions are various, having many of our shows and games. The paternal and school influence in China is similar to that of the Romans and Hindoos,—perhaps excessive, but salutary and wise. There we find no undutiful children, no fighting boys, no youthful candidates for the gallows, no stubborn scholars, no rebellions in schools and colleges, no disrespect to teachers.

The Chinese laws are just and wise. They are severe for crimes, but otherwise mild and equal. The custom-house duties and regulations are better than those of England. The trading nations of Europe have always shown a disposition to intrude and impose their own whimsical laws of trade to foreign nations, with their monopolies, excessive duties, prohibitions, etc. Yet they dare to complain of the Chinese regulations. They desire to do in China as they have done in India, Turkey, Java, America, etc. They ought to beware, and remember their fate in Japan, where they are now almost excluded. The Chinese have been compelled to concentrate their trade in Canton and Macao. If they should be compelled by wanton aggressions and open smuggling to exclude all foreign trade, what should we do for teas and Chinese luxuries? Let the Americans be wise, and if the English should be excluded, they may secure the whole tea trade.

Japan is an old Chinese insular Colony, a flourishing state of 25 millions population, once receiving kindly all nations, and trading themselves afar. By the pride, intolerance, and intrigues of the Portuguese and their converts, it has been compelled to exclude all foreigners except the Chinese and Dutch, while its own trade is confined to the dependencies from the Kurile Islands to the Luchu Islands.

That beautiful country is now an anomaly in the social system of the world, from which it has almost seceded. Meantime 25 millions of human beings dwell there in peace, plenty, and happiness, since 200 years. Thunberg,[8] the learned Swedish traveler, who went there as a Dutch surgeon, has said, that it offers the most extr[a]ordinary social phenomenon. The laws are strict, unviolated, and conducive to social happiness. Japan has no wars, no diplomacy, no parties, no distress, no discontent, no strife, no discord, no lawyers, no law-suits, no dearth, no emigration, no need of foreign commerce, no corporations, no monopolies, no slavery, no intolerance (except against the intolerant Christians,) no intemperance, no grog-shops, no throne, no crown, no royal foppery, no waste lands, no cavalry, no wheel carriages, no officers unfit for their office, no public debt, no paper currency, and no taxes: all the public revenue arising from a land rent paid in kinds, and a few customs.

This is also the case in China, where out of a revenue of 350 millions of dollars, a tithe in kind on the produce of the land, forms the main income. But there is no additional tithe for the clergy, as in England. They are provided for in lands, as well as the soldiers or militia: thus both the priests, monks, and soldiers are also land owners or cultivators. The other taxes of China are a tithe on manufactures, the customs, the mines, the salt monopoly, etc., but there are no inland taxes, no excise, no game laws, no poor laws, and no paupers. Each family, or the emperor, supporting the old and poor. The emperor receives also tributes from afar, presents, and national gifts. No one in China is allowed to keep waste land: he is compelled to let it on shares to the poor, and to pay the tithes.

The freedom of the press has never been impaired in China nor Japan, while it is yet doubtful in many parts of Europe. The effectual restraint on its abuses, is provided by making the printer, seller, and reader of libels liable to the bastinado. In Japan, the Catholic books and images are forbidden, owing to the rebellions they have occasioned; but all the religions and sects of China are tolerated, although there is a national religion, the *Sinto*, a branch of the *Tao* of China.

The Chinese soldiers, while in actual service, are made useful, as among the Romans, by being employed as guards, watchmen, jailers, constables, police-men, couriers, collectors, etc. After wars, they form agricultural settlements and colonies.

The population of China is immense. All writers agree in this, although they differ widely in their calculations, which vary from 175 to 400 millions. Macartney and Barrow[9] state it at 333 millions in China Proper, having followed the original Chinese statistics. Many have doubted their statement; but it was true, notwithstanding, and later accounts increase the number to 360 millions. This population is liable to fluctuation; but has ever been on the increase, since 200 years of peace after the conquest by the Manchus's.

All these facts have been doubted by the revilers of China, and European pride, blushing to find wiser men in the East! Many assertions of these traducers are truly ridiculous, and hardly deserving of notice, although they are believed by many to this day.

Among all the paradoxes upon the Chinese, none are more singular than those deeming them all one people, speaking the same language, and having a set of characters which can only be read by themselves, and not by their neighbors. While on the contrary, the reverse is the case in both instances. See Duhalde,[10] Leyden,[11] Klaproth, etc.

It is well ascertained that every province of China has a peculiar language, or dialect, and often many. Secondly, that all these provinces, and all the neighbors of China, can and do read the Chinese symbols in their own respective languages. As these facts are of high importance in our future intercourse with these nations, they deserve to be examined carefully and in detail.

In a late review of Gutzlaff's work on China, published in the [American] Quarterly Review[12] of Philadelphia, the critic, who appears to have a tolerable knowledge of the Chinese, has fallen into these errors, and has praised Dr. Duponceau for having made the mighty discovery that *the Chinese characters must be read phonical[l]y*, or in other words, that you must learn to speak Chinese before you can learn to read their letters. Yet the critic has qualified this assertion, by confessing, that various Chinese nations of the provinces can read them in their own dialects. This allows that there are many dialects in China, and that those who speak them can read the written Chinese without learning Mandarin dialect of the learned; which explodes at once the new, erroneous opinion. But it remains to be shown that the neighbors of China can also read them in their own languages. Nay, it may be further proved, that they can be read in almost all languages, and thus the Chinese books might become universal, if their symbols were adopted every where, and even in Europe.

This was the general belief, before the Philadelphia savant thought otherwise, and published his mistaken impressions, as truth, in the Transactions of the American Philosophical Society ["Vol. III, new series, p. 69"[13]], and in the Philosophical Magazine, for 1829.[14] He thought that if they had a different grammar, they could not use the Chinese written grammar or style in their respective languages. But these two facts ought first to have been proved: 1. Have they a different grammar? 2. Cannot the Chinese style and syntax be adapted to their speech? This was not done, as it was too troublesome to investigate: but whoever will inquire, will find that they have a grammar nearly alike, and that the Chinese style is quite familiar to them.

If these critics of Chinese facts had taken the trouble to make an experiment, they would soon have seen their error. We have done so, and are quite satisfied that the Chinese characters and symbols are pasigraphic, or universal, and may be read in all languages, and, particularly, very well in English, which has the same grammatical simplicity in verbs, with few inflections, and adjectives always prefixed to substantives. They are equally adapted to those languages which put them either before or after, as in Italian, and often drop the inflections of the verbs in the dialects: while languages like the Arabic and Malay, having the qualifications after the nouns, may yet understand the transposition. As to inflected verbs, it is well known that the Creole dialects drop all the inflections, as in the *lingua-franca* dialect of Italy, and thus use the Chinese style.

We have seen the Chinese symbols read with perfect ease, either in English or in Italian, and also in Spanish, by a learned Mexican, Mr. Nasera,[15] who acquired in six months the written Chinese, although he could not speak any Chinese dialect.

If this is the case with our European languages, how much more will it be with those of Japan, Corea, Siam, Anam, etc., so nearly akin to Chinese in grammar and syntax, idiomatic expressions and ideas. It is well known that one of the acts of subservience to China in the tributary states, is to receive the Chinese calendar, printed by the Board of Astronomy, and annually sent to them to regulate time. This is received and read in the respective languages, the tributary rulers not speaking Chinese.

Japan, being independent, does not receive the Chinese calendar, but the Dairi, or Pope, issues a peculiar calendar, based upon the Chinese, written in Chinese characters, but read in Japanese. The Chinese characters and style are used also in Japan for all official decrees, although written for the Japanese, and read in that language ["see Kempfer,[16] Thunberg, Golownin,[17] etc."]. Yet the Japanese has a peculiar syntax, putting the adjectives after the substantives, as in Malay, and all the Polynesian languages. Chinese books are an article of trade and import in Japan, where they are bought and read by the learned, who know the written Chinese; although the Japanese have three syllabic alphabets of their own, each of fifty letters, as figured by K[a]empfer, in which they write letters and print books. But the Chinese traders in Japan must use interpreters, as few are able to write; nor do the merchants they deal with, read Chinese.

The same happens in Siam, and all places where the Chinese now trade. They must every where employ interpreters to deal with the natives, unless both dealers are learned in written Chinese. Duhalde, who wrote from the memoirs of French Jesuits dwelling all over China, asserts positively, in his history of China, vol. xii. p. 392 of English translations,[18] that the people of Japan, Anam, and Tonkin, use Chinese books, although they cannot express themselves in oral Chinese, nor understand each other's languages. This is confirmed by nearly all the writers who have been in China, or in the neighborhood,—Guignes, Barrow, Hall,[19] Ellis,[20] Morrison,[21] Thunberg, Gutzlaff, Staunton,[22] etc. This last, who knew the Chinese well, says: "Almost all the countries bordering on the Chinese Sea, or Eastern Asia, understand and use the written Chinese, although *not the oral Chinese!*"

This fact was well known, and never doubted, till Dr. Duponceau, without having been in China, or speaking the Chinese dialects and languages, fancied, and would not but believe, that many nations could use the same grammar and syntax. This learned philologist has become famous for his paradoxical opinions on many other languages. He has even fallen into the egregious error of omitting two sounds of the English language, in his work[23] on English Phonology,—the French mute E in *receive, believe*, etc., and the soft Portuguese LH in *billiards, steelyards*. He has denied to America any monosyllabic languages, though the Othomi of Mexico, and Guarani of South America, are such. He has stated that all the American languages are alike in structure, and unlike those out of America: while both assertions are evidently erroneous; as is well known to all who have studied many languages. Yet these paradoxes are becoming current among us, and many believe them upon trust.

The correction of his mistakes in regard to the Chinese, is of the utmost importance, because if it is the fact that 300 or 400 millions of Chinese and neighbors can read the Chinese books, it follows that the translations of our Bible, and books of science, can be made accessible to one third of mankind, without separate expensive translations in one hundred languages and dialects of China or the neighborhood.

Our written Chinese translations may thus be read by the Japanese, Coreans, Manchus's, Mongols, Thibetans, Siamese, etc., whenever they have learnt to read the written Chinese, which is the only writing now of all the Chinese nations, except in Japan, Siam, Thibet, Manchusia, etc., where syllabic alphabets have also been introduced.

It is a general complaint with the missionaries in China, and repeated by Gutzlaff, that they have experienced the greatest difficulty in acquiring the different oral languages and dialects of China, in order to preach and instruct [see also Duhalde, vol. ii, p. 405]. Besides the *Kiang-nan* language of Eastern China, which has become the learned dialect since about 600 years, when the court was held in Nankin, he says that *every province, every city*, and even large villages, have a peculiar dialect. The women and common people do not understand any other. After three or four years' study, a missionary wants an interpreter to speak to strangers, if he travels out of the limits of his dialect! Some of the dialects are real languages, as different from the *Kiang-nan* as the English is from the Russian and German.

The language of Fokien, in S. E. China, is spoken by forty millions. It has the sounds of B, D, R, Z, which are lacking in the Kiang-nan and many other languages; but are also found in Japanese. We know very little as yet of these dialects, because neglected in the grammar; but we know the *Kong*, or language of Canton, that of *Peking*, and half a dozen more, spoken by the sailors, or in Chinese colonies. We still lack the dialects of *Yunan, Honan, Shensi, Sechuen*, and fifty other inland dialects.

Even the *Kong* is very different from the *Kiang-nan*, and both from the Fokien, of which we give some instances:

	Kiang-nan	*Kong*		*Kiang-nan*	*Fokien*
Man	Jin	Yun	*Father*	Tu	Pe
One	Y	Yut	*Mother*	Mu	Bowo
Fish	Yu	Ngu	*Woman*	Niu	Lu
Sea	Yang	Hoy	*Sun or Day*	Ji	Mit
Tree	Shu	Sut	*Moon*	Yue	Guar

Thus there are double difficulties in China both in learning the written and the oral languages; but they are worthy to be overcome, in order to put ourselves in better communication with 400 millions of human beings, to become acquainted with their books and learning, translate the best works, and to impart to them our own knowledge by translating ours in the written Chinese.

The difficulties of learning their characters are well known. They consist chiefly in the great number and intricacy of those now employed. But there is an easy key to them; they having elements and roots, whereby the whole are formed and known.

Six strokes form all the letters by their repetition. Duhalde has figured them. Only 214 radical symbols form all the others by their combination. Although as many as 80,000 characters are in the great dictionary, yet the small Chinese dictionary contains only 10,000, which are sufficient for common use and common books. A man is learned when he knows 20,000, and but few know 40,000; because 60,000 out of the whole are obsolete, or antiquated synonyms, or relate to peculiar arts and sciences.

If all the signs and symbols used by us for writing, cyphering, printing, astronomy, geography, algebra, botany, alchemy, music, etc., were collected and calculated,

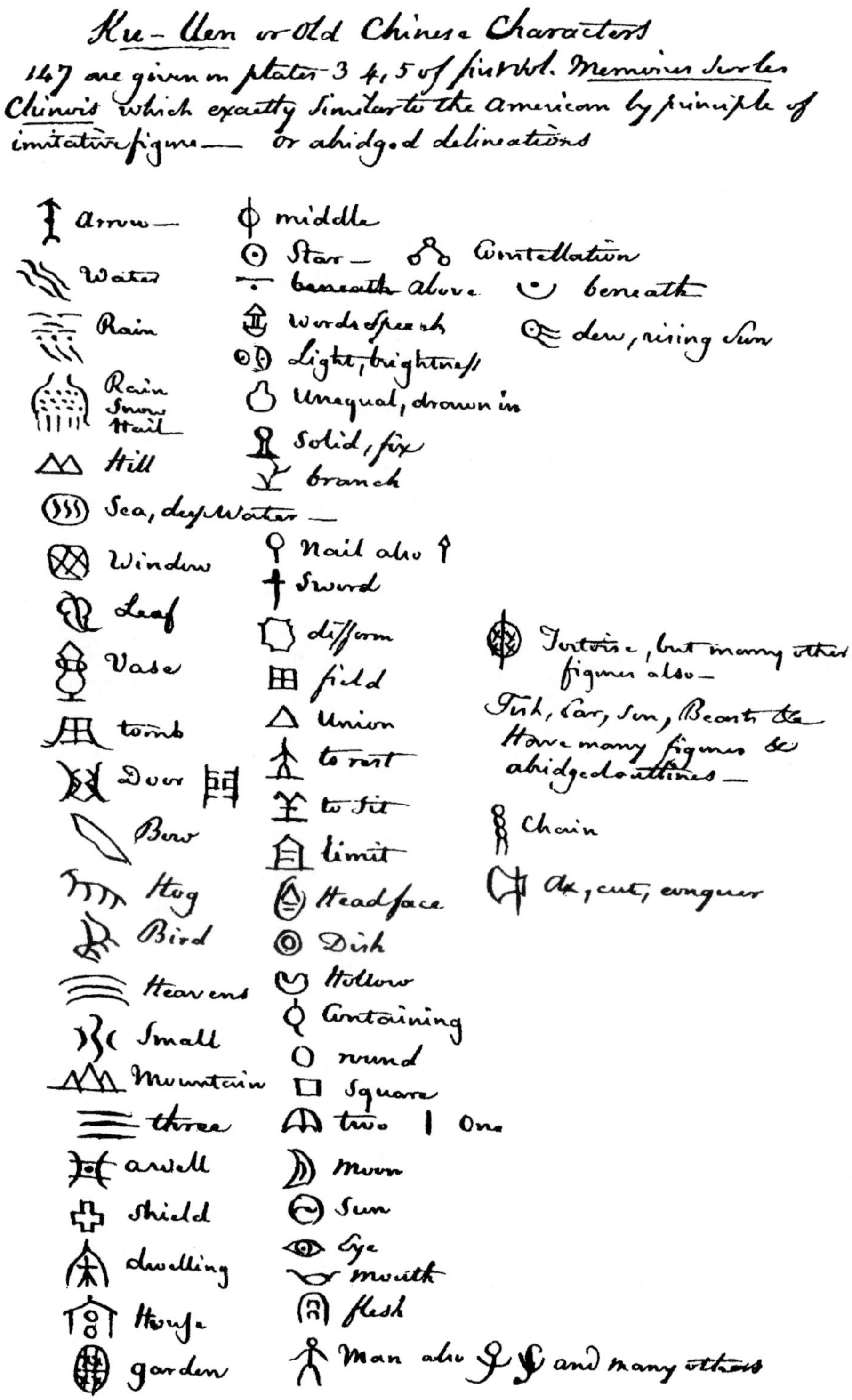

Ku-Uen or Old Chinese Characters

147 are given on plates 3, 4, 5 of first vol. Memoires sur les Chinois which exactly similar to the american by principle of imitative figures — or abridged delineations

arrow —
Water
Rain
Rain Snow Hail
Hill
Sea, deep Water —
Window
Leaf
Vase
tomb
Door
Bow
Hog
Bird
Heavens
Small
Mountain
three
a well
shield
dwelling
House
garden

middle
Star — Constellation
beneath above beneath
words speech dew, rising Sun
Light, brightness
Unequal, drawn in
Solid, fix
branch
Nail also
Sword
diform
field
Union
to rest
to sit
limit
Head face
Dish
Hollow
Containing
round
Square
two One
moon
Sun
Eye
mouth
flesh
Man also and many others

Tortoise, but many other figures also —
Fish, Car, Sun, Beasts &c Have many figures & abridged outlines —
Chain
Ax, cut, conquer

Figure 7. Ideograms in a Rafinesque manuscript at the American Philosophical Society.

they would amount to many thousands. Our letters alone in their various forms of capitals, print, script, manuscript, italics, gothic, calligraphic, ornamental, etc., amount to nearly 1000. Thus the single letter A, can be written or printed by us with forty or fifty different variations: this, however, we all know. It may not be more difficult, therefore, to learn the Chinese signs, than our own; particularly the 214 roots; and when we know how to combine them well, we know the whole of the ideas they express, without learning their names and sounds. It is well known to be easier to learn to read than to speak Chinese.

There were formerly in China sixteen kinds of primitive characters, which may be seen in Kircher.[24] They are now as obscure to the actual Chinese as the hieroglyphics of Egypt are to us. The first set, the lung or dragon symbols of Fohi, is antediluvian. Another set, the turtle symbols of Yao, is over 4000 years old. These sixteen sets formed 540 symbols, that continued in use till 840 years before Christ. When the modern mode of writing came into practice, many successive changes took place in it, until 200 years after Christ, when the modern straight lined letters became of general use. The written Chinese has become crowded with obsolete synonymical signs, divided into six classes.

Meantime the *Kiang-nan* language had only 330 oral monosyl[l]ables to express all these characters, which were multiplied to about 2000 oral words by accents and tones, often difficult to acquire and express in our letters; but in fact not more difficult than our long and short vowels, aspirations, and accents. But these words and tones vary in all the Chinese dialects, whence the second great difficulty of oral speech.

We are not aware that these variations have as yet been properly sated in our Chinese grammars and dictionaries, which speak more to the eye than to the ear. Even the latest by Remurat [Rémusat] and Morrison are deficient in this respect. They merely give the grammar and sounds of the *Kuan-hoa*, or style of the learned, in the *Kiang-nan* language. There are four classes of Chinese styles: 1. Old styles. 2. Poetic styles. 3. Learned styles. 4. Vulgar styles. Yet these do not differ so much in syntax as in the choice of synonyms in writing or speaking, which ought, however, to be given.

There are also homonyms of symbols. The sun was formerly written by a circle with a small bar in the centre; now it is a square with a bar; but both are called GA, as in the old language, while the actual languages give many different names to the sun.

The Chinese grammar is perfectly simple and regular. It has none of our anomalies nor irregularities, therefore it is easily read and translated into any language reduced to its simplest form.*

**Dr. Duponceau, in his letter to Capt. Basil Hall [cited in note 14] in the fifth volume of Philosophical Annals, 1829, although confessing himself* ignorant of the Chinese, *has labored hard to disprove this simple fact, and to prove that all foreign nations who use the Chinese characters* must have *learned the spoken Chinese, (not that they do,) before they use them. It would be much easier to prove that they need not.*

If the people of Anam or Cochin China use some Chinese characters in a different sense from the Chinese, it is of course as written anomalies, well understood, as the French use the three letters SON *to write five very different ideas,* bran, his, her, its, sound; *without impairing their discourse, the connecting words showing the meaning of the written homonymy. In English,* sound *is written for three French words, son, canal, sain, expressing very different ideas; yet we make no mistake when we say, the sound of a bell, Long Island* sound, *and a* sound *head.*

The Chinese idiomatic forms are common to all the neighbors, (although very different from the European

Men, meaning *many* and *more*, forms all the plurals; and the verbs have no inflexions, but receive particles to modify the tenses. Thus verbs are construed pretty much as in English with auxiliary particles.

Remusat has well shown, that the oral Chinese is not purely monosyllabic, since it admits of two words being expressed by a combined character, and there are about twenty particles without separate meaning, that take a mark in writing to express this, while another mark shows that characters must be read phonically to express foreign names.

Since there are more symbols than words and accents, it must happen that several symbols bear the same name. This has been stated as a great inconvenience; but a similar homonymy is found more or less all languages, and particularly in English and French, without great inconvenience. Several words of very different meanings seem pronounced nearly alike in English, such as:

Where, Were, Ware, Wear, Weir, pronounced WER.
Reed, Read, Rid, " RID.
Vain, Vane, Vein, " VEN.
I, Aye, Eye, Ay, High, " AY.

It is as bad in French, where, for instance, to say *five holy seals*: although written *cinque saints seings*, these are pronounced SEN SEN SEN, and require a paraphrase for explanation, as in Chinese. It follows that every apparent difficulty and anomaly of the Chinese languages, both written and spoken, has its equivalent in our idioms, and need not surprise us. After all, the written Chinese is more easy to acquire than is generally supposed, and when acquired, it may be made a medium of imparting ideas, without the multitude of oral languages, which so much perplex and impede the intercourse of mankind.

R.

Published in *The Knickerbocker* magazine, 5 (May 1835), 365–76.

1. This is a woman named Afong Moy, who was exhibited mostly in New York in 1834 but also at Peale's Museum in Philadelphia. She is said to be the first Chinese woman to appear in the United States, where she amazed onlookers not only by her tiny feet but also by her ability to feed herself with chopsticks.
2. Karl Friedrich August Gützlaff (1802–1851), *A Sketch of Chinese History, Ancient and Modern: Comprising a Retrospect of the Foreign Intercourse and Trade with China: Illustrated by a New and Corrected Map of the Empire* (New York, 1834).
3. Karl Friedrich August Gützlaff, *Journal of Three Voyages along the Coast of China, in 1831, 1832 & 1833 with Notices of Siam, Corea, and the Loo-Choo Islands* (London, 1834).
4. David Abeel (1804–1846), a Dutch Reformed Church missionary: *Journal of a Residence in China and the Neighboring Countries from 1829 to 1833* (New York, 1834).

---

idioms,) and even if they were not, they might easily adopt them as synonymical phrases. The Italian comme state, *(how are you,) is understood in English and in French as a synonym of* how do you do, *or* comment vous portez vous,*—idioms widely different.*

The Roy, *or Chinese dialect of Japan, is by no means general; it is confined to very few, else no interpreters would be wanted. The* Yomi, *or real Japanese, can be written in Chinese characters as easily as the Italian, by using the Chinese idiom and syntax.*

If Remusat and Champollion have partly fallen into the erroneous belief of Duponceau, it is to be regretted as giving a currency to error. They must have been misled by not reflecting on the possibility of a simple syntax and idiom becoming almost universal.

5. Probably:

Joseph de Guignes (1721–1800), *Le Chou-king, un des livres sacrés des Chinois, qui renferme les fondements de leur ancienne histoire, les principes de leur gouvernement & de leur morale* (Paris, 1770).

Abel Rémusat (1788–1832), *Recherches sur l'origine et la formation de l'écriture chinoise; Remarques sur quelques écritures syllabiques tirées des caractères chinois, et sur le passage de l'écriture figurative à l'écriture alphabétique* ... (Paris, 1827).

Julius von Klaproth (1783–1835), *Notice de l'encyclopédie littéraire de Ma touan lin, intitulée ... Wen hian thoung k'hao* (Paris, 1832).

6. Waln, Robert (1794–1825), *China ... with Remarks on the European Embassies to China; and the Policy of Sending a Mission from the United States to the Court of Pekin ... To which Is Added, a Commercial Appendix ... of the Trade ...in China and India; and a Full Description of the American Trade to Canton* (Philadelphia, 1823). At Rafinesque's request, Waln wrote a long letter about the ginseng trade with China. Rafinesque published it for the guidance of those Kentuckians who gathered the root for the market. The first installment appeared in Lexington's *Western Monitor* newspaper, 16 April 1822, and was concluded there in its 23 April 1822 issue.

7. Sir John Barrow (1764–1848), *Travels in China, Containing Descriptions, Observations, and Comparisons, Made and Collected in the Course of a Short Residence at the Imperial Palace of Yuen-min-yuen, and on a Subsequent Journey through the Country from Pekin to Canton* (Philadelphia, 1805).

8. Carl Peter Thunberg (1743–1828), *Travels in Europe, Africa, and Asia, made between the years 1770 and 1779* (4 vols., London, 1793–1795). First published 1788–1793, 4 vols., in Swedish, a language Rafinesque probably did not read.

9. Sir John Barrow (cited above, fn 7) was the private secretary to Britain's ambassador to China, Earl George Macartney (1737–1806).

10. J[ean]-B[aptiste] Du Halde (1674–1743), *The General History of China, Containing a Geographical, Historical, Chronological, Political and Physical Description of the Empire of China, Chinese-Tartary, Corea and Thibet, Including an Exact and Particular Account of Their Customs, Manners, Ceremonies, Religion, Arts and Sciences* ... (4 vols., London, 1736).

11. John Leyden (1775–1811). Possibly his *Comparative Vocabulary of the Barma, Maláyu and T'hái Languages*, a mission publication (Serampore, 1810); but more likely his *Malay Annals*, translated by Sir Thomas Raffles and published in London in 1821.

12. The anonymous review of Gützlaff's *Sketch of Chinese History* occupies pp. 100–43 of volume 17 (March 1835) of the *American Quarterly Review*.

13. Appears as a footnote in the original. Rafinesque is referring to "The Translator's Preface," pp. 65–96, which Du Ponceau inserted before his "Grammar of the Language of the Lenni Lenape or Delaware Indians. Translated from the German Manuscript of the Late Rev. David Zeisberger" which follows, in pp. 97–251, of the *Transactions of the American Philosophical Society*, n.s. 3 (1830).

14. "Letter from Captain Basil Hall ... Inclosing a Communication from Peter S. Du Ponceau ... on Some Points Connected with the Nature of the Chinese Language," *Philosophical Magazine or Annals of Chemistry, Mathematics, Astronomy, Natural History, and General Science*, 5 (January 1829), 15–24. Du Ponceau's letter is dated 7 July 1828, to which he added a postscript dated 14 July 1828.

15. Manuel de San Juan Crisóstomo Nájera (1803–1853). In his autobiography, Rafinesque calls him a Mexican philologist.

16. Engelbert Kaempfer (1651–1716), *The History of Japan, Giving an Account of the Ancient and Present State and Government of that Empire* ... (2 vols., London, 1727).

17. Appears as a footnote in the original. Vasilii Mikhailovich Golovnin (1776–1831), *Memoirs of a Captivity in Japan During the Years 1811, 1812, and 1813, with Observations on the Country and the People* (3 vols., London, 1829).

18. The English translation of Du Halde's book consists only of four volumes. The reference to "vol. xii" is a misprint for vol. ii.

19. Basil Hall (1788–1844), *Voyage to Loo-choo and Other Places in the Eastern Seas, in the Year 1816* ... (Edinburgh, 1826).

20. Sir Henry Ellis (1777–1869), *Journal of the Proceedings of the Late Embassy to China, Comprising a Correct Narrative of the Public Transactions of the Embassy, of the Voyage to and from China, and of the Journey from the Mouth of the Pei-Ho to the Return to Canton* (London, 1817).

21. Robert Morrison (1782–1834), *A view of China, for Philological Purposes Containing a Sketch of Chinese Chronology, Geography, Government, Religion & Customs. Designed for the Use of Persons Who Study the Chinese Language* (London, 1817), or his *Memoir of the Principal Occurrences During an Embassy from the British Government to the Court of China in the Year 1816* (London, 1820).

22. Sir George Staunton (1737–1801), et al., *An Authentic Account of an Embassy from the King of Great Britain to the Emperor of China: Including Cursory Observations Made, and Information*

Obtained, in Travelling through that Ancient Empire, and a Small Part of Chinese Tartary (2 vols., London, 1797).

23. Peter Stephen Du Ponceau, "English Phonology or an Essay towards an Analysis and Description of the Component Sounds of the English Language," *Transactions of the American Philosophical Society*, n.s. 1 (1818), 228–64.

24. Athanasius Kircher (1602–1680), et al., *La Chine d'Athanase Kirchere de la Compagnie de Jesus: illustrée de plusieurs monuments tant sacrés que profanes, et de quantité de recherchés de la nature & de l'art : à quoy on a adjousté de nouveau les questions curieuses que le Serenissime Grand Duc de Toscane a fait dépuis peu au P. Jean Grubere touchant ce grand empire: avec un dictionnaire chinois & françois, lequel est tres-rare, & qui n'a pas encores paru au jour* (Amsterdam, 1670).

5

Ruter Reviewed

Rafinesque and Ruter were both honored at the graduation exercises of Transylvania University in 1822—Rafinesque with the university's honorary M.A. and Ruter its honorary D.D. Martin Ruter (1785–1838), a Methodist minister, was in charge of the Methodist Book Concern in Cincinnati at this time. There he published school texts in arithmetic, spelling, and more substantial books such as a martyrology abstracted from John Foxe, as well as the little book reviewed here. After eight years in Ohio he returned to Kentucky as the first president of Augusta College; later he was president of Allegheny College in Pennsylvania. In 1836 he moved to the new Republic of Texas, where he traveled more than two thousand miles on horseback organizing congregations. Apparently his exposure to the elements during this work hastened his death.—*Editor*

* * *

Review

An easy entrance into the Sacred Language, being a concise Hebrew Grammar without points, compiled for the use and encouragement of learners, and adapted to such as have not the aid of a teacher—By Martin Ruter, D.D. *Cincinnati, published for the Methodist Episcopal Church, Morgan and Lodge Printers*, 1824; I vol. 12mo. 96 pages.

The publication of such a work in the Western States, where no ancient languages are taught as yet, except Latin and Greek, must be both a matter of surprize and congratulation. The Oriental languages had not even attracted much attention till lately in the Atlantic States; and but few, if any, good grammars had been published there: while this short introduction to the Hebrew Language is equal, if not superior to most of them.

We are indebted to the Methodist Episcopal Church, which has lately done so much for the cause of piety, benevolence and instruction, for this new manual of Oriental and Biblical study; and to a worthy, learned and modest Divine of that church, for the compilation and perspicuous arrangement of its pages. We have seen with pleasure such an effort, considering it as the first dawn of Oriental literature in these distant occidental regions. The concise, analytical and easy style of this new grammar ought to be commended and appreciated, and we have no doubt that it will recommend itself by its utility and small bulk, to all the students of the Sepher or Biblical text in the west, if not elsewhere.

This useful production, deserves therefore our unqualified approbation, and the immaterial blemishes which we have discovered in it, relating more to omissions than to inaccuracy, do not in the least invalidate its value.

It is almost a phenomenon to have such a learned work published among us; and still more so, to see it compiled and condensed with such perspicuity and attention. We perceive with satisfaction that the *exuberant points* of the Mazoretic dialect, which increase so much the difficulty of the Hebrew grammar, have not been employed in this manual. The division of the alphabet into radical, servile, final and dilated letters is to be praised. The remarks on the figures and Chaldaisms of the Hebrew, are useful additions, as well as the literal translation of the first chapter of Genesis, some prophecies, &c.

We should have wished to have seen another addition; that of the principal elementary roots of the language; altho' it might have increased the bulk of this introduction. The utility of such a vocabulary is obvious, since it is a kind of substitute for many bulky Dictionaries. If the Rev. Dr. Ruter ever gives a second edition of his grammar, we recommend this improvement; and if it is in his power to procure the celebrated work published in 1815 by D[']olivet[1] on the radical structure of the ancient Hebraic, we advise him to consult it by all means, as the best and most comprehensive work on Hebrew Grammar and elementary roots. This work, which is scarce in America, and has not yet been properly studied nor consulted by our Hebrew scholars, contains treasures of biblical erudition. The second volume includes besides, a new and radical translation of the first ten Chapters of Genesis, in English and French, with copious notes, in which all the most obscure points of philology & history are illustrated.

Another improvement which we would venture to recommend to our worthy author, would be the introduction of the simultaneous use of English and Hebrew letters in all instances, and whenever any Hebrew word occurs. This is now done in the best Oriental Grammars published on the Continent of Europe. It obviates the painful attention, often producing disgust, which is requisite in the first study of the Hebrew characters and writing, and it has the advantage of accustoming the eyes, by the continual recurrence of the two kinds of letters, to annex definite and peculiar sound to the Hebrew characters, which is a great facilitation and requisite, in the acquisition of this or any other language.

The correct pronunciation of Oriental languages, is very difficult to be attained; but it is not always an indispensable requisite in the study of dead languages. However we should wish that more attention could be paid to the enunciation of vowels at least. The Rev. Dr. Ruter has of course followed the erroneous pronunciation of the English scholars, which is different from that of nearly all the remainder of the world. That the Continental or European pronunciation is the correct one, can be easily proved by the living or spoken dialects of the Hebrew, such as the Syriac, Maronite, Carait, German-Hebrew, Polish-Hebrew, Portuguese-Hebrew, &c.; and likewise by the other spoken Oriental languages, such as Arabic, Persian, Coptic, &c. Therefore the vowel *A*, ought to be pronounced as in *Far* and *Map*; *E*, as in *Bell* and *Rest*; *I*, as in *Mill* and *Lip*; and *U*, as *oo* in *Root* and *Book*. The incorrect abuse of giving to these vowels the English or Saxon power of letters, pervades the American schools and books, in servile imitation of an obsolete Oxford theory, but will probably give way in time to the correct practice, not only in the case of Hebrew, but also of Latin and Greek, where it is still more ridiculous, since the modern Latin or Italian and the modern Greek or Romaic, both living languages, are altogether against the English practice.

We have been sorry to perceive that our learned author has copied another collateral error of many Hebrew scholars, by wishing to exalt the Hebrew language above all others. He says at the very outset of his preface, that *the Hebrew has a higher claim to antiquity than any other language now existing*. We know that this has been the opinion of many Orientalists; but we cannot assent to it, while we are aware that this false view has repeatedly been confuted by enlightened Philologists. Since there are yet, some authors, who appear unaware of this, and their high standing or holy character might contribute to disseminate this error, we shall attempt to disprove it, by a few plain and evident statements: in doing which we have no other aim than to serve the cause of truth and learning, being persuaded that the Hebrew cannot lose its importance by being proved to be no older than many other ancient languages.

We know at least fifty languages as old or older than the Hebrew of the Bible, of which [the] real name is Syro-Chaldaic; among these languages the Chinese, Sanscrit, Arabic, Celtic, Latin, Greek, Zend, &c., are conspicuous, and all living yet in their spoken dialects. We shall not investigate in detail, the relative antiquity of these various languages; this would carry us too far; but we shall merely compare the relative priority and age of the Latin and Hebrew, which may be proved to have originated at a contemporaneous period, and to be equidistant from the *Primitive Language*, the common parent of all subsequent Dialects or languages: being both removed five steps from that *Primitive Language*, and having both therefore four successive parents older than themselves.

A tabular view of this succession or genealogy may serve to bring the subject under a regular shape, and it will be still better illustrated by a comparative table of the successive changes of *one radical word* in those gradual dialects of each other.—This word will be *Earth*.

Primitive Language.

		(Earth or Land.)
1*st* step or dialect,	*Iranic*	Zar?—Tila.
2*nd* d[itt]o	*Aramic*	Ar.—Tira.
3*rd* d[itt]o	*Pallistin* or *Canaanit*	Tar?—Arz.
4*th* d[itt]o	*Sepheric, Mosaic* or *Hebraic*	Eartz.—Artz.
5*th* d[itt]o	*Hebrew* or *Syrochaldaic*	Arez, Earetz, or Adamat.

Primitive Language.

1*st* step or dialect,	*Iranic*	Zar?—Tila.
2*nd* d[itt]o	*Gomerian*	Zur?—Tira.
3*rd* d[itt]o	*Ausonian* or *Oscan*	Tur.—Tera.
4*th* d[itt]o	*Salian* or *Old Latin*	Aru.—Tela.
5*th* d[itt]o	*Latin* or *Roman*	Arvis, Tella, Tellus or Humus.

The *Iranic* was one of the first branches of human speech, from which the Hebrew and Latin have both sprung. In the *Chinese*, another branch as old as the *Aramic*, the earth is called Ti. From the *Iranic* or *Aramic* have been derived many of the ancient Oriental dialects, such for instance as

	(Earth or *Land.)*
Pelhavi or Assyrian	Arta—Damik. Akhe.
Chaldee	Ara—Dameh.
Arabic	Ardi—Arz.
Syrian	Aro—Kam? &c. &c.

In which the root of this word is always *Ar*. All these are older than the Hebrew. The *Aramic* was the language of Aram and of Abram. The posterity of Abram adopted in Palestine the dialect of that country which was somewhat mixt with Egyptian. When residing in Egypt under the Palli Dynasty, their Goshen or Mosaic or Sepheric language was formed, which is the old Hebrew, and became more akin to the Egyptian. D[']olivet contends that the Sepheric or Hebraic was almost pure Old Egyptian. But this is merely a presumption, while the actual loss of this Hebraic language is a positive fact; it became extinct about 500 years before Christ, or 2325 years ago by changing into Syrochaldaic or vulgar Hebrew: which is the language of our Bible, none of the Bible manuscripts being older. The Samaritan dialect is another Hebraic dialect mixt with Ph[o]enecian and Chaldaic, and of the same age as the Syrochaldaic. The Rabbinic and Mazoretic dialects are younger still, as likewise the Syriac.

Thus the actual or Bible Hebrew or Syrochaldaic is no more than 2325 years old. See Adelung and Vater on Languages.[2] The Sanscrit is two thousand years older at least; in that language the earth is called *Stira* or *Dhara* or *Bumi*, the analogy of which is striking in *Terra* and *Humus* in Latin.

The *Gomerian* became *Celtic* in the Alps, *Cantabrian* in the Pyrenees, *Ausonian* in the Apennines. The *Ausonian* was divided into many dialects[:] *Ombrian, Eugubian, Etruscan, Salian,* &c. From the Salian was formed he Latin or Roman language, towards the beginning of Roman History, or about 2500 years ago, therefore somewhat earlier than the actual Hebrew or Syrochaldaic. This has been proved by ancient inscriptions discovered in Italy. Let us show, by the same word *Earth*, how some of the ancient dialects (earlier than the Latin) were formed from the *Gomerian* and *Ausonian*.

Ancient Dialects.

	(Earth or Land.)
Cantabrian	Lar?—Lurre, Erri.
Pelasgic or Ancient Greek	Gaya—Era.
Celtic	Talu—Itala.
Teutonic	Cord—Arth.
Cimbric	Duar—Nore.
Etruscan (In Italy)	Ther?—Thira.
Ombrian (ditto)	Fri—Ferri. Humi.
Eugubian (ditto)	Herna—Herra.
Egyptian (Peculiar language near to Pelasgic)	Kahi—Dum?

These disquisitions may we trust demonstrate that the Hebrew or Syrochaldaic has not a claim to very remote antiquity although it is a valuable ancient language.

We had omitted to mention that the letter *Tsadi* called *Ts* by our author and *Tz*, by many others, has no consimilar sound in the English language; but it has in many Oriental, Celtic or Latin dialects, for instance the modern Italian, in which it is written Z or Zz, as in *Zia* or *Razza.* It is a peculiar sound, impossible to be expressed, but easily taught, and which has even two modifications, like the *th* in English words *thy* and *thigh. Ts* and *Tz* may represent them although *Zh* and *Zz* might be less liable to ambiguity. But this belongs more particularly to the Phonology of languages.

We now dismiss the author with our best wishes for his first attempt, and in a confident hope that if he applies himself to Oriental literature in general, he will be able to enlarge his views and usefulness in this philological department.

C. S. R.

Transylvania University.

Published in the *Cincinnati Literary Gazette*, 1 (22 May 1824), 161–62.

1. Antoine Fabre d'Olivet (1768–1825); like Rafinesque an autodidact who dabbled in dozens of fields of knowledge, he published *La langue hébraïque restituée,* (Paris, 1815–16).

2. Johann Christoph Adelung, *Mithridates oder Allgemeine Sprachenkunde* (1806–1817). Volumes 2–4 were edited from Adelung's MSS., with additions, by Johann Severin Vater.

6

Hebrew Studies

Rafinesque's reflections on Hebrew began at Transylvania University, in whose library he probably first read Fabre d'Olivet, for he included a book by the latter in the bibliography of his *Ancient History, or Annals of Kentucky* (1824). If the book had any relevance at all to the ancient history of Kentucky, it might have been in regard to what Fabre d'Olivet had to say about the "ten lost tribes of Israel" theory, which Rafinesque always opposed. It was at Transylvania too that he met Martin Ruter and later wrote a review of Ruter's book on Hebrew grammar. His interest in Hebrew was like his interest in Indian languages—as a tool to decipher the mysteries of remote prehistory, and both were consonant with his fascination with word puzzles. The book he published in 1838—*Genius and Spirit of the Hebrew Bible*—treats Hebrew in the tradition of Kabala studies of ancient secret wisdom embedded in the structure of the language itself. Modern Kabalaists do not seem to know of the book; others have accepted the judgment of Rafinesque's biographer, R. E. Call, who wrote that it is "a book without a single redeeming literary feature" (p. 120) and added for good measure that is "without the least value from any possible standpoint" (p. 202). Whatever one's opinion of this "key" to esoteric wisdom, still of interest are Rafinesque's discussion of the structure of the Hebrew language, of its pronunciation, and his proposal for transposing the characters of its alphabet. These are the portions transcribed below.

No printer's name appears on the book, but the workmanship, trim size, and type font are similar to those of Rafinesque's *Flora Telluriana*, manufactured by a job printer on North Fourth Street in Philadelphia named Probasco. Whoever was responsible for the printing must have had trouble reading the author's manuscript and, as customary with him, Rafinesque squandered little time on proofreading. Because the compositor was unable to puzzle out how many loops were intended, the word *vowel* habitually came out as "wowel" and the points customarily used to distinguish Hebrew vowels were called "nutations." Since neither word makes sense in this context they have been amended here to *vowel* and *mutation*; obvious misprints have been silently corrected, as has the compositor's confusion of ע with ץ. Other necessary changes have been bracketed.—*Editor*

* * *

OBRI
or Hebrew Alphabet.
Reduced to English letters and rectified.
Invariable Signs and Sounds.

Names	*Xaldi*	*Sounds*
1 Alef	א	A—Ah as in Father
2 Beh	ב	B—B
3 Gimel	ג	G—G harsh as in God
4 Daleh	ד	D—D
5 Œe	ה	E—Eh as in Belt
6 Uau	ו	U—U as in Full, O in Do
7 Zain	ז	Z—Z
8 Œeh	ח	Œ—as U in Fur, I in Bird
9 Teh	ט	T—T
10 Iod	י	I—I as in Bill
11 Xaf	כ	X—Kh, aspirated
12 Lamed	ל	L—L
13 Mem	מ	M—M
14 Nun	נ	N—N
15 Samex	ס	S as in initials
16 Oin	ע	O—O as in For
17 Fe	פ	F—F or Ph
18 Yadi	ץ	Y—Tz, as the German and Italian Z
19 Kof	ק	K—K or harsh C
20 Rec	ר	R—R
21 Cin	ש	C—Sh, Fr. *Ch*, Germ. *Sch*
22 Hau	ת	H—Th, or the Greek Θ

Remarks on the Alphabet

This Improved and rectified alphabet for the Hebrew Language has no need to employ double letters for single sounds, and has no equivocal letters nor sounds like the Xaldi Alphabet. It has long been needed and it is strange that it never was thought of before: although premiums had been offered for writing the Oriental Languages in our letters, and we have still 5 unemployed letters for the Oriental sounds not in the Hebrew, J for J or Dj, P, V, W, Q, Æ, besides accents.

In selecting English signs for the equivocal Xaldi letters of the Hebrew, I have been led by analogies. The Y employed for Tz is quite like the Xaldi sign, altho' now widely different with us, being made similar to French I in sound it was the French U of the Greeks. X for Kh is exactly the Greek and Spanish letter and sound. F is the real sound of Ph, pronounced between F and P in Greek and Oriental dialects. C[,] quite a useless letter with us, may very properly represent the sound of Sh, as it partly did in the Italic languages although commonly modified now in Tsh.—H for the English TH is a novelty, I might have prefer[r]ed to introduce the Greek Θ, if I had not meant strictly to employ the English Alphabet: in that case H might have stood for the 8th letter; but its sound is certainly the vocal sound of Œ as in Latin and French who write it now *Eu*, *Heu*, *Œu*, and this was probably the exact sound of the Greek H, taken from the Ph[o]enician and Hebrew, and not at all our mild aspiration of H similar to the French mute E, which was the *Sheva* of the Hebrew, or soft breathing put between consonants to prevent a clashing of hard sounds, by producing a gentle hiatus.

Lastly each letter must be fully sounded; none are ever mute or silent, and the phonic utterance is always invariable. By attending to these few directions, we shall be able to pronounce the Hebrew Language exactly as it was done 3000 years ago.

As to the mutations or points of the Masoretic Jews, they are entirely useless, nay pernicious; because they stand for the changes and additions in the words of later dialects, whereby we distort and lose the real original words, obtaining instead another Language. It is as if when writing Latin we were to write instead[,] in modern Italian Language, *Uomo* for HOMO, *Albero* for ARBOR &c. These difficult vowel points being rejected altogether, make the Hebrew Language much easier to attain and by no means equivocal as the Rab[b]is pretend, since the points rather produce many more equivocations. And in fact they do not employ them when writing the modern Rab[b]inic or Syriac Dialects, nor in the Talmud.

By this reduction to the real elementary sounds, the OBRI, is easily attained, the words shortened and rendered quite concise: whereby we could if we liked write the pure old OBRI as easily as we write English, and in half the space. We should find it a very philosophical Language, able to express nearly all our ideas and even modern terms, by adopting them, as the Rab[b]is adopted many foreign words in later times.

By discarding altogether the ugly and equivocal Xaldi letters we shall lessen the expense and trouble of printing the Hebrew, and render it attainable to those who are disgusted or puzzled by these Xaldi letters, and the useless old mode of writing and printing and reading from right to left—which I also dismiss altogether[,] adopting our usual mode of left to right in all cases.

It is well known that the Hebrews did not employ the Xaldi letters till after their return from the captivity, and that the Samaritan letters are supposed to be their real ancient letters, unless they had an earlier Alphabet now lost, akin to the Demotic Egyptian and Ph[o]enician, such for instance as is found on the Rocks of Mt. Sinai. Therefore the Xaldi signs being foreign and spurious, are by no means essential to the Hebrew in any way, and the old Hebrew or OBRI could be written as well in Greek as it was once by many Jews, or in Arabic letters as now done by some of them, but most conveniently in our Roman or European letters, now generally in use by nearly all the civilized nations, even the Germans and Russians beginning to employ them.

Of the 22 signs used for the sounds of the OBRI, 6 represent vocal sounds or vowels A, E, U, Œ, I, O—10 are consonants B, G, D, T, X, L, M, N, K, R—and 6 are sibilants Z, S, F, Y, C, H.

Relative Applications of Those Signs or Letters.

The two main applications of the 22 OBRI signs are to express a classification of ideas, and numeration or designations of numeral values. Our numerals are now distinct from our letters, in OBRI they were not, and thus they were employed for arithmetical signs, as well as painting ideas.

In our actual languages, words are so multiplied that they must be classified, which is often attended with great difficulty. In OBRI the words were fewer, all derived from Roots of 2 letters chiefly, forming Generas of Ideas, while each single sign was a kind of order or class of ideas.—In the biliteral Roots the first letter had commonly the preponderating influence. The triliteral Roots are often formed by double roots, whereof the two joining letter[s] are similar and therefore blended into one. All the words of 4 letters are either compounds of double roots or derivations by affixed letters.

This peculiarity makes the OBRI Language very philosophical, and analytical, while it proves to a certainty that the pure names alone ought to be used, discarding all the accessories of Masoretic accents and mutations, with all superadditions and expletions.

Many of the Hebrew Books being poetical, admit of such additions or amplifications of words to suit the measure or harmony of the verse. While they admit also for the same purpose of elisions or abbreviations; the most frequent being when two similar letters come together, which are almost always reduced to one. Whenever two similar letter are united or made double, they imply an intensity of the meaning. Triplication[s,] which are very rare[,] designate the utmost intensity or energy of meaning.

As there are also words for numbers beside the literal numbers, these often imply a complication of meaning and numeral energy, which may be distinguished by the annexed context; but is one of the most obscure part[s] of explanations.

Table

Of the value and meanings of the 22 Signs.

Value	*Letters*	*General meanings or class of ideas.*
1	A	Man, Unity, Stability, Centre, Power &c.
2	B	Open, Paternity, Visibility, Action, Dwelling &c.
3	G	Throat, Canal, Organs &c.
4	D	Bosom, Division, Square, Plenty &c.
5	E	Breath, Life, Entity, Spirit, Self &c.
6	U	Eye, Vision, Light, Creation, Passage &c.
7	Z	Arrow, Demonstration, Image, Refraction &c.
8	Œ	Field, Labor, Work, Law, Elements &c.
9	T	Roof, Shield, Protection, Resistance, Strength.
10	I	Hand, Power, Fluidity, Potential, Manifestation.
20	X	Hollow, Mould, Assimilation.
30	L	Arm, Wing, Expansion, Possession.
40	M	Woman, Mother, Passive, Plastic...
50	N	Child, Extension, Production...
60	S	Ball, Circular, Spiral...
70	O	Body, Form, Material, Bad, False...
80	F	Mouth, Speech, Voice, Face...
90	Y	Air, Wind, End, Term, Flowing...
100	K	Sound, Cry, Cutting, Compression...
200	R	Fire, Ray, Head, Motion, Change...
300	C	Celestial, Justice, Propensity...
400	H	Soul, Influence, Mutual, True.

It will be easily perceived that these meanings embrace two series of ideas chiefly human, material and intellectual: besides often a third, physical or potential.

Yet they are all connected philosophical[l]y and rational[l]y, thus affording a beautiful evidence of ideal associations, and intellectual philosophy. Some have supposed that this Language being at once so analytical and synthetical, must have been made on purpose as a sacred vehicle of knowledge; but its affinities with spoken Languages preclude this supposition as in the case of the Sanscrit.

These literal meanings were probably rather applied to the Sounds they represented than the Alphabetical Signs themselves, as we have no positive evidence that the MKRE or OBRI Bible was written in letters before the Xaldi were adopted. Some contend that the Samaritan letters were the original OBRI; but we have lately acquired some evidence that the OBRI had another Alphabet, quite Syllabic like the Sanscrit, or rather the Old Chinese, each letter being a monosyl[l]abic Word and Root. Inscriptions on Rocks and Old Buildings have been found in that old style, from Mt. Sinai to Hauran East of Damascus, and also in Palestine (see Bur[c]khard[t], Buckingham &c)[1] which it will be as interesting to study, unravel and decypher as the Egyptian letters, that were on the same plan, but not the same model, being commonly symbolic.

* * *

The Xaldi Letters are yet employed by us, without any necessity: those Letters are ugly, uncouth, difficult to distinguish, print and read. Those standing for A, Y, C, L, T, F, K are alone perfectly distinct, all the others more or less alike and dubious, B and X – G and N – D and R – E and Œ – U and Z – are so similar as to be hardly distinguishable and very perplexing, hurting the sight (see the Alphabet), while in ours these and all Capital Letters are perfectly distinct. I have therefore attempted to reduce the whole Hebrew Bible to our Letters, and I wish the whole may be thus printed. I have used now large Capitals only in order to make them more conspicuous; but hereafter less bulky letters may be employed.

* * *

Pronunciation of OBRI Letters.

Many opinions have been entertained on that score, but all those based on the theory of points are false, applying to changes of dialects; as in our English language for instance where we have the *written sounds of Yore*, which we pronounce quite differently and variously in the modern English. What difference between *Plough*, *Cow*, *Cough*, *Sea*, *Fowl*, *Bee*, *to do*, *Face*, *Rice*, *Iron* &c now pronounced *Plau*, *Kau*, *Kof*, *Si*, *Faul*, *Bi*, *tu du*, *Fes*, *Rais*, *Ayorn*! The table of Sounds given with the Alphabet is the most correct, and liable to but few objections, A is stated by some to be sounded broad as in *Ball*, *Fall*, and is accordingly written Æ by D'Olivet,[2] those who hold that opinion, presume that the Sheva or hiatus between clashing consonants was the short pure A instead of the French E mute; but this is not proved, and would change the roots, therefore improbable, although A may have had sometimes the broad sound, as in fact all the vowels may also, and the broad E, O, U, have quite as probably existed in modulations and music. Y is often deemed Ts instead of Tz, and may have these two modifications that exist yet in the Italian. U and I are stated to have become V and J when initials, but this was only in subsequent dialects. F which is so often deemed P or PH latterly had probably the sounds of F, V, P in dialects. Meantime the radical sounds were as stated, nay may have been still less in number, if the oldest Alphabet had only 16 letters from A to

O, the other letters and sounds from F to H being blended with the others as in the oldest Alphabets of Ph[o]enicia, Greece, Etruria &c.

XALDI AND *ACURI.*

They are identical synonyms; whenever I use the name of *Xaldi*, I mean by it our Chaldean or the learned language of Assyria or ACUR[;] the Xaldi letters are called Acuri in the Bible. I have commonly used this as an anglicized adjective without mutations nor plural, instead of Xaldic, Xaldean &c, just like OBRI instead of Obric, Hebrew, Hebraic. But it may be noticed that in OBRI the language was feminine called OBRIH, and the Hebrews had a plural OBRIM; to admit these gram[m]atical forms or *Hebraisms* in the English language would be awkward, and thus OBRI, XALDI, ACURI &c stand for the singular and plural, in all cases indeclinable as all our adjectives even when expressing nations.

Pages 5–13, 16, 227–28 of *Genius and Spirit of the Hebrew Bible* (Philadelphia, 1838).

1. John Lewis Burckhardt (1784–1817), Swiss traveler who mastered Arabic and became so deeply conversant with the Quran and associated literature of Islam that he was accepted by Muslims as an Islamic scholar. Among his publications were *Travels in Nubia* (1819), *Travels in Syria and the Holy Land* (1822), *Travels in Arabia* (1829), and *Notes on the Bedouins and Wahábys* (1830). James Silk Buckingham (1786–1855), English journalist who established the *Calcutta Journal* in India and the *Oriental Herald* in England. Among his books were *Travels in Palestine* (1821), *Travels among the Arab Tribes Inhabiting the Countries East of Syria and Palestine* (1825), *Travels in Mesopotamia* (1827), and *Travels in Assyria, Media, and Persia* (1829).

2. Antoine Fabre d'Olivet (1768–1825); like Rafinesque an autodidact who dabbled in dozens of fields of knowledge, he published *La langue hébraïque restituée,* (1815–1816), a major source of inspiration for Rafinesque's *Genius and Spirit of the Hebrew Bible.*

III.

Society

7

Life in Lexington

There is no reason to doubt that the French original of the following letter was dispatched to Bory de St. Vincent, and there must have been other letters in connection with Rafinesque articles that did appear in a scientific journal edited by Bory. However, Rafinesque figured so little in Bory's varied life that no letter by him appears in the *Correspondance de Bory de St-Vincent* published in 1908 by Philippe Lauzun.

This letter did have a profound effect in Lexington, where, even before its publication in Rafinesque's new magazine, the *Western Minerva*, it evidently was the cause for the magazine itself to be suppressed by the printer, "at the request of some secret foes of mine," Rafinesque wrote in 1836, "who probably paid him for it." Rafinesque went on to say that he had only saved three copies. It may be one of these—though only in the form of proof sheets, some of which bear the author's corrections—which was discovered eighty years later in the library of the Academy of Natural Sciences of Philadelphia and finally published by photo-offset in 1949.

Characteristically, Rafinesque himself believed his magazine was "too learned" for Lexington, as he wrote at the time to Thomas Jefferson. But if the officious men of Lexington who were the boosters of the community—and especially of Transylvania University—were distressed by being called "paltry Owls or rather *Whip-poor-wills*, which live in the dark and now and then utter their lamentable and dismal cries," their snooty wives would have been no less pleased by the observation that their parties had "a degree of formality and monotony, which makes them become tiresome after awhile," according to "Lavinia," whose letter follows. "Lavinia" may have been chosen as a pen name for the same reason Rafinesque used "Benjamin Franklin" as author of two philosophical essays—to make it appear that the first issue of his magazine had contributors other than the editor. Nevertheless, like his mouthpiece Lavinia, Rafinesque had incisive opinions about Lexington's "Belles and Beaux" as shown in an unpublished personal letter, included here to round out this account of Lexington's social life.—*Editor*

* * *

Fragments of a Letter to Mr. BORY [DE] ST. VINCENT[1] at Paris (member of the Academy of Sciences of Paris, author of Travels in the Islands of Africa, &c.[2]) on various subjects.—Translated from the French [by CSR].

LEXINGTON, 7th January 1821.

I send you via New Orleans, many of the objects which you have asked of me, and particularly a large collection of plants from this state, and shells from our streams. I cannot send you as many of these last as you request, because I have been asked one thousand specimens for the museums of Liverpool, London and Paris, and I must endeavour to comply with the request.

The edition of the first and second volume of the Western Review and Miscellaneous Magazine is exhausted, and cannot be had; if a second edition is printed you shall receive them.

I send you at last my Ichthyology of the River Ohio, containing as many as one hundred and eleven species, of which I have myself personally seen, described and figured about 90. This branch of Natural History has lately received great increase in this continent; Dr. [Samuel Latham] Mitchell [*sic*] has ascertained about 100 new species of atlantic fishes, Mr. [Charles Alexandre] Lesueur over 150 species of fish from our great Lakes, to which I have added as many from various quarters; 400 new species discovered within five years!

Figure 8. Baron Bory de Saint-Vincent, frontispiece portrait from his *Correspondance* (1908).

There is here as elsewhere a set of unfortunate individuals, who have two eyes; but cannot see: their minds are deprived of the sense of perception: they are astonished and amazed at my discoveries, are inclined to put them in doubt and even to scoff at them. The art of distinction is entirely unknown to them; they are like the uncivilized savages who call cabins all our various buildings, let them be huts, cottages, log-houses, brick-houses, stone-houses, barns, churches, palaces, jails, colleges, capitols, &c. Thus our Cat-fishes, eels, shads, sturgeons, &c are for them mere fish to fill their stomach! and moreover they are all of European breed, and were carried here by Noah's flood direct from the Thames, the Seine and the Rhine!—I let them rail to their heart's content, and I laugh at them as much as the members of the French Institute and the British Royal Society, do at the marabouts of Morocco and Arabia, who contend to this day that all knowledge is in the Alcoran! and of course in their own brains.

I have sent you during last year and by various opportunities, 21 memoirs for your general Annals of Physical Sciences; they were

No. 1. Principles of the System of the Universe.
2. Monography of the Bivalve Shells of the Ohio, with figures.
3. Prodromus of a Monography of the Roses of North America.
4. Prodromus of the Fossil Turbinolites of Kentucky, by Mr. John D. Clifford and myself.
5. Remarks on the genus *Houstonia.*
6. Ditto on the G. *Eustachya.*
7. Analytical view of the Orders, families and genera of the natural class *Endogynia,* sub class *Corisantheria.*
8. Remarks on the natural affinities of the genera *Viscum, Viburnum* and *Samolus.*
9. Remarks on the Polystomous and Porostomous Animals.
10. New notices of Natural History.
11. Remarks on some new hybrid animals.
12. Remarks on some ichthyological errors.
13. On a new species of *Manis. M. ceonyx.*
14. Correction of the genus *Lysimachia.*
15. Synandrical Nomenclature.
16. Description of a new Spider, *Tessarops maritima* with fig.
17. On two new genera of fossils from Kentucky[,] *Endolobus* and *Sutorites.*
18. On a new species of mole, *Talpa sericea.*
19. Remarks on the family of Convolvulaceous.
20. Descriptions of two new genera of Cephalopodes, *Anisoctus* and *Todarus.*
21. Classification of the Antipedes among the Cephalopodes.
I now add thereto seven other memoirs, which are
22. Analytical classification of Vegetable Odors.
23. Designations of the durations of Plants.
24. Florula Mandanensis, or description of 70 new or rare plants, collected near the Mandans on the Missouri, by Mr. [John] Bradbury.
25. Classification of the Sharks and Skates, with four new species and fig.
26. Descriptions and figures of twelve new Atlantic fishes.
27. Illustrations of twelve new genera of fresh water fishes from the United States, with figure.
28. Descriptions of many new spiral shells from New York.[3]

I hope to send you this year about twenty additional memoirs and tracts for your Annals, several of which are already partly prepared, such as the natural history of our Salamanders, the descriptions of our western spiral shells, both from our land or fresh waters, the natural classification of all the American genera of plants of doubtful affinities, &c.

It is only in Europe that my labors and discoveries may be fully appreciated:

here I am like *Bacon* and *Galileo* somewhat ahead of the age and my neighbors; but a time will come, and perhaps within a short period, when such labors as mine will meet with general approbation, even here, as they may now do in Europe. I am however happy to perceive that this apathy and reluctance for scientific researches is very far from being general: we have already at this early period of existence of these western states, as many enlightened citizens and writers as in any part of Poland and Russia of equal extent, already more than in our southern states, and will soon rival and surpass the middle or eastern states.

We have lost this year two of the most zealous friends of knowledge in Lexington, both worthy contributors to our Western Review, Mr. John D. Clifford[4] an eminent antiquarian, geologist, &c. and the Rev. Mr. Birge,[5] a young clergyman of the most promising talents as a scholar and a writer.

You wish to know who are our writers: we have very few as yet, and since most of them wish to keep their names concealed, it would not be well to drag them forth before the public; they shrink from the light as yet, and they are perhaps in the right, since as soon as a writer becomes somewhat eminent or displays the least glimpses of genius, he is assailed by a host of revilers and croaking frogs.

The Western Review has been attacked by the most contemptible sophisters, aristarchs and moles; but thrives and goes on without minding them. The Western Minerva has been threatened before her birth; but the [A]Egis of Wisdom will shield her against all these paltry Owls or rather *Whip-poor-wills*, which live in the dark and now and then utter their lamentable and dismal cries.

I have not been spared of course, and so much has lately been said against my labors, &c. that they would endeavor to make me believe that I am a great man, since they put me on a level with our most eminent philosophers, such as Franklin, Rush, Jefferson, Clinton, Mitchill, &c. who have been equally assailed and slandered by turns. But the good sense of the people makes amends for this slight inconvenience: fools laugh and grin, while the wise shrug their shoulders and smile at them. The American nation has been accused of vanity, and this vice has been called our national sin; but if we have a national vice, we have also a national virtue, this is *good sense*, which cures all evils by degrees, and to which an appeal is seldom made in vain.

The state of Kentucky[,] which is the oldest and most populous in the western states, has taken the lead in western Literature; and Lexington the seat of the University of the West, (the only one as yet in complete operation west of the mountains,) is become the central focus of it. The state of Ohio, which is the next child of the west, has not yet been able to rival us, and all the attempts made there have been ineffectual; but they try to mimic us in every thing. Our medical school was hardly established when they attempted another at Cincinnati, within eighty miles of this town, which has begun to go in operation with about twenty students, while we have over ninety. They have three or four embryos of Universities which are mere grammar schools! I wish them well with all my heart, but I wish also that they knew how to go about it and would not mislead and deceive. Many of their puffs are mere tricks, for instance, they have established a Museum,[6] which has issued proposals of exchange; but when applied to, they had nothing to give, but were very greedy to receive! How different has our Museum behaved! It has sent during last year with my assistance, over two thousand specimens of natural productions to England, France, Holland, Switzerland, Baltimore, New York, &c in anticipation of proposed exchanges.

You will perhaps be glad to hear that there are two writers of some talent in Ohio: Dr. [Daniel] Drake of Cincinnati and Mr. [Caleb] Atwater of Circleville. The former has shown himself an author of capability in his first work called *Picture of Cincinnati*;[7] although that work is not free from defects and even errors: but he has not published any thing since, except small rhetorical pamphlets: he aims at knowledge however, and if he does not know how to reach it, it is perhaps because he has a share of the unfortunate short sight.

Mr. Atwater is an antiquarian and geographer: he has lately begun to explore and describe the Alleghawian antiquities of his state in the transactions of the American Antiquarian Society, and he has in cont[e]mplation a general description of the state of Ohio.[8] He is an able man; but a diffuse writer, his style being deprived of order, perspicuity and elegance.

It is a pity that these gentlemen should appear to labor under the moral diseases called selfishness and conceit, adding thereto a proportionate share of jealousy of each other, and every body else who may attempt to be on a par with them. Each would wish to be the only writer of eminence between the Alleghanies and the Mississippi! How deplorable! If they should acquire and join the happy gifts of benevolence, liberality, am[i]ability and correct perceptions, to their zealous endeavors, they might soon become the pride of their state.

There are throughout the western states, many enlightened and learned men, who study nature or cultivate the Muses, without ever publishing any thing. I have met them at Pittsburgh, on the banks of the Ohio, the Wabash and the Kentucky. I wish they could be induced to communicate their researches or effusions to the public: they would soon begin the first era of our Literature. I hope they will be inclined to join us in our endeavors; every thing must have a beginning, and their appearance shall be hailed by all the friends of letters and knowledge.

But let them beware to prostitute their talents to the cause of Ignorance, by satirical or ironical momentary effusions, which are sure proofs of bad taste and the harbinger of malignity. Those poetasters, sneaking critics, and burlesque scrawlers, which are now the bane of literature, will disappear gradually before them as the fog before the blaze of a bright summer sun.

I will conclude by two anecdotes which will prove how easy it is to dispel these mists of literature, and how completely their authors may be hoaxed. I have lately had the pleasure to exercise over some of them, the most perfect *mystifications*; they fell into the snares like turkey buzzards, and I have enjoyed the fun to quiz the quizzers!

In the first instance I was challenged by a champion of dead languages to write him a letter in one of them: I sent him merely two lines in one of the Oriental languages (the _____ language,)[9] and unable even to know what it was, he took it for gibberish, and made a fool of himself in an unintelligible answer. When I explained the lines, he became ashamed of himself and has been shy ever since.

The second instance is still more ludicrous. I had often been attacked in the Cincinnati papers,[10] and some one said once that it would be more useful to introduce a new article of industry, (meaning probably some new patent frying pan) than to discover one hundred new fishes fit to fry. In order to quiz these fishmongers and fish eaters, I threw out a small bait, properly mixed with two other ingredients of unexceptionable appearance. The bait was so tempting and so well disguised that the fishes were caught readily! It consisted in a proposal to teach the inhabitants of the banks of the Ohio, the art of compelling their huge muscles to produce pearls.[11] The fishmongers and musclemongers of Cincinnati and elsewhere thought that I was in earnest and hoaxed themselves completely, turning in ridicule my proposals &c. The fools are such ignorant folks, that they do not even know that this process is as well known in Europe and Asia, as the art of making and mending shoes; but seldom produces enough to feed the cob[b]ler. Yet Linn[a]eus was made a knight of the Polar Star by the king of Sweden for having revived the process. I do not expect such honor here, as we have no knights except Templars and Quixotes; but we have a crowd of Squires and Sanchos, some of which would be more unfit to govern Barataria than the Spanish clowns. I hope however these Cincinnati and Ohio Sanchos, will elect me President of their learned club, in order that I may give them a lecture on the arts of mending shoes, fishing pearls, carving cameos, baiting and frying fish, getting patents for new frying pans, inclined planes, water wheels, museums, and so forth. Yet I am afraid that they would not believe me now, even if I was to tell them that Cincinnati is a fine city in Ohio; they might probably say that it is not fine, but superfine, that it is a town and not a city, nor is it in Ohio, but in the state of Ohio or on the banks of the river Ohio! Oh Wiseacres and silly Sanchos! You are welcome to have your way, bad as it is.

C. S. Rafinesque

Printed in *Western Minerva* (Lexington, 1821), pp. 72–76; published 1949.

1. Jean Baptiste Georges Marie Bory de Saint-Vincent (1778–1846), an army officer much of his life, saw action in the battle of Austerlitz. Because of his dislike of the Bourbons he had to live in disguise from 1816 to 1820. When not on active duty he pursued natural history; rejecting the idea of the fixity of species (like Rafinesque), he believed that species change under the influence of the environment. From 1822 to 1831 he edited the *Dictionnaire classique de l'histoire naturelle* in Paris. He was also one of the editors of the *Annales Générales des Sciences Physiques*, printed in Brussels, where several of Rafinesque's articles appeared.

2. *Voyage dans les quatre principales îles des mers d'Afrique, fait par ordre du gouvernement, pendant les années neuf et dix de la République* [1801 & 1802], *avec l'histoire de la traversée du capitaine Baudin jusqu'au Port-Louis de l'île Maurice* (3 vols.; Paris, 1804).

3. Of these 28 essays, only those numbered 2, 3, 4, 5, 6, 7, 8, 9, 11, 12, 13, 14, and 16 can be identified in the pages of Bory's periodical, the *Annales Générales des Sciences Physiques*. Those numbered 10 and 15 may have contributed, under different titles, to other Rafinesque essays published there but not named here.

4. Rafinesque's best friend. They first met in 1802, when Clifford visited Livorno as supercargo on his family's ship the *Philadelphia*. About the year 1808, he established a branch of the Clifford Brothers firm in Lexington, where he married. He later became an independent merchant there, where he helped to build the Episcopal church, served as a trustee of Transylvania University, and established an athenaeum and museum. He stocked the museum with collections resulting from his interests in geology, natural history, and Indian relics. He was instrumental in getting Rafinesque's appointment at Transylvania. He died suddenly, according to Rafinesque of a "fit of gout in the stomach, which proved fatal in a few days."

5. Benjamin Birge (b. 1797), nephew of the Rev. John Ward, first rector of Lexington's Episcopal church. Ward was married to John D. Clifford's daughter. When he says that "we have lost this year" Clifford (b. 1779) and Birge, Rafinesque means during the twelvemonth, for Birge died March 29, 1820, and Clifford died May 8, 1820.

6. The Western Museum Society, established under the leadership of Daniel Drake.

7. Daniel Drake, *Natural and Statistical View, or Picture of Cincinnati and the Miami Country, Illustrated by Maps, with an Appendix Containing Observations on the Late Earth Quakes, the Aurora Borealis and Southwest Winds* (Cincinnati, 1815).

8. When Rafinesque wrote, Atwater had published his "Description of the Antiquities Discovered in the State of Ohio and other Western States," *Archaeologia Americana: Transactions and Collections of the American Antiquarian Society*, 1 (1820), 105–267. His *History of the State of Ohio* was published in Cincinnati in 1838.

9. A series of ten articles had appeared in a Lexington newspaper, the *Kentucky Reporter*, sarcastically critical of the *Western Review* magazine and its contributors, most of whom were connected with Transylvania University. The critics, who probably were students, flaunted their classical learning by printing their essays under the title "Peritimatist," meaning an appraisal of everything. The essays remained anonymous but were signed Zoilus & Co. in allusion to the 3rd century B.C. grammarian who bitterly criticized Homer. These articles were answered in kind by a series of equally pedantic essays under the title "Antenclematist" (counter-accusation) and signed Dikaiosune (Righteousness), to which Rafinesque probably contributed. When Zoilus & Co. finally attacked him by name—calling him, among other snide epithets, "the stone on which our wits were sharpened"—Rafinesque struck back over his own signature, challenging them to translate TON NI-ON ES-RHAM KIN-BAW STE-SO-MEN VAN LIS-IB LIS-MAR PHOER-APH ES-THAM KIN-MOTH. When they failed to do so, he supplied this translation (*Kentucky Reporter*, 9 August 1820): *When fools become similar to barking dogs and set themselves against their wise masters, they must be beaten and even killed, least* [*sic*] *they should next become mad-dogs and bite.* The name of the language, he said, could be ascertained by consulting Barthélemy d'Herbelot de Molainville, *Bibliothèque orientale; ou, Dictionnaire universel* or Antoine Court de Gebelin, *Monde primitif.* Because a diligent search of both books has failed to reveal the language involved, there may be even more of a hoax here than that claimed by Rafinesque.

10. Such attacks in the Cincinnati press prior to the date of this letter have not been found, but there is no reason to doubt they occurred. Shortly afterward, on 23 June 1821, an amusing versified attack was published in Cincinnati's *Western Spy & Literary Cadet* by "Horace in Cincinnati" (Thomas Peirce). An ode addressed to "Professor Muscleshellorum, of Transylvania University," it had this among its nine stanzas:

> 'Tis not my present business to bemoan
> The loss to letters and botanic science,
> When all your papers on the waves were thrown,
> Of countless "new discoveries" in defiance;
> But still, I'm glad your memory has assisted
> To name so many plants which ne'er existed.

Rafinesque also was the butt of many a joke in the *Cincinnati Literary Gazette* as late as 1824.

11. Under the title "Three Notices of Natural History," Rafinesque published in Lexington's *Kentucky Reporter* (6 September 1820) a note on the "American jaguar," another on his discovery near Lexington of *Scutellaria lateriflora*, and the third his offer, for a share of the profits, to communicate the secret of causing Ohio River mussels to form pearls. The note on *Scutellaria lateriflora*, or skullcap, was widely reprinted in the region since this herb was believed to cure hydrophobia.

* * *

Fragments of Letters from Lexington. by a Lady.

Western Athens—Society—Parties—Balls—Belles and Beaux.

I have been greatly astonished by the number of periodical Journals published in this town.—This must be ascribed to the fact of being the seat of a University.—There is a magical spell in this name, which calls forth into existence literary attempts.—And although a want of perseverance or other casualties often blast them at an early period; yet there is something cheering in the sight of these youthful children of fancy.—I cannot forbear to hope that some of them will escape premature death and reach at last a vigorous maturity.

The surname of *Athens of the West* has already been given to this town, and methinks on very plausible and reasonable grounds. There is certainly not a single town west of the mountains that can rival with it on that score. Pittsburgh, Cincinnati and Louisville are mere commercial towns, and all the attempts to establish there permanent seats of learning have failed.—Let Pittsburgh become the Manchester of the Western States, Cincinnati their Liverpool and Louisville their Bristol; but Lexington must be their Edinburgh.

Lexington contains less of 6000 inhabitants and supports four newspapers and four literary Journals.[1]—It occurs to me that if Philadelphia, New-York or Boston, which contend for the title of American Athens wanted to deserve it, they ought to support a proportionate number of Journals, according to their respective population.—You know that in this age of periodical criticism, this ought to be considered as a fair argument.—For instance Philadelphia which reckons over 100,000 inhabitants ought to muster by an easy rule of three, about 66 newspapers and as many literary Journals, while only 20 newspapers are published there and about 10 literary Journals, if I remember right. The conclusion would be obvious and the proud Philadelphia ought to give up the palm of periodical literature to Lexington. The same remark applies to Boston and New-York.

There are here besides, as many or more learned and cultivated minds (in proportion always) than in any atlantic cities. Even our sex vies with men in acquirements—Your ideas of Kentucky are yet connected with deep woods, barrens and buffaloe hunters.—But how mistaken!—The scene is totally changed, and in this town at least nothing is seen but elegance, refinement and accomplishments.—I mean among the well bred and educated classes, although there are yet occasionally faint glimpses of roughness and clownishness even among them, but these exceptions are few and are far from being approved of.

The Society is excellent, we enjoy a delightful social intercourse, and some puritans might deem it inclining towards dissipation. We have often large Parties and Balls.—It is customary here to give them on several occasions, to a bride, to a stranger, on changing lodgings, on our national days, in return for such favors.—These parties are commonly very agreeable; but they are not deprived of a degree of formality and monotony, which makes them become tiresome after awhile.—I will attempt to describe you one of them, and as they are all alike, the only variety consisting in the number of faces rather than their diversity; since, by the bye, we almost always meet the same persons at the whole of them, it will convey a tolerable idea of the whole.

This party was given by Mrs....[2]—It is always the lady who invites and of course presides.—A crowd of one hundred human faces or more were collected, the sexes being nearly equally divided.—The ladies were ushered in the rooms by a couple of gentlemen ushers, and comfortably seated along the walls, to be gazed at by the flock of gentlemen standing in the middle of the rooms, and to enjoy the conversation of their neighbors, and such gentlemen as condescended to pay their respects or are introduced to them.—Meantime the waiters brought the Coffee (although these parties are called Tea-parties, Tea is very seldom offered) with the usual appendages, cakes and relishes. These poor waiters found it rather difficult to go through the crowd.—After the coffee came the sweetmeats, next the wines and liquors, after which followed the des[s]ert, consisting in fresh and dry fruits.—Will you believe me when I tell you that we have had sometimes even Dates brought from Mogodor all the way to Lexington; but Oranges are great rarities.—Thus there is no lack of good things, and those who are fond of light food, may even cram themselves at pleasure.—After the meal (it is indeed a better one than many heavy Dinners) we had some musick. Some young ladies played on the piano and Mrs.... sung beautifully. Those who were fond of harmony crowded round the singers, and afterwards the company divided itself in small groups or *Coteries* to enjoy a kind of private conversation.—But no one can shine to advantage in these circles, since it is ten chances to one that the best *bon-mot* will only be heard by a couple of neighbours; those however who can say nothing but stupid and dull things, enjoy the advantage to have few witnesses.—The whole pleasure, if pleasure it is, lasted only three hours, beginning late and ending early, towards 10 o'clock.—We retired after bowing to the good lady who had the high gratification to entertain us, at the expense of her pantry and....

There are some other lesser kinds of parties, which are less formal and more animated, they are called private parties, happen with or without invitation, and are principally intended for the gratification of young folks, who amuse themselves innocently in twenty different ways.—Card parties are not fashionable here.

The Balls would be well enough, if the fiddlers knew how to accord, and if some variety was introduced among the dances.—These eternal *Cotillions* have the sway here as with you, nothing else is taught or heard of here, and they make the Balls too dull by far for me.—It has often occurred to me to ask myself why the pretty English country dances are gone out of fashion? and why the lively German, Polish, Russian, Spanish and Italian dances are not yet come in fashion?—And I can only find the solution of this fact, by considering that all our dancing masters—I beg their pardon, *Professors of Dancing*—are petty French teachers, such as might perhaps teach in their country, the village clowns; but have perhaps never ventured in an elegant Ball or polite party.

The other amusements of this town are occasional Concerts, Performances of Conjurors, and sometimes the Theatre.—Yes, there is a Theatre here, but it opens only for a few weeks or months in the year, and is never well attended.—Many Presbyterians and Baptists scruple to go there.—The Concerts are generally pitiful both for singing and performing; but they please those who have seen no better.

You have asked me to give you my opinion of the fashionable Belles and Beaux of this town.—I might have a great deal to say about them; but I must now speak of them in general terms.—I may perhaps descend to particulars in a future letter.—[3] Upon the whole the Belles of Lexington (and in Kentucky in general) are pretty and amiable; but deficient in grace and blandishments.—None can claim to possess a high share of personal beauty; but on the other part none can be called homely and very few are plain. Although they do not always acquire all the most needful accomplishments, yet many are found without deficiency on that score.—But what must be stated, and is greatly to their honor, consists in their minds being often adorned with secret beauties, not easily detected at the first glance; but which reveal themselves on a more intimate acquaintance. I do not mean to say that they are all such. There are even a few coquettes among them; and some fewer, who have the misfortune of a shallow mind.—But they are few, and exceptions rather than fair examples.—Many are modest, unassuming, and withall gifted with the quickest perceptions and happiest conceptions.—I know several who are scholars, have studied several ancient and modern languages, acquired some knowledge of history, literature and sciences, can perform on the piano, sing, draw, &c.[,] are fond of reading, have a good memory, and possess besides all the most enviable and peculiar qualifications of our sex, crowning the whole with a gentle temper and the best feelings.

It is natural that in a town, where so many young men resort for classical instruction, the young ladies should strive to rival them in acquirements. This emulation is very useful to both sexes. We know well that men do not like to have wives, wiser or more enlightened than themselves; whence elsewhere our sex is compelled to keep rather beneath than above men.—But here where no such danger exists and where learning begins to be fashionable, the ladies are not afraid to overstep or overreach their future partners: whence results the happy and mutual inclination of improving the mind.

Figure 9. Rafinesque's sketch of a Kentucky Belle, from his *Kentucky Friends* (1936).

The young Beaux are with us pretty much as elsewhere.—We have here fops as usual to match our coquettes, and even a couple of real fools; but we have also some very clever and amiable young men, able to converse upon almost every subject and who would be deemed ornaments of society, any where.—I am however sorry to perceive that they do not constitute the majority: real worth appears to be in the minority as yet among the male sex, while in ours it forms already the majority. We meet in company with some very insignificant beings, and although they are somewhat reserved in our presence, I understand that they are often very rude and uncivil out of the immediate sight of their Belles. Some of them indulge in dissipation and those dirty practices of chewing Tobacco, &c. I wish that a general conspiracy against this vile and baneful habit, could be formed by our sex, and I should have no objection to be at the head of it.

I cannot answer now your enquiries respecting Transylvania University. It is thus that they call the University of this town.—A very awkward name I always thought.—It means beyond the woods. A name meaning in the middle of the woods would have been better, or *Transmontana* which means beyond the mountains.—But I perceive that I become a learned critic; this lot must be left to the great Critics and Reviewers of the age, who have assumed the sway of the modern Republic of Letters.—I shall however write you soon, another letter on the Literary Institutions and Personages of this *Western Athens*, but I hope that you will not divulge my letter, or at least what may be rather too pointed.—I do not want to quarrel with our great men and little folks.

LAVINIA.

Lexington,———1820.

Printed in *Western Minerva* (Lexington, 1821), pp. 76–79; published 1949.

1. Elsewhere in the same publication (p. 49) Rafinesque listed these periodicals as of December 1820 in the following paragraph:

Periodical Works published in Lexington. 3 weekly gazettes, the *Reporter*, the *Kentucky Gazette*, and the *Monitor*: one semi-weekly, the *Lexington Advertiser*. In the beginning of the year four literary journals were published, two weekly, the *Castigator* and the *Censor*, one semi-monthly, the *Journal of Belles Lettres*, and the *Western Review* monthly: these have been suspended by the death or departure of the Editors, except the last.

2. Here and later the ellipsis sign is Rafinesque's.

3. Though not intended for publication, a future letter of Rafinesque's did in fact discuss "fashionable belles and beaux." Written to Dr. E. L. Briggs, the hitherto unpublished letter belongs to the College of Physicians of Philadelphia with whose permission it is now reproduced as a diplomatic transcript.

[Letter from C.S. Rafinesque to Dr. E. L. Briggs[1]]

Lexington Kent^{y}. 7th. Octr. 1823—

My good friend—

I have not forgotten ~~your request~~ <the request that you made> before your departure, nor the promise that I made in consequence to write to you: if I have not sooner done it, my rambles and yours during this Summer may partly account for my delay. I was told of your actual residence some time ago, but informed that you contemplate returning soon to Natchez. I hope that your actual good prospects and the pestilence of Natchez will induce you to settle in Missouri, but wherever you may go my best wishes will attend you.—I have been rambling a good deal myself this year, altho' not gone out of Kentucky, my principal journey in May and June was a tour of 500 miles in the Barrens and as far as the Cumberland and Tennessee rivers, another in Septr. of 300 miles in East Kentucky and to the falls of the River Cumberland. These journeys have as usual been very useful to me, I have increased my knowledge and my Collections, made many physical and historical discoveries, relieved and improved my mind, strengthened my health and fortitude &c. I went in a carriage in company with Mr. Ficklin[2] as far as Bowlinggreen, in another with Mr. Morehead[3] to Russel[l]ville &c, the remainder was mostly performed on my legs, which are yet pretty active. I saw many acquaintances in the Green river Country and made many more. I was treated very friendly by the Campbells of Hopkinsville and introduced to all the pretty girls of that town, some of whom were very agreeable, but I did not see any equal to my select Lexington favorites! altho' they were superior to some of our reputed belles.

I saw Dr. Hardin in Russel[l]ville, he is gone into partnership with Dr Heard,[4] and will I hope do well in time, there are many reports afloat about him, some say that he is courting his cousin Miss Payne, and is shortly to be married to her; but others expect him in Lexington this month, where *a lady* will again have a chance of saying yes or no!

All your family are happy and well: your mother has a house full of boarders, about a dozen, Mrs.& Miss Hunt, two Mrs.Warham and a young Miss W.[,] Dr.

Drake, his wife and two little girls, besides Mr. Martin and the two brothers Dunn.[5]—These two last are the only ones that I am sorry to see there, they are hardly reputable boarders, they went there after being sent away from the College building by Mr. Holley, for improper and immoral conduct! let this be between us, I hope they may not stay long.

Our friend Dr. Henry Miller[6] was also there lately, but is now gone home after resigning his appointment of Demonstrator of Anatomy. Dr Dudley did not like such an adjunct and he has contrived to disgust him beforehand I suppose. I know the whole story, but it is too long to write. I am sorry for this on many accounts, and many of his friends must feel as I do, altho' it is likely that he will do very well again in Glasgow.

Your sisters pretend to be indifferent about Drs. M. and H, but I reckon that they cannot be totally so, you know how long I have preached <to them> sincerity in vain! Your sister Margaret is as pretty, lively and sensible as ever and really improves; but your sister Juliet[7] altho' ~~as~~ <still> lovely and fascinating ~~as ever~~, will not yet take the advice of her best friends, she will not be sincere, and must flirt when opportunities offer even with those that she cannot esteem, I could say a great deal about her, but must forbear, she would not like that I should tell you all: I once despaired of her reform, and she was angry with me because I told her my mind boldly and bluntly; but however she has lately shown a small appearance of amendment, and even once said that she ~~would return~~ <was returning> to the ~~path of …~~[8] <good path>. I wish it, but do not believe her yet, she has been too long used to deception and abuse of power.

I have one hundred news to tell you, but have hardly room, many you may see in our papers, such as Marriages and deaths; but I will add such details as do not meet the public eyes.—

There have been 20 marriages in town or neighbd since you left us among which the following may interest you.—

Miss Eliza Warfield Coleman to Mr. Irvine of Richmond, they will live on a farm in Madison County.

Miss Emily Satherwhite to Mr. Whitney, who is studying Medic.

Miss Amanda Leavy[9] to Ch. Morehead, they ~~are~~ <went> to live in Russel[l]ville, but she disliked the place and *is sorry* to have married, altho' she has such a good and clever husband, if ~~he~~ <you> knew her as I do, you would not wonder at this, she is now here and Morehead is in Russel[l]ville attending the courts, he will come for her in 3 months!

Miss Henrietta Hunt and Miss America Higgins have married two twin brothers[,] Mess. Morgan of Huntsville Alabama, who are neither handsome, nor intelligent nor rich, *mere country bucks* who came here fortune hunting with false titles of Colonel and Major and puffing themselves, they were thought rich, but there has been a sad reverse of the medal. *There is no knowing what women will do* after this and so forth.—

The other marriages to take place shortly or talked of are Miss Catherine Dumesnil[10] with Mr. M'Ilvaine—a good match[.] Miss Rebecca Warfield with Charlton Hunt—she has some objections! Miss Emily New with Mr. … [,] Clark of the Court at Elkton.—

Then you see that the chances are daily lessening for the candidates for the smiles of beauty, among which I still reckon myself, altho' I am no longer a candidate for matrimony, I am too unlucky, whenever I begin to feel something more than friendship for a lady, she turns out to be unworthy of it, besides you know that I have no heart to spare? ~~As~~ When ~~as~~ *an event* will take place, I shall think seriously of leaving Lexington, as soon as I can make a good bargain elsewhere, I have some prospects in Baltimore, Virginia, Carolina & Huntsville, but have neglected them till now: it is time to think of settling in the lap of Science, since beauty's lap is too dear to buy.

This scrawl hardly deserves an answer, and I do not mean to exact one; but when you write to your family, put in a few lines for me, and let me know how does the world fare with you in the distant region of the West; meantime I send you my warmest wishes for y^r^. health and wealth, besides love if so inclined. May the choicest blessings of heaven be yours, these are the sentiments of

Your sincere friend

C. S. Rafinesque

PS. I have visited twice the Ayres family[11] in Danville since their removal there, the girls regret Lexington very much. Mrs. Widow Gatewood is returned here from thence, she is no longer in partnership with the half crazy Widow. Miss Sophia Mellen is still here with her mother, Mr. Prentiss is in no hurry to take them away—All are well—Mr. Bainroth is now here selling some goods, your sisters are rather cool towards him I think.—

---

1. All that is known with certainty about E[dmund] L[loyd] Briggs is that the 1818 Lexington city directory listed him as a physician. Then, after taking three courses of lectures instead of the usual two, he was awarded the M.D. degree in 1823 by Transylvania University for which he submitted a twenty-five page manuscript "Inaugural Thesis Designed to Disprove the Doctrine of Direct Debility." The thesis is extant. All the rest is inferential. No record of him has been found in Natchez newspapers or cemetery records, nor does his name appear in the index to the 1830 federal census for Missouri. Rafinesque's letter indicates that Briggs had moved "west" and implies that he was related to others of that name who lived in the vicinity of Bowling Green. No record of him has been found in the extant newspapers of western Kentucky.
2. Joseph Ficklin, Lexington's postmaster.
3. Charles Slaughter Morehead (1802–1868), Kentucky governor, later U.S. Congressman. He took his bachelor's degree in 1820 from Transylvania and his law degree in 1822; Rafinesque said that he was once Morehead's teacher.
4. Starting in May 1823, the Russellville newspaper carried a notice in its weekly issues that Doctors D. M. Heard and R. Harding were conducting a joint practice. During this period, Russellville, a town of only 1,500, had at least eight physicians
5. Alexander Montgomery Dunn and John Adamson Dunn, from Amity County, Mississippi, both took bachelor's degrees at Transylvania in 1825.
6. Henry M. Miller (1800–1874) was awarded the Transylvania M.D. in 1822. After leaving Transylvania, he was in private practice for a time in Glasgow, Kentucky, later in joint practice with Dr. Christopher C. Graham in Harrodsburg. In 1837 he became professor of obstetrics in Louisville, where he pioneered in the use of anesthesia during childbirth. He was president of the American Medical Association in 1859.
7. Among the portrait sketches drawn by Rafinesque is one titled "Juliet." It depicts a young woman in left profile; arching over her is an emblematic weeping willow and behind her a bitter-sweet plant, while below the sketch appear these lines:

   I knew her in the prime of her beauty and youth:
   When she was the chaste emblem of candor and truth.
   But alas! what a change!....

   The drawing is reproduced in *Rafinesque's Kentucky Friends*, ed. by Harry B. Weiss (Highland Park, 1936) and appears also on p. 335 of *Profiles of Rafinesque* (Knoxville, 2003).
8. The ellipsis sign here, and later on, is Rafinesque's own.

9. Daughter of the Lexington merchant William A. Leavy, Amanda Leavy was married to Charles Slaughter Morehead on 10 July 1823. After her death Morehead married her sister, Margaret Leavy, in 1831.

10. Catherine Dumesnil probably was the daughter of one of Lexington's numerous French residents, the watchmaker and jeweller Anthony Dumesnil.

11. One of the sketches in *Rafinesque's Kentucky Friends* is of Samuel Ayres, said to be Kentucky's first silversmith. Beneath the picture Rafinesque penned:

A worthy man, religious good & kind
His equals or as good, we seldom find.

Samuel Ayres was noted in Lexington for his many charitable endeavors and had been in business there since at least 1787. Late in life he removed with his family to Danville, where he died 20 September 1824.

8

Utopian Society

Like other social theorists of his time, Rafinesque believed that most human misery results from deficiencies in society. In search of a community that would promote happiness, he queried his sister in France about the theories of the socialist Claude Saint-Simon. Writing to the Rappites in Pennsylvania, he admitted that "since my younger days I have dreamt of a plan & society like your own, the crimes, vices and misfortunes that I have seen arising every where from personal exclusive property." When the Rappites moved to Harmonie on the Wabash, he visited them there. After they sold out to Robert Owen, who renamed the settlement New Harmony, Rafinesque considered living there. He was fond of both Moravians and Shakers, and enjoyed participating in their way of life on occasional visits.

Yet, in every instance there were restrictions—either of thought or action—that he was unwilling to accept. Meanwhile, having developed his own "new" economic system and even patenting it as the Divitial Invention, he offered it to William Maclure as a means toward perfecting what Robert Owen had begun. His description, which follows, was published in the *New Harmony Gazette*, but as far as we know nothing more came of it.

Having also written extensively on subjects as diverse as horticulture and animal husbandry, colleges and education, it was reasonable for Rafinesque finally to devise his own ideal society. The providential appearance in his life of the wealthy Charles Wetherill made this seem possible. Their joint plans for a Midwest colony foundered as a result of the Panic of 1837 and Wetherill's death, but the scheme itself remains of interest.

At the conclusion of his autobiography, *A Life of Travels*, Rafinesque wrote that he wished he had space to describe his "beneficial plan" whereby he "could make 1,000 families happy and rich, on 10,000 acres of land, by useful plans of peculiar economy and partitions." He attempted to do so in the second essay, a pamphlet, which follows. He wrote elsewhere that in 1837 he had issued five numbers of *New Aurora*. Only a single copy is known today. It is the one reprinted here, and it is labeled "Second Edition, 1st June 1837."

In the year of his death Rafinesque took a final stab at promoting Utopia, in another publication called *The Pleasures and Duties of Wealth*, "Printed for the Eleutherium of Knowledge"—which probably means that whatever remained of Wetherill's investment in Aurora paid the printing cost. With no editor to restrain his exuberant use of capital letters and italic fonts, the appearance of his prose may reflect the fanatic he likely had become on the subject. A few paragraphs from this 32-page pamphlet are included to complete the story.—*Editor*

* * *

OUTLINES
Of a Plan for Cooperative Associations and Mutual Societies.
ADDRESSED TO WM. MACLURE.

The zeal of many philanthropists, and the speedy diffusion of knowledge upon the subject of human happiness, are exerting their influence on the minds of a multitude of individuals, scattered all over the United States, and preparing them for a change, from a state of selfishness and private warfare, to the desirable existence of friendly cooperation and reciprocal labors.

Mr. Robert Owen[1] has had the influence (if not the glory) of giving the impulse, to many minds partly prepared for such a change—which has led to the establishment of many friendly associations, and cooperative societies in several states of the Union. Although the constitutions of those societies differ widely in detail; yet they are all based upon the benevolent plan of inducing individuals and families to congregate for the purpose of mutual cooperation, and exerting their mental or muscular faculties for each other's good and happiness. This appears to be the main principle or link of all the communities and societies already established, or now forming, for practicing the social or mutual system of co-operation.—The metaphysical dogma of fatality, or the abstract notion of absolute equality, or the gradual abolition of slavery, which has in some instances been annexed thereto, are by no means essential to mutual operations, since men can cooperate, whatever be their religious, moral, and metaphysical ideas.

But there are every where many individuals for whom the transition from the actual social intercourse and dealings, to a better state of amicable cooperation, appears too sudden and too great; because their minds are not prepared for it, or because their interests may be injured. They must be instructed and conciliated; since it can easily be shewn that the mutual system is not at variance with their rights nor interest, when applied to the more special and essential purpose of producing greater happiness and cooperation.

The legal, political, and financial difficulties and restraints that the laws, manners, and opinions have offered, or will soon offer, against this benevolent scheme, can only be obviated by organizing the mutual societies in accordance with our laws; and by conforming to the actual feelings of virtuous men:—wherefore, men of legal attainments and benevolent minds ought to be consulted, and nothing done to disgust or irritate our fellow-beings, and the authorities in power, when they do not exceed their duty or right.

If it were not for those difficulties, and want of knowledge, many other societies would have been established, and the actual ones would have less to contend with. It is beyond my purpose to enumerate these checks; they are more easily felt than described; but without entering into metaphysical or legal discussions, it is merely my wish to suggest and inculcate some practical principles or operations that may obviate many objections; and thereby enable persons so disposed to multiply these benevolent institutions, or prepare the members for a better change still, and such future improvements in the social system as knowledge, time, and circumstances will allow or produce.

My plan is not perfect, nor such as could be practiced if unclogged by bad laws and habits; but it is such as our actual laws or habits will admit of any where, in a large city as well as a sequestered spot of land, and will not hurt the prejudices, the notions or intercourse of any sect, community, or town. It is not the ultimate plan of equal communities, but a plan for auxiliary societies, to help them or to form centers of cooperation for actual and future associations. This plan can be modified in any way to suit circumstances and events; while others cannot: it will leave to every one a home and the fruit of his exertions; while it will remove all the evils for which the social system is a remedy. It is in fact a plan of preparation and amalgamation, such as Mr. Owen might have proposed or adopted; but he preferred separating himself altogether from general society; which transition was too great, and may be resisted by many who have too much at stake there. I shall now, without further preamble, offer the outlines of my plan, and conclude by stating its advantages.

First Part of the Plan.
Principles to be adopted by the societies.

1. All the principles of Mr. Owen, and Mr. Gray,[2] except the metaphysical dogmas and tenets.

2. An increase of human happiness by mutual cooperation every where.

3. Labors, both of the mind and body, exerted at pleasure, ought to provide for all the wants and comforts of life, as well as property, at all times and places.

4. Our comforts are to be proportionate and commensurate with our exertions.

5. Whatever has an exchangeable value, such as property, rents, wages, labor, &c. may be used for mutual exchanges and cooperation at all times.

6. Money is no longer to be a medium of exchange in the society; but stocks, or the representation of the above exchangeable values, rendered divisible at pleasure, and *ad libitum*, according to the principle of my *Patent Divitial Invention.*

7. The members of the mutual societies will pledge themselves to use no other medium among themselves, and thus be enabled to create a medium commensurate with their labour to supply all their wants.—Money will only be needed with strangers.

Second Part of the Plan.

Practical operations of the mutual associations.

8. Any number of individuals or families, from 5 to 5000, may form themselves into associations any where, for the purpose of mutual cooperation and interchanging among themselves their property or labor, whenever wanted; adopting the name of Mutual Institutions or Associations, with a local appellation.

9. They will select three, or five, or seven, worthy members, among those who shall put most property or labor into the stock or mutual exchangeable capital, as trustees or managers of the deposited stock or sums invested, and responsible agents of the association.

10. These managers shall provide rooms and markets for the deposited stock, and transact the business of exchanges. Also stores and factories, land and gardens, carts, wagons, &c.

11. Every kind of property will be received in deposite; moveable property, such as tools and materials of arts and sciences, merchandize, manufactures, &c. will be deposited in the general store of the institution: while land or houses will be deposited by assignment or transfer of titles, so as to be available by other members; and the whole property valued.

12. The product of material labor will also be received in store and in kind, whatever be their nature; while mental or permanent labors or services done, will be estimated according to their nature, purpose and utility to the association. Nothing will be refused however small or large.

13. Money, stocks, rents, wages, and every other kind of income shall also be received, in stock or deposite at the value or nominal import.

14. Appraisers will be appointed to decide on the exchangeable value of every thing.

15. As soon as any value is deposited, the managers shall give to the depositor, not by name, but as a bearer, or to this order, a certificate or certificates of the same, upon the principle of the *Patent Divitial Invention*, divisible into any required amount, and exchangeable into any other required amount, transferable and available by the bearer, for their nominal value in dollars and cents. These tickets shall be signed by all the managers.

16. When money, stocks, rentable property, or any other profitable value is deposited, the certificates thereof will bear an interest of 4 or 5 per cent, payable quarterly to the bearer.

17. Any owner or bearer of these exchangeable certificates, will be entitled, at any time, to exchange them at the stores and markets of the institution for any commodity he may want, and they will be received in payment of rents, wages, or any other articles, or services, or materials.

18. Lands, tenements, &c. held in deposite by the institution, will be let to members or any one else, and the deposite notes received in payment of rents or shares.

19. Money and stocks deposited will be lent to members who have dealings with strangers, at the interest of 6 per cent. upon adequate security. The money not lent will be employed in purchasing articles of assortment for the general store.

20. If any articles are not exchanged after several years, (or perishable articles

in less,) they may be sent to auction for disposal. If any loss accrues, it will be made up by the profit fund of overplus, interest, and commissions.

21. Every thing will be exchanged at Cost; but a Commission of from 2 to 5 per cent. shall be deducted on every deposite of property or labor, to meet expenses and form a fund to meet losses, pay insurance against fire and other contingencies, or useful institutions of the Society.

22. Every year the profits and losses will be balanced; and the balance of profit will be used as donations to the sick, infirm, and other members incapable of labor.

23. Instruction in arts and sciences, amusements of a rational nature, and other recreation, will be provided for the members, who will pay for it in deposite tickets.

24. A friendly and social intercourse, conferences, discussions, and calls for peculiar labors, will be encouraged among the members, who will consider themselves as a great family. Schools, Lectures, and Moral meetings will be provided.

25. All the books of the members, will be deposited in a public library, accessible to all, and even lent out except valuable and rare books, and their value given to each member: duplicates will be sold.

Advantages of This Plan.

1. Every man, woman or child will be allowed to exert their talents, industry, or skill in any way, and receive at once an adequate compensation for the same in a medium of exchange.

2. There will be no longer any poor, since any one being able to obtain materials to work with at the common store, will do some work at home and provide for themselves.

3. The rich will find a good and benevolent investment of their property, whereby they can also obtain whatever they may want for comfort: and thus be reconciled to co-operation.

4. Money will be rendered partly useless; and the deposite tokens of the Institution will soon become current even beyond the society, by their exchangeable value.

5. The association will not be deemed a partnership at law, the members being nameless, and each working for himself, and interchanging their labors or the value thereof.

6. No great capital is wanted, hands and tools are sufficient. It will not be needful to spend much money in buildings, nor to go to a vast expense at a great distance, in order to enter or form a Community. Freedom of action will be preserved, &c.

7. Members will be allowed to have a home, or may unite several families under one roof: or form large boarding houses and hotels for cheaper living; or build new villages as they please.

8. The vicious, intemperate, or dissolute will have no claim nor influence on the society (and as soon as known as such may be expelled) while their labors alone will entitle them to exchanges.

Such are the outlines of the great plan which I propose to the serious consideration of all the friends of cooperative industry. If any one can devise a better one, or improve upon this, he will receive my thanks and support. Altho' the main spring of this scheme is my *Divitial Invention*, which I have patented, in order to give to it a higher legal claim, it is my intention to allow these societies to use it at such a trifling rate, as benevolent institutions, that I hope no selfish views will be ascribed to me on that score.

C. S. RAFINESQUE, *Ph.D.*
Patentee of the Divitial Invention.

LEXINGTON, (Ky.) 18th April, 1826.

Published in the first volume of *The New-Harmony Gazette* on 17 May 1826.

1. Robert Owen (1771–1858), social reformer who, after amassing a fortune in an English cotton mill, established a communitarian society at New Harmony, Indiana (1824–1829).

2. John Gray (1799–1883), British Owenite theoretician whose pamphlet *A Lecture on Human Happiness* (1825) was reprinted in Philadelphia in 1826, where it was enlarged by the addition of the "Preamble and Constitution of the Friendly Association for Mutual Interests, Located at Valley Forge"—a group Rafinesque hoped to join when he left Lexington.

* * *

PLAN
Of the Philadelphia Land Company of Aurora, Intending to build a City, and four Towns on 66,640 Acres of Western Land. Capital $96,000 in 1600 Shares of $60. Each entitling to a Farm of 40 Acres, a City Lot of one-fourth Acre, and a Town Lot of one Acre.

Wealth and Happiness.

Several patriotic and benevolent citizens of Philadelphia and elsewhere, wishing to provide wealth and happiness for their families and children, by becoming the founders of an Inland Colony in the Western lands of the United States North of the River Ohio; but having in view at the same time to proceed on the steps of Roger Williams and William Penn, upon the most liberal principles, they have formed themselves into a Company for this purpose, under the above name, and invite every one to join them in this patriotic and profitable undertaking.

The advantages they offer are entirely novel and very liberal, since any one owning a Share of **$60** in this Company is entitled to a farm of 40 acres, a City Lot, and a town lot at cost ... which shall be improved for them, or money lent for the purpose.[1]

It is contemplated to purchase at once or gradually as the funds will allow, a tract to be called AURORA, of 100 square miles of good land being 64,000 acres, and to divide them into 1600 farms of 40 acres, besides adding thereto 640 acres for a City to be called AGATHOPOLIS meaning the Good City, with 2000 acres whereon to build 4 towns to be called INDUSTRY, HONESTY, BENEVOLENCE, and TOLERANCE, each upon 500 acres of land.

The capital stock of the company will be of $96,000—whereof $80,000 will be required for the purchase of the land at $1.25 per acre, and $16,000 for the city and towns lots, public improvements, surveys and expenses: the surplus with the premium on the choice of lots, will form the capital to be loaned for improvements.

The City of AGATHOPOLIS, will be divided into 1600 lots of quarters of an acre, leaving 240 acres for the public squares and walks; streets, public buildings and institutions; the college, churches, &c., or any other useful purposes, factories and needful establishments receiving free lots.

Each town shall be divided into 400 lots of one acre, and 100 acres laid out in squares, streets, walks, with schools and public buildings, factories, &c.

Each farmer owning a share and farm of 40 acres, shall have a City lot, and a town lot.—If he owns several shares and farms, he shall have as many city and town lots. The public roads shall be taken out of the front of the farms.

Thus for $60, any man can provide for himself or each child of his, a good farm and two valuable lots that may become worth $1000 within five or ten years.

Many mechanics, laborers and emigrants are often unable to purchase small farms and lots at cost, or must be at great expense to choose their land, which will be obviated by this plan.

To accommodate them still more, they will be allowed to pay the $60 by quarterly instalments of $10, the first in advance, and if in need of help, will be furnished with money for their removal, tools, seeds and store goods for their improvements.

$16,000 were subscribed at the outset by the promoters of this Colony, and they have left $80,000 to be subscribed and paid by any suitable person willing to unite with them, during the months of May and June 1837, and the Books of the company will be left open to all until the first of July.

The choice of lots and farms will be allowed for a small premium, which shall be employed to establish a store, hotel, mills, schools and public buildings, besides being used by the SAVINGS BANK OF AGATHOPOLIS, to loan to improvers and farmers. Those who do not choose to pay any premium, will draw for the other farms and lots, by numbers put in a wheel.

In the month of July 1837, after the complete organization of the Company, one or two Trustees shall be elected and sent, either from Philadelphia, or from New-York, to go and locate the land on a healthy spot, making the preliminary surveys and arrangements.

At a general meeting of the share holders, the regulations of the Company will be revised and confirmed. Meantime the promoters of this patriotic Colony, being anxious to secure the contemplated advantages of *Wealth* and *Happiness*, have agreed that no one shall be allowed to join the Company, unless he approves of the following essential conditions and regulations.

1. To avoid quarrels and trouble, no Drunkards nor Lawyers shall be allowed to join the Company and settle in AURORA. Temperance is recommended to all the members, and the disuse of tobacco. No spirituous liquor, nor tobacco shall be sold at the stores nor hotels, nor in AURORA; but only pure wines, cider and beer, with other healthy liquors.

2. Every dispute, quarrel or lawsuit shall be settled in AURORA by Arbitrators

chosen by the Company, who must decide within one month every difficulty between members, or even strangers if they agree to it.

3. The cultivation of grape vines to make American wines, of mulberries and silkworms for American silks, of bees for wax and honey, and of beets for beet sugar, shall be undertaken as a public concern, encouraged and fostered.

4. Agriculture, pasturage, and the produce of wool, cattle, &c., shall be deemed the main support of the Colony; unless other resources are found in the soil. Manufactures shall afterwards be gradually attempted.

5. The great expense of fences, pailings, walls and ditches to divide the farms shall be avoided; no fencing shall be wanted except around gardens or on the outskirts. The farms shall be divided by single furrows or hedges, or marked trees. Cattle, sheep and hogs shall be kept in pens at night and pastures by day, watched by herdsmen, and paid by all the owners of cattle.

6. In clearing the lands, maple-trees and fine trees shall be spared, the timber shall be cut into lumber, and the refuse made into potash.

7. Whenever an improver or farmer shall want laborers and help, his neighbors shall help him by turns at a fair compensation. All the members shall unite for mutual help. Resident members will be paid by the absent to improve their lots by degrees. No member shall be compelled to become resident.

8. The Savings Bank and Store shall furnish funds and goods; but never on credit, only upon evidence of utility, shares paid for, labor performed and deposites of produce.

9. The circulating medium between the members and settlers shall not be bank notes; but cash or certificates of deposited funds.

10. If squatters are found on the land, they shall be conciliated, paid for improvements and allowed to settle if good people, but expelled if drunkards and quarrelsome, by interdiction of all trade and intercourse rather than force.

11. Stockholders of the Company, may sell, lease, or let any one of their farms or lots to any Emigrant or stranger, provided he is approved of as a settler, otherwise the Company will buy up those farms and lots offered for sale at valuation, and must have the refusal of all bargains.

12. Any share holder impairing the happiness and peace of the community, will also be invited to retire and compelled to sell off and remove.

13. Public free schools shall be established in each town by the Company, and a free College in the City of AGATHOPOLIS.

14. They shall also establish reading rooms, libraries, museums and public lectures on knowledge of all kind.

15. Every religion shall be admitted, every opinion tolerated, provided it leads to no evil deeds. Free Churches and meeting houses may be built, open to all sects and preachers, who must be tolerant liberal men; none other shall be allowed to preach, in order to prevent religious disputes.

16. Public festivals and amusements shall be encouraged; but supported by private subscriptions. Every kind of trade, exertions and industry fostered and perfectly free.

17. Every settler or traveller committing a misdeed shall be expelled, or handed to the Courts if needful.

18. The State laws shall be obeyed; except the militia laws by those who are

Christians or followers of peace, as no one ought to be compelled to fight or learn to fight.

19. The Company and Colony shall be managed at first by a President, Secretary and Board of Trustees.—When the Colony shall be established, a Committee of Trustees and Censors on the spot shall help them.

20. In the affairs of this Company, every members shall have a vote for every share he or his children holds, and for every member of his family settled in the Colony; but no one shall be taxed without his consent.

21. Yet it is agreed that in case of need, a mutual contribution of not exceeding half a Dollar per Annum on each farm and lot may be laid, by a majority of two-thirds of votes, and after ten years or 1847, it may be increased to one Dollar per Annum on each farm and lot. The whole of which is to be used for improvements, roads, canals, mills, education, instruction, &c.

22. Needy settlers will be helpt by the funds of the company, houses built for them, some acres cleared, &c., for which they will pay gradually with interest. When they undertake further improvements, every facility shall be afforded to them. While the State taxes of the absent members shall be paid by the residents, and charged to their account.

23. A free College or rather ELEUTHERIUM OF KNOWLEDGE shall be built in Agathopolis between 1840 and 1850, where all the arts and sciences shall be taught gradually, and all opinions tolerated. It shall be established and supported by allotments of land, (a purchase of 66,640 acres is intitled to a bounty of about 2000 acres education lands) contributions, manual labor and moderate fees. The President of the Company will be the patron of it.

24. If any of the share holders, should neglect his engagements, by not paying his instalments, it is agreed that he will forfeit those he may have paid, and his farm and lot may be sold for the Company; but that industrious settlers may be allowed to pay them in work.

25. If any shares should be unsold on the 1st of July 1837; they shall be held by the Company for sale to Emigrants and settlers at $100 instead of $60, and the advance of $40 added to the fund for improving the land.

26. If any share holder shall do any injury or detrimental act to the Company, he shall be liable to be fined or expelled, and any sale made by him to improper persons shall be null and void.

27. If he shall neglect the regulations of the Company, he may either be fined or compelled to leave his lots and farms for sale on his account by the Company to better settlers.

28. All fines, forfeitures, donations and contributions, advances and profits of the Company, shall be employed to improve the Land, City and Towns, and help the actual settlers.

Therefore this Land Company is a Beneficial Mutual Association, to procure and enjoy wealth and happiness for the contributors or their children, and any one who may join them in time with similar views: besides providing comfort and labor for industry, and funds to co-operate with.

The Books of the Company are to be opened on Monday the first of May 1837, for subscriptions of shares, and are to remain thus freely open to the public and all applicants willing to conform to this plan, for two entire months until the first of July 1837.

The books will be deposited at the office of the DIVITIAL INSTITUTION OF NORTH AMERICA, Nr. 172 Vine-street, in behalf of the Trustees, Agents and Members of the Company, where is to be also the Office of this Land Company.[2]

The Secretary of the Company, C. S. RAFINESQUE, is authorized to receive the subscriptions, with instalments, and to issue the Scrip, entitling to a deed in fee simple of the freehold; but subject to these rules, and forfeitures as will be signed by the Articles.

The Share holders will meet on Monday the 10th of July to organize the Company, revise the regulations and elect the Officers, and select the State where the Land shall be bought, whether Ohio, Indiana, Illinois, Michigan or Wisconsin.[3]

Philadelphia, 15th April 1837.

CHARLES WETHERILL,[4] *President, 21 Arch st.*
C. S. RAFINESQUE, *Secretary, at the Office.*

At a meeting of the Trustees held on the 11th May 1837, in consequence of the general suspension of Specie payments,[5] it was resolved to continue to receive the first instalment in Philadelphia Bank Notes; but that the other 5 Instalments due in October 1837—January, April, July, October 1838, must be paid in Specie or such Currency as will be received in payment of Land by the United States. The premiums will also be received in Philadelphia money, and the unsold Shares may be sold for any good Currency for $100 after the first of July.—By order of the Trustees.

C. S. RAFINESQUE, *Sec'ry.*

---

1. The ellipsis sign appears in the original.
2. By this time, the Divitial Institution was the small workingmen's bank Rafinesque had founded, and the address is that of the rented house which was also his residence.
3. A site in the vicinity of Bloomington, Illinois, was chosen. In a letter of 20th March 1838 to John Torrey, Rafinesque wrote: "I must go this Summer to Illinois to survey & lay out a City with the foundations of 5 Colleges to form a Central University." Six weeks later he remarked in a letter to A. P. de Candolle that he planned to deposit specimens of all his discoveries of new plants in the "Museum of the Central University of North America which is going to be started at Bloomington in Illinois between 1840 & 1850, and of which I am one of the founders." Of course, he did not make the trip, probably because his co-founder died sometime that year, and he himself died two years later.
4. Philadelphian Charles Wetherill (1798–1838) was a wealthy manufacturer of paint. He had retired by the time of his association with Rafinesque, when he probably was suffering from lead poisoning.
5. There could not have been a worse moment than the spring of 1837 to launch the scheme, for this was the beginning of the financial Panic of 1837. It is not known what land, if any, actually was purchased by the Aurora Company, but other Illinois holdings owned by Philadelphians were abandoned when the owners could not even pay their taxes.

* * *

The Pleasures and Duties of Wealth.

1. Ever since *Cupidity* became a human passion, the desire of acquiring *Wealth* has more or less prevailed among all Nations and Individuals; often becoming their banes when the proper uses of *Wealth* have been forgotten.

2. Ever since *Property* was introduced as a reward of *Industry*, the wish to

obtain, increase and secure its accumulation by Individuals, families, corporations, tribes or Nations, has become paramount to all others; and this passion has often been gratified by very improper means, *Injustice*, *Oppression*, *Strife* and even *Bloodshed*...[1] thro' Cunning, Force, War or Bad Laws: whether this desirable property was Land, Chattels, Estates, Money or Services.

3. In the greedy pursuit after *Wealth* and *Property*, the human sinful Vices of *Cupidity* and *Avarice*, have been unfolded and gratified; nay indulged and encouraged by Social Laws, in spite of the precepts of *Morality*, *Honesty* and *Religion*. Instead of *Wealth* being the reward of Industry, Sobriety and Prudence, it is oftener the consequence of Extortion, Monopolies, Speculation, Inheritance or even Despoiling ... disguised under the terms of fair trade, good luck, legal rights, &c.

4. This has been perceived by the best Legislators, Moralists and Founders of Religions: they endeavored to restrain the abuses and evils introduced by *Cupidity*, and the allied passions of *Ambition*, *Avarice*, *Lust*, *Coveting*, &c. ... but often in vain, the sinful propensities of mankind nearly overcoming all the nominal restraints, unless coerced by direct laws, or guided by the holy Spirit of Charity.

* * *

28. The *Apostolic* or *Evangelical* plan of improving Mankind, is perhaps the very best, since it is the very Religion we profess to believe. Deemed of Divine Origin, venerable by Age and good deeds, pure and holy when first promulgated and undefiled by human passions, it however admits and tolerates in our times of all the sinful practices of Paganism: of Wars, Quarrels, Lawsuits, Pride, Cupidity, Vices, Crimes, &c. but fortunately it admits also and predicts a different *Era* to come, called a 2d Golden Age, or *Millen[n]ium*, when all these evils shall cease, and Justice prevail with Love and Charity, producing Happiness.

29. The majority of nominal *Xristians* do not follow this *Evangelical* Religion, but are immersed in Sinful practices. There are however some Churches, Societies or Individuals, who aware of the Sins of the worldly throng, abstain from these evils, and dwell in peace, in or out of the Social crowd. Such are the Friends or Quakers, the Believers or Shakers, the Moravians, Harmonites, Zoarites, Russian Quakers, and many others, who commonly live in communities as did the *Apostles* and follow the tenets of primitive Evangelical Religion.

* * *

36. As if to shame us, the *Social* plan of Owen and others, has lately been promulgated; altho' revived from the dreams of Plato, or the Cretan Laws, or the Pythagorean Schools, or the Hindu and Peruvian Communities; nay published 100 years ago in France. The *Socialists or Owenists* now claiming to be reformers also, (altho' discarding Religion and Moral principles or even the merits of Skill and Industry,) have spread in England and America, by turns successful or desponding, they still cling to their infatuation, by adopting the principles of Social Harmony; which are in fact the practical part of the first Apostolic Communities, and the Essenians and their progenitors.

37. Community of property has always been deemed a more perfect mode of life, and when the *Nazarens* or *Evangelists* became gradually Sinful *Xristians* admitting Cupidity and other Vices, they yet venerated the Therapeuts or Monks and Nuns, that kept the practical tenet of common goods and labor. In order to restore the purity and further *improve* this very old Apostolic Dogma, it is only needful to

add thereto the important improvement of adequate and *proportionate reward of Wealth, Knowledge, Talents and Labor!* acting in concord. Thus the baneful social (or Anti-Social) tenets of Owen and others, will be banished, by the pure light and practice of *Evangelical Concord.*

38. This may be done by the *Mutual* plan, proposed in France by St. Simon, Fourier, Considerant[2] and others; which they have unfortunately promulgated chiefly in theory, and drowned in long intricate details, whereby it is often difficult to ascertain their meaning. Fourier was so Sanguine as to suppose that the whole world would adopt at once his Scheme of *Social or Mutual Phalanx*; but he died before a single one could be fairly tried. He has left however zealous disciples, some of whom have reached America, and a *Phalanx* is said to be under trial near Paris: while the St. Simonians attempting a New Sect, abolishing family ties, have been persecuted and dispersed.

39. The pith of this new theory (if new it is) appears to be that *all the human passions* are to be made subservient to useful purposes by their gratification; a dangerous principle if some baneful passions are not excepted, and evil propensities not control[l]ed when they inflict pain on others—However the utmost tolerance is inculcated which is right, and each *Mutualist* or member of a Phalanx, is to be lodged, fed and treated in proportion to his capital, skill, talents, industry and activity.

40. Such an ultimate plan is of course excellent and quite practicable, since it appeals at once to the *interested passions* of all, even *Cupidity, Vanity, Ambition,* &c. It admits also of as many separate and intricate bodies as our individual Society Systems, since as many *Clubs* or *Associations* for any purpose whatever can be formed within the *Phalanx* or Mutual Social Body, by any one who chooses or can.

41. Fourier contemplated a Hierarchy and Political System in his plan, that may not suit us; but may do in Europe. He said that Nobles, Princes and even Kings might be members of his Social System, and become wealthier and more powerful thereby. We may have instead Legislators and Judges, Priests and Teachers, Generals (of peace) and Presidents, and this may satisfy the ambitious.

42. But to be the *Founder* of a *Mutual* Cluster of Happy beings (which may be called an *Eden, Elysium, Agathon, Arcadia, Olbion* or *Eutopia*) will gratify the highest ambition; and to do so, it is only required for a *Wealthy man* to set the example, and say, *I wish to be happy my making the happiness of 1000 fellow beings,* without losing a Dollar by it, since I may give them my Estate or money in perpetuity, for which they will pay me a good annuity, or a perpetual income if I have heirs to gratify!

43. The inducements for Laborers under this Scheme is their mutual comfort and support, cheaper food and expenses by clubbing for cooking and every thing else, besides a proposal quite peculiar *to make Labor as easy and pleasant as a festival.* Ploughing, reaping, manual labor, &c. are to be performed with songs, music and dances, just as if going to a wedding—This is certainly something new, (unless martial music for soldiers is a pattern) and altho' it may not suit the good followers of Penn, it will suit nine-tenths of Laborers and working men: four hours daily labor or feasting[?] will support any one and leave him twelve hours for study, reading, recreation, extra labor, exercise, meals, &c. besides 8 for sleep. Competition is also to be abolished and the hardest work the best paid.

44. Altho' the *mutualities* were contemplated to be chiefly rural Communities, in order to raise the needful food; they might also be attempted in Cities, which are Centralities of Knowledge and Capital. A few houses put in common stock, some tools, shops, trades, &c. might be sufficient to begin with; but rents and profits that absorb nearly all the fruits of Labor must be avoided as much as wages and extra expenses. By all being put in Common Stock for a term or forever, food, clothing, and Dividends might be procured at any rate for all concerned.

45. Such *mutual* plan if duly combined with religious Tolerance and Love to all men, might be still more successful any where. There will be no doubt many pious men who may give it a trial ere long, and if they do not split upon the rock of intolerance, or imposing their tenets on all their associates, it must eventually succeed on a large scale, but only partially if peculiar dogmas are to be dictated and assented to, or a monkish life inculcated.

Published in *The Pleasures and Duties of Wealth* (Philadelphia, 1840), pp. 3–4; 11; 13–17.

1. Here and elsewhere, the ellipsis sign is Rafinesque's.
2. Claude-Henri de Rouvroy St. Simon (1760–1825); François Charles Marie Fourier (1772–1837); Victor Prosper Considerant (1808–1893).

IV.

Education

9

Colleges

In 1833, when Rafinesque wrote a 5,000 word essay on higher education for his *Atlantic Journal* magazine, he was having trouble paying his printer's bills and could not afford to print it. As a consequence of his financial position, all *Atlantic Journal* articles are much shorter than this. After he and the retired paint manufacturer Charles Wetherill became associates, however, his publishing activities blossomed, and this is but one of several manuscripts he dusted off and printed—very likely with Wetherill's money, even though Wetherill himself died in 1838. As an offering of the "Eleutherium of Knowledge," the pamphlet was appropriately "Distributed Gratuitously," as it says on the cover. Only three copies have been listed by American libraries.

During 1833, Rafinesque did find an outlet in the pages of a New York State magazine for a shorter piece on the need for agricultural schools. Though actually printed five years earlier than the Eleutherium pamphlet, it follows here as an elaboration on one recommendation made in that pamphlet. The dateline at the end shows that Rafinesque was a house guest of the hospitable Amos Eaton at the time.

See also the *Plan of the Philadelphia Land Company of Aurora* for Rafinesque's additional thoughts about educational institutions.—*Editor*

* * *

Improvements of Universities, Colleges, and other Seats of Learning or Education in North America. Philadelphia: Printed for the Eleutherium of Knowledge, 1839.

[Rafinesque's note:]

The following remarks were written in 1833 for my ATLANTIC JOURNAL AND FRIEND OF KNOWLEDGE, but could not be inserted, as they required to be curtailed; they are now printed apart 6 years after, in the series of useful publications of the ELEUTHERIUM OF KNOWLEDGE, AND INSTITUTE OF LEARNING,[1] chiefly for the use of individuals managing State Institutions or contemplating to become founders of new Universities and Colleges, or to promote their gradual improvement.

* * *

A general wish appears to prevail in the United States to foster Education and Collegiate Instruction. In such an extensive and improving country with so many different aims, but ample scopes and means, besides talents employed to support both, we ought to have obtained an efficient and creditable system of learned Education, altho' it might neither be uniform nor perfect; but we have fallen instead into a kind of servile imitation of some of the most imperfect European systems. Yet there was and there will be for a long while the best opportunities in the various institutions of Education begun or contemplated in the old and new States to improve gradually our system, or introduce better features and regulations.

Without entering at large upon the comparative merits of the various higher Schools, such as Academies, Seminaries, Lyceums, Gymnasiums, Normal Schools, Colleges, Universities, &c., gradually established in Europe chiefly in the midst of the dark ages, and but very slowly improved in modern times, it may not be amiss to recapitulate the main features of the 3 systems adopted by the 3 most enlightened Nations of Europe, the French, English and German, in order to appreciate their useful plans and intrinsic defects.

The French system once copied from Italy, and reformed at the revival of learning, was highly extended and purified in later times; it is based on the liberal principle of making the higher branches of Education, and professional learning, accessible to all *free of cost*, the expenses being supported by the State or by Endowments. Students pay only small fees, in some Colleges they even are boarded. The courses last many years, the discipline is mild but rigidly enforced, the examinations severe and studies enlarged, all sciences admitted none excluded, tolerance now ample &c. The Professors and Officers well chosen by merit or open competition, being adequately paid and independent of students. Many peculiar Schools of Natural Sciences, Agriculture, Mining, Engineering &c are supported as auxiliaries to the Universities.

The English system as evinced in the two Monopolies of Universities is far less liberal and efficient, altho' amply endowed or even too wealthy; it is more exclusive, intolerant, shy of reforms and improvements, with obnoxious tests, privileges, abuses, sinecures, &c. The discipline is lax, the cost enormous, poor Students despised, wealthy Students pampered and indulged.—It is now generally allowed that the system is both erroneous and at variance with the age: therefore a new system has necessarily sprung up in Scotland since spread to England and among us. It is a system of dissent and competition; not being so well endowed, the Professors are ill paid or not paid at all, made to depend on the fees of the Students, to whom they must be indulgent, wherefore the discipline and examinations become relaxed. No deep knowledge is imparted except to a few youths of genius; Professors are appointed by patronage or servility, and in the smallest possible number, which 2 or 3 years of study are often deemed adequate for degrees, in order to lessen the expenses of the Students, and invite them by cheapness and indulgence.

The German and Italian systems are nearly alike; owing to the number of Independent States of the same speech wishing to retain their youths at home, or invite neighbor students, they are also based on competition, and offers of advantages, *free instruction*, cheap living, fine libraries, eminent teachers—Professors are however seldom adequately paid, but allowed to receive private pupils or classes, paying for more attentive instruction; they are appointed in various ways, sometimes fairly by competition, elsewhere by recommendations, or state patronage. Many modifications exist of course in Prussia, Austria, Bavaria, Italy, &c.[;] in some seats of learning the discipline is very lax and students often riotous, in others they are kept under proper control, they seldom dwell or board with their teachers, but the indigent are not neglected, nay enabled in many ways to go thro' their studies. Every where a spirit of religious tolerance is introduced (except in Italy); in the Protestant Universities liberal principles prevail. All over Germany chairs are multiplied, new attractions devised, learned pursuits encouraged, and new paths of knowledge opened.

Therefore we shall find in all these national systems something to imitate or to reject, to select or improve upon, whenever and wherever it shall become desirable to maintain in the U. States a national or rational Education, with high attainments accessible to all. The French plan would be best, if we could combine with it the advantages of competition and excellence of the Germanic improvements, discarding the abuses of English Monopolies and Scotch Parsimony. But we might yet improve over them all, by the vast resources which our rich soil, education lands and value of labor combine to unfold.

Our actual system in North America has many defects borrowed from abroad, and as yet but few peculiar features of its own. As we admit competition and improvements in all instances, we ought to apply this process to purify and elevate our standards of attainments by opening a free scope of emulation, and prospects of rewards and preferment in all those who study and teach.

Some practical experience and deep studies, in the difficult arts of learning and teaching by all the various modes usually employed or lately devised, have enabled me to perceive many defects, superfluities and wants of our Institutions. I will therefore venture to suggest some remedies, or propose some better substitutes.

1. Our Institutions of Education and Learning, Colleges and Universities, are generally not sufficiently endowed, and thus made dependent on Tuition fees and the number of Students for success, as if they were private schools—competent endowments either by the States or generous Individuals are the obvious remedy. Congress has been very liberal to the New States on that score, and it is only by misapplying the education lands, or neglecting their duty, that they can impair the contemplated liberality.

2. Religious sects often strive to obtain the control of these Institutions, so as to enlarge their influence, spread their tenets, grasp the funds, and make them subservient to sectarian views. Thus in almost every State those that cannot obtain control of the main University, start Colleges in opposition, or to secure therefore at least the patronage of their own sect. If this was merely done in friendly competition, or more instruction imparted, it might be well; but it is seldom so: on the contrary[,] hostility is evinced, less knowledge taught and but too often more facilities of misbehaviour, or lenient indulgence in studies and examinations afforded to the students.—The plain remedy for this tendency is for each state to establish

free Colleges and Universities, where nothing shall be paid by Students as in France, at least for Tuition; as such can fear no competition of rivals, unless equally largely endowed and also made free to all. Next to admit of the most ample tolerance and liberality in the boards of Trustees or other controlling officers, also in Professors and teachers, Students and the acquirement of degrees, leaving their religious instruction in the hands of the respective denominations of Religious communities.

3. In forming the boards of control of these state Institutions none but men of liberal, tolerant and benevolent or enlightened principles should be selected, who[,] fully aware of the important charge entrusted to their care, will faithfully perform their duties, and supply vacancies by similar men, make proper appointments, never waste nor misapply the funds or endowments—this has not always been the case in all the states, and we have ample cause to regret that different men have got into these boards, men of little knowledge or devoted to some sect, sometimes even Clergymen of intolerant sects; who of course did lean towards their own party or opinions, trying to make them prevail. Other unsuitable men misapplied the funds or wasted them in expensive buildings, yet always *liable to burn down*, as they usually are by turns, by having wooden stairs, floors and roofs. When Presidents and Professors of favorite Sects have been appointed, they have sometimes introduced religious intolerance or improper discrimination into our Colleges, which ought to be confined to the Theological Seminaries.

4. Too much money or other ample means are commonly wasted in bricks and mortar, or stones and lime, or marble and pillars, and still worse on boards and shingles liable to repeated conflagrations; while too little is spent on Iron and Zinc, Stucco and Tiles that might prevent them. Else not enough on the tools of knowledge that ought to fill these Halls, such as Instruments, Apparatus, Models, Books, Engravings, Objects of Study &c., for the Libraries, Museums, Gardens &c.—altho' they teach as well or better than the usual routine of Oral Lectures, Recitations, and tiresome Examinations, by introducing instead pleasing Demonstrations, Recreations, Dictations, Perambulations &c—the remedy is obvious, let us have both or all of them if it is wished and we are able; but not too much of the first or usual kind, and at any rate as many books as possible rather than splendid empty Halls.

5. It is often deemed sufficient for a College to have halls, chapels & boarding houses; but these last are mostly useless, as Students ought to dwell and live apart with their parents, guardians, or protectors, although under a proper collegiate control as to studies and decent behaviour. It is very important to have also Libraries, Museums, Gymnasia, Gardens; nay also Printing Presses and Labor Schools, either Mechanical or Agricultural, where indigent Students might maintain themselves and learn useful professions, while under the Collegiate course—this would obviate one of the main objection[s] made by the bulk of the people to Colleges, that they are only suited to the rich, and in fact it is made so in many of ours; since while labor is high, Students are prevented from laboring, and made to pay high for tuition and board, besides fees.

6. At any rate nothing should be charged for tuition, or only small fees asked for matriculation, use of Libraries and other conveniencies, besides Diplomas—no uniform boarding prices ought to be fixed, except in the labor school, where labor

should be received for money. Thus each Student might live according to his means, and if allowed to labor *as a recreation* if he likes, or paid for it, if he needs it, this improved system will be more conductive to health than continual sedentary studies: all may vie with the rich[,] even if poor; may even indigent Orphans be entered in all the schools, and supported by their own labor.

7. The appointment of Professors and Teachers is quite important, since on them mainly depend the utility and reputation of Universities. It has been too often the custom to appoint thro' recommendations of Friends or Clergymen, instead of taking into accounts the talents, abilities and reputation of the applicants or suitable individuals to be invited. It would not be proper to bribe them (as in Germany or as Churches do with us) by higher salaries to leave one College for another; but competencies and some inducements might be held out for applications of suitable officers or high[ly] gifted men: such as higher preferment, or a more acceptable Professorship, or less laborious duties, outfits for removal, traveling expences, purchase of their Libraries or M[anuscri]pts. collections of Science, offers to print for them.

8. To send to Europe for Professors as was done in the University of Virginia was rather unjust and antinational: while there were so many in our country in want of preferment. It was as bad as if we were to send for Judges to England instead of appointing our best Lawyers.—Teachers, Authors and Men of Learning have so few rewards and preferments in perspective that to deprive them of those that may offer is quite an injustice. Clergymen ought to be promoted in their Churches and Dignities, but not to control Colleges. Teachers of all kinds ought to have the same chances in their profession, by merit and assiduity or talents, evinced by their labors and works. They ought never to be discarded while they have been faithful servants to the public, but rather promoted as they increase in knowledge and experience.

9. When there is a lack of suitable applications, why not invite a concurrence of applicants and a public examination of their respective merits, evinced by a lecture and written Thesis? as done in France, Germany, Italy, &c. Nay it is done there habitual[l]y in several Universities, and in most instances Authors are examined on their published works only if they ask it. This is called a *Concours* by the French, *Concorso* in Italy; it is a stated rule to enable worthy Students or learned men to obtain situations, offices and emoluments; securing them to the most worthy if no abuse nor predilection creeps in[to] the examination, to impair the impartial decision of the men appointed to decide: who ought to be also learned men quite impartial, and neither rivals nor patrons of the applicants—this open and just mode of proceeding is of very ancient origin and extensive application every where except in England and N. America, since it is the great fundamental principle of all literary and civil appointments in China since the most remote ages. It cannot be too earnestly recommended to the attention of our Colleges and Institutions.

10. One of the inducements prof[f]ered to Teachers and Professors is often to allow them rooms and board, or dwelling houses if they have families, when they always prefer of course to dwell and live by themselves; but to provide houses for all will be found expensive if not impracticable in many instances, while if their emoluments are adequate, and the locality a city or town, they can easily provide suitable lodgings. It is generally preferable for Teachers and Students to dwell apart

on many accounts, which will occur to any one who has been either, and too familiar intercourse often lessens the respect which Teachers ought to inspire—what may be deemed adequate emoluments, will depend on the locality, rents, prices &c.; $2000 in one place are hardly equal to $1000 in another: unhealthy situations or far removed from good Society, Libraries, Literary news &c will be less acceptable to men of worth and talents than those deemed healthy[,] sociable and well stored with Books, or food for the mind as well as the body.

11. It is not professors that are alone to be appointed, but also Presidents and Deans, Counselors, Demonstrators, Teachers, Librarians, Curators, Provosts, Janitors, Monitors &c. These graduations offer scales of employment and preferment, such as faithful service may deserve or entitle to. Even when they are superan[n]uated after long and useful services, they ought not to be ungratefully removed and discarded, but retained as honorary professors, or worthy Counselors, on a small allowance for a few years of retired life—honorary professors might also be appointed with or without emoluments, when distinguished men[,] or Lecturers may wish to become enrolled in a faculty of some great State Institution, and may deserve it by giving at least a course of Lectures or Demonstrations; by such men and their names being enrolled in the Universities, they rather confer than receive an honor.

12. To dictate to the faculties or the members thereof, the modes in which to impart knowledge, or to restrain them in enlarging the sphere of it, is not merely illiberal but injurious, and ought to be avoided or rather never attempted. Men of talents and experience are the best judges of the means to teach what they know so well, and they may by turns use conversations, lectures or dictations, demonstrations, interrogations, disputation, printing &c—printing and circulating their

Figure 10. Principal building at Transylvania, where Rafinesque lived in two rooms in close proximity with the students.

works, opinions, observations, lectures &c., is one of the most efficient mode[s], too much neglected in our Colleges, altho' it scatters knowledge far beyond the narrow halls of the Schools, imparting eminence and high reputation to these learned Institutions.

13. They also know best how to inspire respect and enforce obedience, when they have the control of the faculties & students. The Government of the Students ought to be *paternal*, based upon a delegated authority from the Parents or Guardians. Disrespect to the Teachers ought to be deemed one of the highest offence[s], *just like an insult to a Father!* To compel Teachers to be looked upon as accusers or stern judges is not only invidious but deplorable; they ought rather to be deemed invested with a share of fatherly authority. Some Students have been so much indulged or spoiled at home that they assume too much when at school, these must be restrained by severity, else they will give more trouble than profit: yet when it is of importance to obtain or retain the fees of many Students, as in unendowed school, they are often overlooked or indulged, to the great detriment and discomfort of all the schools. Reprimands, degradations and expulsions ought not to be spared in case of need; but at the same time leniency offered to repentance, reinstal[l]ation on apologies &c.

14. Whenever Students are allowed to combine, bind themselves not to reveal names of delinquents in annoyances and riots, form secret societies and conspiracies...[2] it is impossible that a proper discipline can be maintained. Thus in Germany and America they will often annoy and baffle their teachers—yet it is very easy to obviate all this by strict prohibitions, and binding each Student to honor or bond to abide by the regulations of the University, as long as he belongs thereto. The faculties and control[l]ing board ought to have the right to enact such rules as will preserve order and good will among all. The prohibition of vicious indulgences, strong liquors, tobacco, cards or games of chance and other practices equal[l]y injurious to youths, their health or morals, ought to be adopted in all Institutions.

15. The Monitorial plan, forming the Students into classes with monitors, allowing each to emulate and supersede each other by turns, through vigilance and good conduct, might have the best effect if introduced in our higher schools—also if in each class the Students be allowed to examine and criticize each other, as in schools for teachers. Lastly it will be found that to restrain the bad, and reward the good, in all instances, must avail there as elsewhere. Rewards, premiums, medals, badges, special diplomas specifying the acquirements and conduct while at College, would prevent and obviate all the actual results of indiscipline and loose control, that have so often disgraced some of our Colleges and Seats of Learning, because depending upon the Students for their main support.

16. Many modifications and improvements in our actual faculties, professorships, classes & degrees of Students, might gradually be introduced to great advantage. None more so than in the degrees usually confer[r]ed, which are even now all different in the 4 admitted faculties. It would be difficult to account for the loose, inaccurate and obsolete degrees of Bachelor and Master of arts, without any licentiate nor doctorate as of old; also for the terms now employed for graduates of Theology and Law: while doctorates are given at random not to Students, but to emine[n]t men, to whom doctorates of Philosophy or Science are denied, although

they are granted in Germany and Italy—and this while doctorates in medicine are profusely granted *for mere fees* to any Student of two years standing only, who is not a mere dunce, but refused to many learned men skilled in medical sciences.

17. Our 4 faculties are based upon the old obsolete notions that Philosophy, Divinity and Law, were the 3 great classes of learning and professional pursuits; while medicine was added rather lat[t]erly. The actual state of general and professional learning now appears to require at least 5 faculties if not more—1 arts and sciences—2 philosophy and belles lettres—3 medical sciences—4 legal sciences—5 spiritual sciences including the Intellectual—while the historical sciences must be protem added to literature, and many other pursuits must necessarily be made auxiliaries, that in peculiar schools are paramount, such are agriculture, mechanics, mining, engineering, navigation, tactics, diplomacy &c.

18. In these faculties will be included almost indefinite numbers of branches or sciences, to become peculiar professorships, according to their generalization or division. This number will always chiefly depend on the means and liberality of the Universities. Their standing, reputation and eminence, depends not so much on the number of the Students, (as usual[l]y deemed by mistake) as the number of their officers, their abilities and acquirements in various pursuits, and great works produced by them. The subsequent eminence of some of their Students, is often to be ascribed to patronage, or wealth, success in life &c, and only imparts some retrospective honor on their *alma mater*; but the talents of the Teachers in any branch, or their contributions to the increase of knowledge impart at once a *Halo* around the seats of Learning, and a lasting reputation.

19. The classification of Students by years of study is both natural and proper; but one year of study does not imply 12 months, in the medical schools it is reduced to 4, and it might with advantage be reduced to 6 in all others, or 2 seasons of 3 months each, Spring and Autumn, as in the Rensselaer School of Teachers. The obsolete and absurd names of *freshmen* (rather freshboys)[,] sophomores, juniors and seniors, given to the usual 4 successive yearly classes, might be dispensed with or modified. In France classes go by numbers, the first being the last attained; elsewhere they go by names of actual studies. The most proper in each faculty should be by numbers of years in College, such as primary, binary, tertiary, quaternary and quintenary or last classes. This would indicate at once, whatever be the years or seasons asked of them, how long they have studied or attended. A diploma of attainments and conduct should be given every year, and the Students classified in 3 series of *Eminent, Proficient* or *Deficient* in the studies attended to.

20. In the 4 or 5 faculties, many arts or pursuits must still be left untaught, unless casual[l]y or imperfectly: whence the necessity of having auxiliary schools connected or unconnected with the Universities. Some of these are already or ought to be national concerns rather than state institutions; but we lack many others, such as schools of Astronomy, Diplomacy and Public Economy, Commerce and Manufactures, Mining and Geology, Navigation and Tactics or the Naval and Military Arts &c; special schools for all these are found in all Countries except our own! with the exception of the West Point Military School. If they were gradual[l]y founded and well supported, we should obtain the needful range to complete the circle of knowledge and pursuits.

21. Agronomy or Agriculture with Horticulture, and the arts and sciences

thereto connected as Botany, Veterinary, &c, are now become so important, and in our country Agriculture being the profession of the bulk of the People, it would be so popular and patriotic to improve it and extend it, that it appears to be the paramount duty of all the States to foster it, with all the auxiliary sciences, by establishing peculiar schools of Agriculture, or annexing Botanical Gardens and experimental Farms to our Universities: which would be still more useful if made Labor Schools for indigent or destitute Students.

22. In many Colleges a peculiar course of studies is prescribed, in others it is left to the choice of Students. Both modes are liable to objections, to prescribe and compel in all cases may be improper, but to advise and afford the means must always be proper. For instance, in the schools of medicine, all are compelled to study Surgery although they may not wish to become Surgeons: in Europe this profession is quite distinct from that of Physician. To compel a Student to learn Latin, or Music, or Chemistry ... for which he has no taste, and to prevent him or not procure him the means to learn Agriculture, or Drawing, or any other study for which he has a taste, appears both unjust and injudicious.—Therefore Students after due advice and consultation with Parents and each Professor, ought to be left free to cho[o]se their course of studies; but their diplomas should state those attended and those omitted, so as to prove their assiduity or neglect, as much as their proficiency in one or more branches, and aptitude for future pursuits through life.

23. The progress of discoveries and human acquirements have gradual[l]y introduced into the European Schools of learning, (but not many of ours as yet) the modern sciences of Chemistry, Botany, Geology, Mineralogy, Zoology, Ethnography, Archeology, Philology &c, with many others, in spite of the reluctant or sturdy opposition they met with at first, just as Natural Philosophy, Astronomy; the Oriental Languages &c, were long opposed at the revival and expansion of modern knowledge. Now we begin to feel the want of Agricultural and Mechanical Schools, of Drawing and Painting, Music and other fine Arts, besides many others late sciences and pursuits. Although *Phrenology* or the *Astrology* of the brain, will probably be discarded just like the *Astrology* of old, and other obsolete vain sciences, yet *Phrenomy* or Mental Philosophy, may well be placed by the side of sublime Astronomy.

24. Lastly, there is a real want, nay absolute need of female Colleges as auxiliaries to the Universities, in order to teach *one half of mankind*, whatever knowledge is appreciated and desirable, or suitable to the female sex. Above all it is an act of justice to provide for female Orphans as well as male, by teaching them some available professions, or qualifying them to become suitable female Teachers, instead of dooming them to servitude. There have been instances in Europe of female professors being appointed to teach men with applause, why not with us? if enabled by talents and studies.

25. It is true that youths of both sexes when inclined and gifted with Genius, may overcome all obstacles, and learn much by themselves, or by reading—yet we nevertheless deem it needful to provide schools for all, in order that the latent germs of talents may meet the needful props and helps—to elevate the character and moral worth of both sexes, it is merely required to put them on a more equal footing, or afford them similar facilities—nay it is not youths alone that require these

means, men and women of good mind, know well that they learn as long as they live, in proportion to their means and opportunities; nay religious minds duly believe that this world and life is but a larger kind of school, to teach us our duties, and point out the paths leading to the HEAVENS, which all may attain who wish it and deserve it, by having qualified in the terrestrial life for the future celestial life.

1. For "Eleutherium" see *Plan of the Philadelphia Land Company of Aurora.* It also is mentioned in a slightly different context in the *First Scientific Circular for 1837.*
2. The ellipsis sign both here and later is by Rafinesque.

* * *

Agricultural Schools.

The friends of agriculture throughout the United States have long hoped to see something done for the instruction of young farmers.

Is it not surprizing that in all the Universities of Europe, even in Spain, Portugal and Italy, nay even in the most despotic and bigoted countries, there is a chair of agriculture; and that useful art is taught as a science, free of any charge to whoever chooses to attend the lectures; while in this country,—a country of farmers, as we might say,—not a single University has ever thought to teach agriculture, and, when thought of, will probably not be done by endowment and free of fee, as in Europe!

We must be sadly deficient in wisdom, and knowledge of our own good, to have so long neglected the duty of enlightening our agricultural population. We have many classical schools and universities, where the liberal professions are taught, and farmers are induced to send their sons there to become lawyers, physicians and clergymen, but agriculture is not deemed a science worth teaching, nor a liberal profession worth instruction, although it supplies one half, at least, of the scholars.

A strong unanimous effort ought to be made by all the enlightened farmers and zealous friends of the various arts and sciences connected with the culture of the earth, such as agronomy, horticulture, floriculture, arboriculture, botany, domestic economy, &c. to produce a better state of things, to make our young farmers better satisfied with their station, and enable them to sain[1] in society, by giving them a scientific agricultural education; by endowing free lectures on those arts and sciences in our universities; but above all, by establishing, either by state laws, or private efforts, *special Agricultural Schools,* with model farms and gardens, where all the practical improvements are to be taught, as well as all the sciences which a liberal farmer ought to know.

This plan is strongly recommended to the public by its absolute necessity; and the need of attending with more care to the real wants of such a large class of the community. The mechanics have already been beforehand, and established institutes and schools, while the farmers, the sinew of the country, are yet idle in the great cause of self instruction or scientific improvement.

Let a Franklin arise among them that will ennoble the art of feeding mankind, and persuade at last our wealthy men and legislators to aid and support them. Much has been done lately by farmers' periodicals, to spread knowledge among this class in general; but it is now that their children must be attended to, and educated so as to honor the profession, and put it on a level with the most liberal arts, as it is already among the most useful.

C. S. RAFINESQUE,
Prof. of Hist. & Nat. Sciences.

Elm Place, Lansingburgh, N.Y. Aug. 15, 1833

Published in the *New-York Farmer and American Gardener's Magazine*, 6 (September 1833), 265–66.

1. The magazine's editor failed to excise this Franglais neologism, apparently intended to mean "to be at ease ... comfortable ... sound," etc.

10

Natural History Surveys

The *First Scientific Circular for 1837* reproduced here is from a copy of the broadside addressed in Rafinesque's hand to the Lieutenant Governor of Massachusetts and preserved at the University of Wisconsin Library, Madison. In his *Bulletin Nr. 7 of the Historical and Natural Sciences* (Philadelphia 1838), Rafinesque said that a *Scientific Circular* for 1838 had been issued, but copies of it have not been found.

The circular was self-serving because at this time Rafinesque was pulling all the strings he could think of to get himself appointed a member of the New York State natural history survey (and perhaps of any other state that would have him). At the same time, having been sent to officials of all the states, it expresses what a seasoned naturalist believed these surveys ought to accomplish and how it should be done.—*Editor*

* * *

First Scientific Circular For 1837 to the Governors, Lieut'ts Governors and Speakers of the Legislatures—Of All the States and Territories in the United States.

BY C. S. RAFINESQUE, Professor of Historical and Natural Sciences in Philadelphia, late Professor in the Transylvania University of Lexington in Kentucky, and in the Franklin Institute of Philadelphia; member of 15 learned Societies in America and Europe; author of 50 works on all the Natural and Historical Sciences; Founder and Promoter of many useful and scientific Institutions, &c.

TO HIS EXCELLENCY [*Here Rafinesque inserted by hand the title of a state official*]

THE WRITER respectfully ventures to address you, on several subjects of some importance, and may try to suggest, or throw some useful hints for further improving your STATE, or acquiring a complete knowledge of its minerals and natural wealth, resources and capabilities of improvements, &c.

THE WRITER has been partly employed ever since the year 1802 in visiting, exploring, and surveying several of the States from Massachusetts to Illinois, Kentucky and Virginia, as a scientific traveller, taking maps, geological sketches, plans of interesting localities, drawings of scenery and antiquities—collecting minerals, geological and botanical specimens, fossil remains, shells and animals, or making drawings of those not easily preserved.

This pursuit has led him throughout the Western States in 1818 to 1826, from Lake Erie to the Cumberland Mts.; while he was professor in Lexington, and from 1826 to 1836 while settled in Philadelphia as a Professor of Sciences, and Public Writer, he has continued to make yearly excursions and surveys in the Alleghanies, Mattawan and Taconick Mts., with the littoral regions, from Lake Champlain to the Potowmak.

He takes the liberty to state this fact, in order to show that he is a practical scientific surveyor, and not a mere theoretical writer. As a Geographer, he has shown his skill in mapping in 1818 the whole course of the River Ohio, in a map 10 feet long, which survey was sold for $100. All his other surveys and researches were made at his own expences, and for the love of science or to promote useful knowledge.

While thus employed, he has perceived with great satisfaction, that of late years, owing to the rapid increase of our population and prosperity, the attention of the public Authorities and State Governments, has been happily bent towards improving the States by public ways and works, roads, canals and railways, which after requiring extensive surveys, has gradually led to feel the need of enquiring into the capabilities of our soil, and mineral wealth, by more regular surveys than the casual or blind operations pursued in the first instance.

This appears to open a new Era in our gradual evolution as a powerful and happy nation: although it is only the first step in scientific surveys. Ever since 1815 the worthy Gov'r. Clinton of New-York, whom I was happy to reckon among my friends, had in contemplation such useful surveys, but deemed best to delay them till our Canal System was fully completed and prized. Meanwhile since this was attained, at the suggestion of several Patriotic Geologists, &c., the Governors of several States have been induced to consider the subject, recommend it to the Legislatures, and many States have already adopted these views, and undertaken partial surveys, chiefly geological and mineralogical, although mainly superficial and often imperfect, except in a few instances.

Meantime the great State of New-York has set the example of more expanded and proper scientific surveys, by dividing the State into Sections, appointing many Geologists, Mineralogists, Chemists, Botanists, Zoologists and Draftsmen to carry it on, for four years. This wise example deserves to be followed by all the other States, as the Canal System begun also by New-York, has been the model for other States.

It is the purpose of this address to inculcate in the proper quarters the necessity and facility of this. To point out the defects of the partial surveys already begun, and to suggest the modes to render them more effective and available.

That the motives of THE WRITER may not be misunderstood, it is proper to state that they are purely patriotic and disinterested. He is not actuated by any jealousy nor hostility against any of the actual Scientific Surveyors, many of whom are his personal friends or worthy men; nor is he seeking to supersede them, or ask for employment. He has never applied for any appointment of the kind, except in the great survey of New-York, and even then, with the sole view of directing by his long experience the course to pursue by inexperienced surveyors.

THE WRITER is now permanently fixed in Philadelphia and connected with vast labors, and useful institutions which claim most of his time. He is endeavoring besides to promote the foundation of some other patriotic Institutions, among which is one contemplated by his friend CHARLES WETHERILL a wealthy philanthropist, who wishes to establish a new institution or ELEUTHERIUM as a Model School for Scientific Teachers and Surveyors, as the future need of a class of competent men to supply the wants of all the public Schools, Institutions, Surveys and Works to be gradually done by our prosperous States, begins to be felt even now, and will be still more hereafter, unless we send to Europe for them, or do this.

In fact some of the surveys begun or contemplated, have already been delayed by the lack of proper Surveyors, or young and inexperienced Geologists and Naturalists have been employed, who have of course performed their duties negligently or not thoroughly, for lack of time to qualify themselves for their duties.

Under these circumstances THE WRITER hopes that the liberty he takes to address you on this public concern will not be deemed too bold, and he now will venture to enter into the details of the suggestions, which his long experience, and zeal for knowledge, prompt him to give.

1. It appears to be the duty of every GOVERNOR of a State or Territory, to consider the subject of Scientific State Surveys, either preliminary or partial, or else minute, complete and perfect: and to recommend the same to the Legislatures from year to year, until begun or achieved.

2. By preliminary survey, is meant such as have taken place in some STATES, by employing a Surveyor to cross them in some peculiar directions, to enquire into agricultural capabilities, kinds of soils and rocks. This is the first step in this kind of survey.

3. By partial survey, is meant either a geological exploration, or an agricultural, mineral and chemical examination of many points in the STATE. Such surveys are now in progress through many States, and are only the second step.

4. A minute survey, would be a gradual examination of every county during many years, so as to leave no section of country unexplored. This ultimate kind of survey has not yet been attempted; but must evidently take place before any STATE is properly surveyed.

5. A complete survey embracing all the above branches and collateral sciences, with botanical, zoological and oryctological researches, is now going on in the State of New-York, and all the States are invited to imitate this useful plan: to which might even be added meteo[ro]logical and medical researches.

6. A perfect survey can only take place, when all these sciences and objects are attended to minutely, county by county, during a series of years, by skil[l]ful scientific Surveyors and learned men. This ought to be always kept in view until it is attainable and effected: and then indeed each State will completely know its own resources, useful productions, mineral wealth, capability of the soil and hidden treasures.

7. The utility of such gradual surveys need not be proved, nor insisted upon, now that the whole country is alive to the wisdom and need of a thorough exploration

of our natural resources, and search after the best modes of bringing them to light, rendering them available, &c.

8. Whenever the principal authorities of the STATES, have become convinced of this, and the GOVERNORS have recommended those surveys to the LEGISLATURES, they have been found ready to grant the needful powers and funds. Therefore it is mainly upon the respective GOVERNORS that these useful explorations depend: as besides the right to appoint surveyors and direct their operations is usually vested in them.

9. It must be urged upon the GOVERNORS of all the States that have not yet begun operations, to delay no longer their enquiries and action on the subject; whenever they become aware of their need and practicability, they ought at once to lay the subject before their LEGISLATURES at their next meetings. If the explorations have been begun, they might enquire into the mode of extending the benefits of it to all parts of their States, and thus perfecting what has been undertaken.

10. Although this address is mainly intended for the GOVERNORS of the States, it is hoped that by directing the attention of the LT. GOVERNORS and SPEAKERS of both Houses to the same subject; the useful scope may meet with more friends and may be influenced by their favorable support. They are earnestly invited to give it their best attention, and to gain friends to such general patriotic measures; handing these suggestions to the most enlightened members of the Legislatures, who may mature a plan to bring forward: whereby our Country will become thoroughly explored in due time, and all our natural resources unfolded to view.

11. Having proceeded so far in general explanations, THE WRITER now ventures upon the delicate ground of the effects of some actual surveys, either begun or gradually in progress: which is done in the full hope that all such results of inexperience at the outset, may be easily avoided in future when pointed out.

12. Some of the best surveys made have been purely geological or mineralogical, neglecting altogether the departments of Agriculture, Chemistry, Botany, Fossil remains, Physical Geography, and Manufactures based upon natural production.—Mines and Quarries, Strata and Rocks, Coals and Salt...[1] are certainly very proper and important subjects of enquiries, but they are not the only ones: although mistaken for the only needful ones at the outset, owing to the employment of mere Geologists, or peculiar interest.

13. These Geologists were generally of the English School, which entertains very narrow and absurd views of the Science; while it is cultivated in greater perfection in France and Germany. There is hardly yet an American School of Geology[,] which ought to be as liberal and expansive as this vast continent; but we may hope yet to have one, and these surveys will surely lead to it.

14. A single Surveyor even with several Assistants, can only make preliminary surveys. His assistants cannot always be depended upon for accuracy and he cannot see all, nor even go over every road and track, much less explore wild tracts, hills and mountains, where the greatest discoveries may be made.

15. All the actual surveys are directed to examine the actual mines and mineral localities, rather than seek for new fields and spots. The exploration is merely of the superficial soil; while in many instances excavations ought to be undertaken, and even in some spots borings attempted to sound the rocks at greater depths than heretofore.

16. Therefore whatever may have been done already by such learned mineralogists as [Parker] Cleaveland, [Gerard] Troost and [Edward] Hitchcock, is nothing compared to what may be done elsewhere, or wherever all the collateral arts and sciences of Chemistry, Iconography, Botany, Physical Geography, &c., will be made available for the purpose any where, as partly done now in New-York.

17. Under this liberal and extensive view, it is evident that several surveyors and assistants ought to be employed by each State, for a series of years, to explore thoroughly THE WHOLE STATE, counties by counties, visiting and exploring every square mile in each; afterwards time must be given them to mature their reports, analyze the minerals, visit some places at leasure, take the plans of mines, and lastly sound the earth or rocks by boring or drilling where required.

18. In the choice of principal Surveyors, it is requisite that competent and experienced scientific men only should be appointed and employed: if instead incompetent and unexperienced men apply, who cannot stand a strict examination, it would be preferable to wait until better Surveyors be found. They will always be procurable by consulting the Writer, or the Professors of Colleges and Universities, who even when not qualified in those sciences, or unable to accept themselves such employment by leaving their own collegiate duties, will be able to point out or recommend the proper persons. It is to be regretted that in all Colleges and Universities, the Natural Sciences have long been neglected, and scientific men seldom encouraged. If they had been, a greater number of qualified Surveyors should be found ready to enter the field, which is now opening for them. Presumptuous young men, or mere students recommended by their teachers are seldom suitable or properly qualified except for assistants.

19. The appointment of assistants has often been left entirely with the Surveyors. It is often proper it should be so; but in some cases young men of little or no worth have offered themselves or been recommended to them, who were merely prompted by the view of emolument, without proper qualifications; or even connected with speculators to ascertain the valuable lands and spots surveyed, so as to base thereon ulterior projects, or take undue advantage of the real owners of the property and lands; this ought to be prevented by all means, and none but suitable clever and trusty men employed as subordinates.

20. Besides the field Surveyors, many learned Chemists and Professors might be employed to advantage, without being taken away from their duties or advocations [*sic*] in Cities and Colleges, for the purpose of analyzing and labelling Minerals, ascertaining and labelling Botanical specimens and fossils, or every object collected, maturing and drawing up the reports of collected facts. This would save a great deal of time and trouble to the Surveyors, and may be done with more advantage by the Sedentary Chemists, Botanists or Naturalists than those who are running over the grounds. In this case learned and eminent men may always be found in Philadelphia, in New-York or elsewhere, willing and able to attend to this duty and give ample satisfaction.

21. It is desirable that a STATE MUSEUM should be formed in each State, to receive the collections and specimens, maps and plans, sections and drawings of the various surveys and explorations, or else that several sets should be collected as in New-York to be deposited in the Museums of the various Colleges in each State.

22. While this useful purpose is contemplated, it would not be amiss to unite such a STATE MUSEUM to the STATE LIBRARY, which almost every STATE has begun to collect for the use of the LEGISLATURE. Respecting those Libraries it is to be regretted that the GOVERNORS OF THE STATES do not oftener draw thereon the attention of the Legislatures. They are generally increasing very slow, and lacking great many historical works, American works, and useful works. If the GOVERNORS of all the STATES were to recommend the purchase of such works, these libraries would soon become larger and more complete: while American Authors would be encouraged.

23. It is one of the duties and views of all the LEGISLATURES to collect and preserve, not only Books, but also Manuscripts or Documents relating to American History, and the History of the States. The Writer begs leave to remind the STATE AUTHORITIES of this fact, and to urge upon them the indispensable need of erecting fire proof buildings, where records, documents and rare Manuscripts or works may be preserved, without the dangers of the constant Conflagrations that are gradually destroying them. This can now be done with great facility and even at a cheap rate, by using the modes recommended by THE WRITER in his pamphlet[2] on Incombustible Architecture published in 1833.

24. It is hoped that a time may come when every STATE, will feel the necessity of establishing a Botanical Garden and Agricultural School, for the purpose of introducing and spreading useful plants and trees, teaching the practical and essential branches of Agriculture, Gardening and Botany. This plan is respectfully recommended to the consideration of the GOVERNORS of each State, and THE WRITER feels happy to have been instrumental by his writings to make better known the cultivation of Grape Vines and Mulberry Trees,[3] leading to the production of American Wines and Silks, that deserve to be fostered by all the State Authorities.

Although THE WRITER is not unknown to the learned and eminent men of the Atlantic and Western States, as an Author, Professor and Zealous Investigator of Nature: he may perhaps be but partially known to those for whom this address is intended. Therefore he deems proper to add the names of several valued friends or eminent men, who are acquainted with his character and writings, in order that if needful, proper inquiries may be made, whereby the suggestions he has ventured upon, may acquire a greater weight by the full assurance of his capacity, activity and zeal: evinced besides by the trouble he takes to print this at is own private expense.

THE WRITER having been in America since 1802, has lost many early friends of high standing now deceased, who could have further vouched for his early zeal and exertions in the cause of science, among which it will be sufficient to mention Thomas Jefferson, Dewitt Clinton, Dr. Benjamin Rush, Colonel [Thomas] Forrest, and Dr. Samuel [Latham] Mitchell [*sic*]; but he can yet boast of the acquaintance or friendship of General [William Henry] Harrison, Henry Clay, Dr. [Samuel] Akerly, Dr. [Charles Wilkins] Short, Dr. [James Haines] M'Culloh, Dr. James Mease, Dr. John Torrey, Prof. [Jacob] Green, Peter A. Browne, Lewis Tarascon, &c., who are acquainted with his labors, and to whom he refers in case of need.

Along with this will be received one of the Bulletins[4] which he prints casually to inform the public and his friends of the works he has published or are in progress,

with his yearly or contemplated ulterior labors, and he refers to it for further information. Any other explanation required, or enquiry made, by any GOVERNOR or STATE AUTHORITY, will be promptly imparted or answered, if the Writer is applied to personally by letters post paid.
Philadelphia, Spring of 1837.

OFFICE AT 172 VINE-STREET

P. S. As the subject of SAFE BANKING is now beginning to draw the attention of the public, and of those who are aware of, or foresee the evils likely to arise out of our actual UNSAFE mode of BANKING, and unsound currency: it is perhaps well to draw the attention of STATE AUTHORITIES to the principles of SAFE BANKING, begun to be unfolded ever since 1825 by THE WRITER; but fully developped [*sic*] and explained in a Pamphlet, called SAFE BANKING, now printing at the expense of the DIVITIAL INSTITUTION OF NORTH AMERICA AND SIX-PER-CENT SAVINGS-BANK at Philadelphia. Therein will be seen how STATE AUTHORITIES, Corporations, Rail-road Companies, Manufacturers, or Individuals, may if they are wise, establish SAFE BANKS and INSTITUTIONS, liable to no risks nor loses; whereby they shall acquire greater security and utility, besides being able to issue a better currency, based on actual values invested or deposited, instead of bank notes or mere promises, often broken, when such Banks fail or are made unsafe. EVERY BANK LIABLE TO SUDDEN CALLS AND LOSSES IS UNSAFE; they shall become SAFE when these risks will be obviated by changing their actual unsafe dealings: besides preventing the practice of gambling in their own stocks, and their depreciation.

---

1. The ellipsis is Rafinesque's.
2. Pages 183–86 of *Atlantic Journal* reprinted as a pamphlet. The only known copy of the pamphlet is at the Library of Congress.
3. *American Manual of the Grape Vines* (Philadelphia, 1830). The only publication on Mulberry Trees known to Rafinesque's bibliographers is his *American Manual of the Mulberry Trees*. The title page of this pamphlet states that it is number 5 of the publications of the Eleutherium of Knowledge and that it was printed in Philadelphia in 1839, so the reference here—in 1837—must be to a publication yet to be discovered.
4. *Bulletin Nr. 3 of the Historical and Natural Sciences* had been published in May 1836, and probably was the one enclosed.

V.

Public Lectures

11

Inaugural Lecture

Proposed in October 1819, the botanical lectures of Rafinesque at Transylvania University did not actually begin until the spring of 1820, because he could not collect a paying audience until then. Written as part of the author's continuing campaign to interest both town and gown in natural science, this lecture makes no pretense at originality of content. Its purpose was to make known both the beauty and utility of botany in order to excite its auditors to want to learn more.

This introductory lecture of the series was edited from a manuscript preserved at the American Philosophical Society and was first published in 1983 as a souvenir of Transylvania's Bicentennial Celebration of the naturalist's birth. Rushed into print, that edition was marred by a number of misprints, which have been corrected here. Since the text was intended for the author's eye only, its many cancellations have not been reproduced; abbreviations have been expanded, punctuation normalized, even paragraphing changed—all to enable the reader to experience the lecture in the theater of the mind as a spoken address. The performance may be further enhanced if one's imagination flavors the words with the noticeable French accent that Rafinesque retained throughout life.—*Editor*

* * *

First Lecture. On Botany.

Ladies and Gentlemen:

I shall undertake to speak of blossoms and flowers. They have always been pleasing objects to contemplate and admire! Should my pictures be equally interesting, their study will not fail to appear amiable and inviting. It will be my aim at least to convince you of it, and to excite a desire of bestowing some attention upon these beautiful ornaments of our gardens, our fields, and our woods.

Let us begin by taking an accurate survey of their study in connection with the other branches of human knowledge.

Whatever be the number of sciences, they may all be reduced to three great classes, according to the three great means they employ in order to reach and obtain the needful certitude in their research of truth. These means are the experience

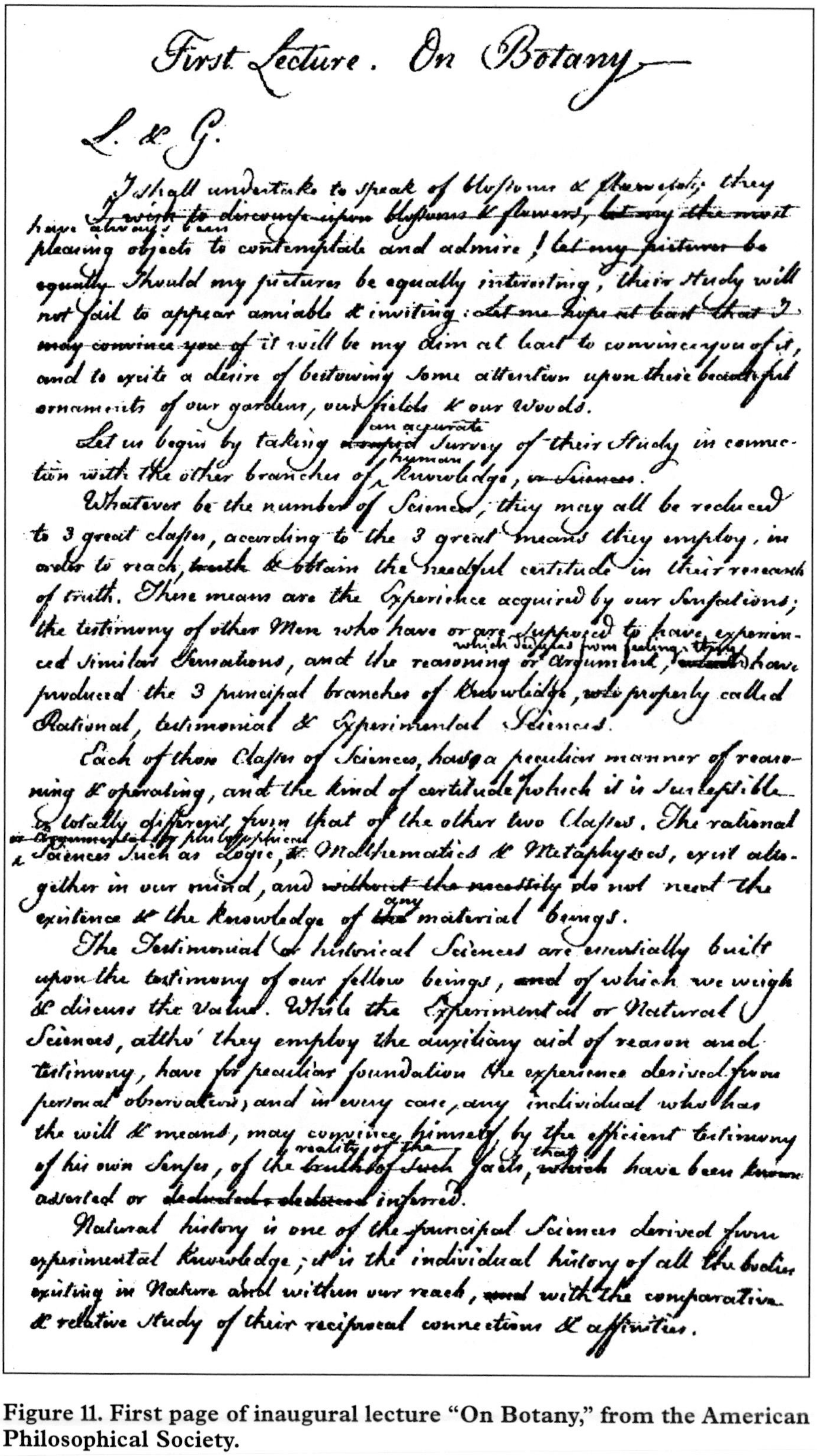

First Lecture. On Botany

L. & G.

I shall undertake to speak of blossoms & flowers; they have always been pleasing objects to contemplate and admire! Should my pictures be equally interesting, their study will not fail to appear amiable & inviting: it will be my aim at least to convince you of it, and to excite a desire of bestowing some attention upon these beautiful ornaments of our gardens, our fields & our Woods.

Let us begin by taking an accurate Survey of their Study in connection with the other branches of human Knowledge.

Whatever be the number of Sciences, they may all be reduced to 3 great classes, according to the 3 great means they employ, in order to reach & obtain the needful certitude in their research of truth. These means are the Experience acquired by our Sensations; the testimony of other Men who have or are supposed to have experienced similar Sensations, and the reasoning or Argument, which differs from feeling; they have produced the 3 principal branches of Knowledge, properly called Rational, testimonial & Experimental Sciences.

Each of those Classes of Sciences, has a peculiar manner of reasoning & operating, and the kind of certitude which it is susceptible is totally different from that of the other two Classes. The rational Sciences such as Logic, Mathematics & Metaphysics, exist altogether in our mind, and do not need the existence & the Knowledge of any material beings.

The Testimonial or historical Sciences are essentially built upon the testimony of our fellow beings, of which we weigh & discuss the Value. While the Experimental or Natural Sciences, altho' they employ the auxiliary aid of reason and testimony, have for peculiar foundation the experience derived from personal observation, and in every case, any individual who has the will & means, may convince himself by the efficient testimony of his own Senses, of the reality of the facts, that have been asserted or inferred.

Natural history is one of the principal Sciences derived from experimental Knowledge; it is the individual history of all the bodies existing in Nature and within our reach, with the comparative & relative study of their reciprocal connections & affinities.

Figure 11. First page of inaugural lecture "On Botany," from the American Philosophical Society.

acquired by our sensations, the testimony of other men who have or are supposed to have experienced similar sensations, and the reasoning or argument which deduces from feeling. They have produced the three principal branches of knowledge, properly called rational, testimonial, and experimental sciences.[1]

Each of those classes of sciences has a peculiar manner of reasoning and operating, and the kind of certitude of which it is susceptible is totally different from that of the other two classes. The rational or philosophical sciences such as Logic, Mathematics, and Metaphysics, exist altogether in our mind and do not need the existence and the knowledge of any material beings.

The testimonial or historical sciences are essentially built upon the testimony of our fellow beings, of which we weigh and discuss the value. While the experimental or natural sciences, although they employ the auxiliary aid of reason and testimony, have for peculiar foundation the experience derived from personal observation, and in every case any individual who has the will and means may convince himself by the efficient testimony of his own senses of the reality of the facts that have been asserted or inferred.

Natural history is one of the principal sciences derived from experimental knowledge; it is the individual history of all the bodies existing in nature and within our reach, with the comparative and relative study of their reciprocal connections and affinities.

Whatever is useful, sublime, and admirable, whatever the heavens, the air, the water, and the bosom of our globe exhibit to our curious and astonished sight—in a word, the whole of material existence—belongs to the dominion of the Naturalist. The atom of sand and the lofty mountain, the huge whale and the imperceptible animalcule, the colossal trees and the inconspicuous mosses and mould are all within the pale and limits of the history of nature, and it extends not only to the investigation of the attributes belonging to the living bodies or animated beings and to the principles of the primitive substances and elements, but to the research of the causes, effects, and affections of matter and life, and of their physical laws and moral actions.

Man himself, that king of the earth, whose powers appear to rise above the level of matter, is yet a dependent on his material existence, and the history of his species is only one fraction of the general history of his fellow beings on earth.

The sciences which he has created, the genius which he has displayed, the arts which he has invented belong to the history of his powers. His civil and moral laws, with the actions of his societies, tribes, and nations, are merely the details of the successive events of his history.

But to facilitate our studies we have divided and subdivided our knowledge and restrained Natural History to the positive acquaintance with the material elements, bodies, and beings.

Among all the branches of this science, that which relates to the living beings inhabiting with us the terrestrial globe and in which number we are compelled to reckon ourselves is certainly not the least interesting. The beings gifted with life surround us with pleasurable sensations and, glowing with spontaneous motions, they throw streams of living powers and actions throughout material existence.

Life is a heavenly spark which shines brightly for a while, appears and disappears successively, but never ceases to assume a perpetuity of shapes. It is the real and exclusive appanage of the organized beings, it distinguishes them from the rough bodies of nature and enables them to resist during a peculiar period the inert properties of matter.

Nutrition and reproduction are the two essential functions of all the living and organized beings; none of them are deprived of it, although a successive increase of different properties and functions are gradually evolved among the various series of those beings. By nutrition the organized bodies are enabled to grow and preserve themselves, while by reproduction they perpetuate themselves, and individuals emanate from one another almost without end. Death is the common lot of individuals; but the types of the various bodies may claim a kind of immortality.

The multitude of vegetables which overspread and adorn the surface of the earth, the crowds of trees, shrubs, plants, and grasses which cover our mountains, fields, and meadows are also organized and living bodies, endowed with peculiar organs and functions. They possess a kind of passive life calculated for the economy of their existence in the scale of nature, and they afford an endless variety of forms, phenomena, and properties, interesting to know and important to appreciate.

All the organized bodies are formed by the reunion of those dissimilar parts called organs, which consist in the heterogeneous aggregation of molecules introduced from within and kept in an active state by the vital force. These bodies acquire continually new molecules and perish when they are unable to continue this acquisition.

The distinction between the two great divisions of living beings, animals and plants, although obvious in the vulgar sense, becomes very difficult in an abstract and accurate sense. Nothing appears more different than a cow and an oak tree; but there are so many connecting links between the lower orders of animals and plants that in fact very few essential dissimilarities or exclusive faculties can be detected between them, as for instance between a polyp and a sponge,[2] between a worm and a conferva, or between a madrepore and a mushroom.

It has therefore become needful to fix their essential distinctions in the invisible faculty which animates all animals, the sensibility, or the faculty by which they perceive the action which external bodies may exercise over them by their qualities or powers. While all the vegetable department of nature is entirely deprived of this faculty, some of them being merely endowed with the irritability, another consimilar faculty by which external bodies may act upon them in some peculiar times or circumstances.

Every other exclusive faculty ascribed to animals or plants has been found to be liable to exceptions. Mobility was, for instance, considered as essentially belonging to animals; but oysters and corals cannot move and some plants travel by their roots or creeping stems.

However, many secondary compound properties and functions, either positive or negative, will enable us to decide whether any organized being must be deemed an animal or a plant. All vegetable bodies absorb their food by pores and roots; but all animals that do the same have spontaneous and collective motions in their whole massive body, while the plants have it restrained to some parts of their body. The plants have no visible mouths, as nearly all the animals have, nor an internal cavity for the reception of food. Their organs of reproduction perish before the individual, while in the animals they last as long, etc.

The vegetables live and die like man and the animals. They appear to be deprived of the sensibility which distinguishes animals, but they have like them the faculty of reproducing themselves and they are furnished with the organs needful to their preservation, and whose motions appear sometimes to be conducted by a kind of instinct. If those beings do not feel, they often act as if they did. They are seen to direct themselves towards the light and to search in the ground or in the air and water whatever may unfold in them the peculiar kind of life spread in all their members. They choose the climates, places, and soils which agree with their constitutions, they inhale or reject the portion of air which is useful or hurtful, and they elaborate each in its manner and within themselves the juices and liquors which maintain them in health and strength.

The name of Botany or Phytology has been given to the entertaining study of all the vegetable bodies, and of all their numberless faculties, properties, and phenomena.

Botany is one of the most useful and amiable sciences. None is more worthy of our early and deep attention. The beings to which it relates embellish and adorn the earth and supply our wants as well as those of the animals. We are indebted to them for our food, our vestments, and our dwellings, besides the remedies which relieve our diseases. These organized beings grow with us, among us, and for us. They afford on every side pleasing pictures full of life and coolness, which relieve and gladden our sight, or produce a deep sensation on our souls. Their sweet scented emanations, their shades and their green carpets invite us to avail ourselves of the pleasures they afford.

If either for our amusement or our instruction we wish to take a nearer examination of these lovely beings and their pretty blossoms, they cannot, like the animals, fly our view; but fixed to the spot which has given them birth and existence they are ever ready to satisfy our desires. None of their forms, nor of their beauties, can escape our sight. We may share with the provident Nature the care of raising them, of presiding over their childhood and watching their growth. Activated by us, they never fail to reward us for our troubles, without reckoning the pleasure which we must feel in seeing the prosperity of our own labours. No man can fail to feel a sensation of delight when he looks over a fine field of wheat, which he has sown with his own hands, or when he rambles over the garden which he cultivates. The beings which have been the object of our cares do not fail to obtain a share of our affections. We love them as our children, and endeavour to perpetuate their existence or their species as long as we can.

The knowledge of the plants is not merely needful to the physicians and the cultivators, but it must interest all those who entertain a taste for the fine arts and the useful arts. None is better adapted to every condition and fortune, nor more suitable to every age and both sexes, or better able to delight our leisure hours and mitigate the daily troubles of life. This study entices us to the country and the healthy fields, it fortifies our bodies by a salutary exercise, driving away sloth and the baneful passions. Those who love that study can never become vicious,[3] they cannot step out of the towns without being surrounded by delightful objects, enticing their attention. In the middle of the tribes of vegetable beings they hold a kind of conversation together: they are interrogated, and they are compelled to reveal their secrets. The enlightened traveller enjoys new and unceasing pleasures; every country adds to his knowledge and to his collections; the farthest he rambles from the human dwellings the greatest becomes his treasures; the wildest countries, the most dreadful regions, are for him fruitful fields, where he reaps a plentiful harvest prepared for him by the hands of Nature.

Such are the sweet and peaceful enjoyments which the study of the plants affords to everybody: it has a great attraction for youth and for the female sex, and how could it be otherwise? How could it not please that gentle sex, which has so much likeness with flowers and has been therefore considered as the flower of mankind? Their hands appear to be made on purpose, as it were, to handle these delicate objects, and to assort blossoms and flowers to adorn and set off their own blossoming charms.

Many ladies of all nations have therefore applied themselves to the study of botany. Some have been contented with reaping the knowledge already acquired, while several have added their share to it and contributed to enlarge the sphere of the science.

But it must not be conceived that botany is a mere study of amusement, or that it serves at utmost to satisfy human curiosity or vanity. Far from it. Although replete with attractive sensations and pleasurable results, it is in fact one of the principal bases of all useful applications of knowledge. We derive from it the needful and the agreeable at the same time.

Every production of the vegetable kind is endowed with some property useful to man or to the other beings and bodies of which he avails himself. The poisons themselves are useful and are not poisons for all the animals. The conium or hemlock is a poison for man, but the goats eat it with pleasure and without danger. The aconite or wolfsbane is a poison for man, dogs, and wolves; but the horses eat it with impunity. It is the best poison for the wolves, which resist or avoid the effect of arsenic. The field mice are easily destroyed by sowing peas and beans previously infused in a decoction of white [h]ellebore or *Veratrum album*;[4] and the catmint or *Nepeta cataria* will entice the wild cats and catamounts into the traps of the hunters.

You must consult botany whenever you wish to know what may suit best your cattle. When you want to distinguish the grasses which they prefer and the plants which poison them or produce most the disorders to which they are liable.[5]

There is a plant which the snakes cannot bear. It is the *Ligusticum*;[6] if it is planted in your gardens or yards no snake will come near, and if you rub your hands with this plant you may handle with safety even the most dangerous snakes, since it appears to throw them in a kind of lethargy or stupor. It is by this means that the *Psylle*, or jugglers of Europe, are enabled to perform many tricks with the vipers. Other plants possess probably a similar property in India, Africa, and South America, since the same results are obtained.

It may be interesting to know that the corol[la] and the bark of the Lombardy poplar drives away the locusts? that bugs are killed by the smoke of the cayenne pepper, the infusion of the *Acorus*[7] or sweet flag, and of the hemp seeds?

In agriculture a crowd of plants and trees could be gradually introduced and exchanged from countries to countries. All our fruit trees came from the East and have now reached America. The apricot tree is native of Armenia, the cherry tree of Asia Minor, the quince tree of the Island of Crete, etc. There was a time when the most cultivated countries of Europe and the United States were woody regions, inhabited by huntsmen. Let us proceed in diffusing the useful productions of the earth from one region to another. Botany will teach us the soil and climate which they may require. The flax is a native of the inundated banks of the Nile; we are therefore certain that the overflowed banks of the Ohio and Mississip[p]i will offer to that plant the most convenient soil, and that it will thrive also in drained marshes and bogs.

The cotton has only lately been cultivated in the United States and has proved a source of wealth.[8] Sugar—and grape vines[9]—begins to be also introduced in such parts as will produce it. Why should we not follow this beneficial plan and enrich our country with the valuable productions of Europe and Asia, such as the olive tree, the tea shrub, the almond tree, the teak wood, the European shumack, the date palm, the pistachoe nut, etc. Let us at least do our best, and do not believe those who may tell you that it will be better and cheaper to go and buy tea in China. They might as well say that it is better for us to go buy our dry cherries in Turkey and our wheat on the shores of the Mediterranean. Let all the nations raise all the productions which their soil and climate will admit of and avail themselves of all their indigenous productions.[10] Commerce will be sufficiently alimentated by those which are necessarily confined to peculiar regions of the earth.

How many native trees and plants of our country deserve yet to be carried from our woods to our fields, orchards, and gardens? Our wild plum trees, our persimmons, our hickories, our pecan trees, our grapes, our whortleberries, etc. ought to be cultivated carefully and they might probably improve as much as the wild pear trees, apple trees, and olive trees of Europe have during a long period of cultivation. The wheat and barley, the cabbage and the carrot, the sugar cane and the cotton, the flax and the beans were once wild weeds, which some worthy benefactors of mankind were induced to recommend to their fellow beings for cultivation. This adoption of vegetables has been the mother of agriculture, and the improvement of the human race was the result of this new art. Who can say where the progress of improvement must stop? We have certainly not reached the utmost limit of this art. Botany is the study by which we may hope therefore to progress in the amelioration and multiplication of vegetables and trees, and through it to increase our individual and national happiness and welfare.

The *Zizania* or wild rice is a native of the lakes and the coldest regions of North America; it could be cultivated all over the United States and in many useless places where the common rice will not grow, particularly on all the overflowed banks of the Ohio and Mississip[p]i. The *Asclepias syriaca* or silkweed deserves to be cultivated with us, as it is already in some parts of Germany and France. It is one of the most common weed[s] along our streams, and grows by millions on the banks of the Ohio. The pods are full of a silky down from which hats and quilts are made; it may even be spun with cotton.[11]

We have wantonly allowed the useful cane, or *Miegia arundinaria*, to disappear from our country. This becomes already a motive of regret, and our cattle are nearly starved in winter. This might have been prevented by the cultivation of the cane and by cutting it in winter instead of allowing cattle to eat it up as fast as it grows and trample upon it. If we should let our cattle eat up the clover as fast as it shoots, we should never be able to make hay with it, and one cow would destroy more than ten cows could eat when gathered. I venture to recommend an early attention to this important subject before the whole of this useful grass is extirpated. It will afford an invaluable perennial and evergreen pasture at all times of the year, when regularly cultivated or merely inclosed. The seeds of the cane are also very useful; they are plump and large, much more than barley, and they contain a fine white meal, which might be used to make bread or cakes, or to feed horses and poultry. The stems of the cane may serve for fishing rods, walking sticks, props, mattresses, looms, partitions, roofs, etc.[12]

The valuable sugar maple disappears also gradually from our woods, being wantonly burned to clear the ground or cut down to make fences and afford fuel. It has been calculated by Dr. [Benjamin] Rush that we could supply ourselves from that tree of all the sugar wanted for home consumption in the United States and have yet a large share left for exportation. If manufactured on a large scale as sugar is in the West Indies, we could afford to sell the maple sugar at 6 or 8 cents the pound.

This tree is so much esteemed in Europe that large plantations have been made of it in France and Germany. In 1803 the Prince of Liechtenstein sent at his private expense to the United States a botanist, Mr. [Joseph] Van der Schott,[13] for the express purpose of collecting the seeds of it. He sent to the prince more than 250 barrels of them, which have been sown in his estates in Germany and Hungary, and whole forests of that valuable tree are now growing from which sugar has already been made.

We ought not to be blind any longer to our own interests; we should respect every maple tree as a holy tree, we ought to enact laws for its preservation, we ought to plant it in our woods and fields instead of the Lombardy poplar, the useless sycamore, and even the less valuable oaks and hickories. If every farm had merely one sugar tree upon each acre, in full produce, we should be independent of foreign countries for the supply of sugar. It will be time to cut them for fuel when they are grown old. Let us consider them as the cows of the vegetable kingdom. With the cow and the maple our country will always flow with milk and honey. The cow gives us milk during her life and feeds us after her death. The maple will afford honey and sugar during its vigor and warm us after its decay.

Botany is then a study of real interest, extensive application, and vast extent. This science applies to all the vegetable productions of the whole earth and even of the bosom of the sea; but it is not necessary for everyone to become acquainted with the whole of them. The most interesting and valuable ought only to claim the attention of every man, together with the local botany of his own country. And even among the plants growing among us, all will not claim an equal share of attention. Although they have each their rank in the general economy of nature and are equally valuable in the eyes of the philosopher and botanist, a certain number only will command the attention of every friend of mankind by their properties and qualities. Nobody ought, therefore, to despair of acquiring this useful knowledge, nor be detected [deflected?] from entering into that study, by the fear of being unable to acquire the needful proficiency.

But they ought not to fall into the common and baneful error of asking merely what a plant is good for and its name without enquiring at all into its ulterior qualities, functions, and structures. What would be thought of a gentleman who, introducing a stranger to a friend, should ask him at once this blunt question? Pray, Sir, what's the name of this man and what is he good for? This question could only be answered if the friend knew all the stranger's circumstances or after he should have enquired into them, and might, upon the whole, be deemed impolite and rude. The rules of society teach us to endeavour to become acquainted with those fellow-citizens or strangers which may happen to be introduced to us by chance, or through friends, by holding a conversation with them, and by indirect enquiries into their characters, professions, circumstances, and acquirements. But should we not understand the language they speak we cannot acquire that knowledge without learning their language or availing ourselves of an interpreter. Plants have also their peculiar language and we can hold with them a kind of mute conversation by which we acquire the knowledge of their structure, functions, qualities, and properties. When we are unacquainted with this language any botanist will interpret for us; but it ought not to be our first question to ask, What is that being good for?, but rather the ultimate and gratifying result of our various enquiries respecting, or directed to, that new acquaintance. As all plants have their names engraved upon them in a peculiar alphabet, it will be also very easy and pleasing to learn them by acquiring this alphabet, and we shall remember them much better than even if we should ask for them repeatedly.

The science of botany divides itself into many branches. The branch which teaches us the alphabet of the science is called taxonomy, meaning the laws of order. Several alphabets have been invented for botany under the name of classifications, nearly as many as the various alphabets of languages used by mankind. But most of them are obsolete and unemployed at present, the Linn[a]ean System or classification having superseded most of them, although it has many anomalies and irregularities, on which account the natural method or classification is now preferred by the best botanists since it is a kind of natural alphabet, in which all the letters and parts have a natural and unvariable meaning.

Glossology is the branch which teaches the technical language of the science, its name meaning the study of language. We are enabled by it to become acquainted with all the terms relating to botany and the plants; and as they are mostly derived from the vulgar language and the other sciences, they are easily acquired.

Nomenclature is another branch, which acquaints us with the names of all the plants. These names, being derived from physical characters always perceptible, or being constantly coupled with the alphabets above mentioned, are easily found out, somewhat in the same manner as you might look for a name into a dictionary, every character of the plants being one letter of that alphabet, and their combination forming the representation of the names. Synonymy is a secondary part of nomenclature; it gives us the knowledge of all the names given by different nations and authors to the same beings, referring them to the scientific and technical botanical names which are thus become common to all nations and to all writers.

The branch called Phytography or description of plants is the grammar of botany; by its mean[s] we are enabled to describe accurately all the plants, and to make them known by employing the technical language of the science, which Glossology has taught.

Botanomy, or the laws of botany, may be compared to the syntax of the language and of this science. By teaching us the rules, principles, and laws of the construction and fabric of botany, it enables us to proceed properly in the study and in the practice of the science.

Besides the above fundamental branches of the science, all the departments of natural knowledge belonging to the natural history of the minerals and animals form also essential parts of botany. They are the historical, physical, analytical or chemical, anatomical, geographical, iconographic, medical, agricultural, and moral departments, which relate to the history, the physiology, the anatomy, the chemistry, the geography, the delineation, the diseases, the cultivation, and the manners of plants. All these departments form an essential and extensive branch of botany which may be called the philosophy of the science.

But another valuable branch of botany consists in the application of the fundamental and philosophical branches to the practical use of mankind, by the investigation of the qualities and properties of the vegetable bodies. This study, which may be called Phytocreny, forms the ultimate result of all previous enquiries and must be divided into four departments.

The first relates to the alimentary properties of plants, and teaches us by practice, experiment, or analogy which of them may afford us food and drink. Among the foods afforded by vegetables are the various kinds of cereals, farinaceous seeds and roots from which bread, pies, and cakes may be made; the roots, stems, leaves, and fruits which may be eaten raw or boiled, roasted, and variously cooked. Their variety and abundance is such that all mankind could be fed by them, without the necessity of resorting to animal food. Among the drinks for which we are indebted to trees and plants may be reckoned the various wines, beers, lemonades, spirits, teas, decoctions, syrups, juices, milks, and oils which we draw from fruits, seeds, or stems, in every climate. In fact all our drinks are derived from the vegetables except milk; and some fruits, such as almonds, cocoa, and the milk tree or cow tree[14] of South America afford likewise a kind of vegetable milk or various milky emulsions.

The economical department of Phytocreny relates to all the uses which we can derive from plants, except the alimentary and medical uses. It is the most extensive of all departments, since we employ plants and trees for our clothing, our buildings, our fuel, our furniture, our weapons, our dyes, our arts, our wants, and our pleasures, in a numberless variety of forms, shapes, and modifications. Each of those uses might afford many interesting details and reflections, which I shall not fail to notice in my lecture on the properties of plants; but there are several other additional uses which hardly fall under any of those descriptions and are yet interesting to know. I shall mention now some of them.

We light fire with tinder and matches of vegetable matters. The cork tree gives us its light and elastic bark, from which corks and floating implements are made. Mistletoe and many other fruits and barks serve to make bird lime and glues. Many plants produce potash and soda, from which soap is made by mixture with oils; and the vegetable ashes serve also to make lyes, wherewith to wash our clothes. Walking canes are obtained from reeds and shrubs. Our brooms and brushes are made with the panicles of many reeds, plants, and palms. Toothpicks are made with the wood of the elder, the umbel of the *visnaga*,[15] and many other plants.

By the medical department we are made acquainted with the proper remedies or palliatives for our disorders. They are enumerated under the many-fold denominations of purgative, emetic, stimulating, tonic, astringent, emollient, narcotic, and poisonous vegetables, and almost every country produces a sufficient proportion of medical plants to cure all the diseases of that peculiar climate. Botany teaches us to distinguish with certainty and accuracy, and excluding dreadful blunders.

The fourth and last department of the study of the vegetable properties is the auxiliary department, which includes the knowledge of the various foods of cattle and domestic animals, of quadrupeds, birds, fishes, bees, silk worms, etc.; the study of the various physical and moral influences of the vegetables acting one upon another, and of the relative influences and harmonies which they bear with light, air, heat, gases, salts, minerals, and animals, and even with our globe.

Such are the sublime and philosophical results of our attentive enquiries into the vegetable world. Let this view of the subject enlarge our ideas of the eminent importance of botany, and entice our attention. We can hardly find an object more deserving to employ our leisure. Whether we mean to acquire the whole knowledge of botany or whether we merely wish to form an idea of the wonderful display of vegetation, our contemplation will be drawn towards beings and bodies amiable and inviting. We shall not be compelled to travel far in search of them: they are always near us, in our gardens and our fields. A single blade of grass, or blossoming plant, will afford us an unremitted enjoyment and might afford us occupation during many years if we were to consider it and study it under all its different points of view.

Ladies and gentlemen, I venture to draw your attention upon these interesting beings. You may make discoveries or add something to our former stock of facts and observations, whenever and wherever you may take up this study, either as an amusement or as a temporary occupation. Let us hope that some among you may be inducted to undertake it with zeal and satisfaction; but let none neglect to do something at least towards it.

---

1. In his introductory "Lecture on Knowledge," delivered November 7, 1820, Rafinesque expanded on his tripartite division and further divided the experimental sciences into two modes, each of them having three degrees.

2. The identification of sponges with either animals or plants was an open question long after the date of this lecture. Rafinesque was writing that "the fibrose sponges" "are real plants" in the year of his death. *The Good Book* (1840).

3. Elsewhere Rafinesque wrote: "Every pure Botanist is a good man, a happy man, and a religious man! He lives with God in his wide temple not made by hands...." *New Flora of North America* (1836).

4. "It is a poison for all insects in decoction, noxious to swine, sheep, geese, fowls; crows [are] intoxicated by steeping corn in it." *Medical Flora*, Vol. 2 (1830).

5. Rafinesque's article "On the Salivation of Horses" (*Western Review*, 1819) advised ridding pastures of *Euphorbia hypericifolia* and *Lobelia inflata* to prevent this disorder in animals which were greatly valued in Bluegrass Kentucky. He added that a cure might be effected by feeding a stricken animal with cabbage leaves.

6. "Lovage, smellage." *Medical Flora*, Vol. 2 (1830).

7. "Cattle will not eat this plant, and it is noxious to insects; the leaves, therefore, may be used to advantage against moths and worms." *Medical Flora*, Vol. 1 (1828).

8. Rafinesque resurrected the late 18th century study of West Indian cottons by J. P. B. van Rohr, on which he based scientific names for 38 species and varieties of *Gossypium* in *Sylva Telluriana* (1838).

9. Rafinesque's *American Manual of the Grape Vines* (1830), which appeared first as a section of Vol. 2, *Medical Flora* (1830), accounted for 300 varieties of 62 species of grapes, 42 of the species said to be native American. The book also deals with the horticulture of vines and with oenology.

10. Rafinesque's *American Manual of the Mulberry Trees* (1839) contributed to the considerable

literature of the time intending to encourage silk production in the United States. He also planned a companion volume on beet sugar, which was never completed.

11. "All the Asclepias [species] are milky; but this [*A. tuberosa*] less than others. They all produce a fine glossy and silky down in the follicles or pods; which has been used for beds, hats, cloth, and paper. This down makes excellent beds and pillows, being elastic, and one pound and a half occupying a cubic foot. Light and soft hats are made with it: the staple is too short to be spun and woven alone; but it may be mixed with flax, cotton, wool, and raw silk. It makes excellent paper, and the stalks of the plants afford it likewise, as in flax and Apocynum. The *A. syriaca* or Silky Swallow-wort producing more of the down, has been cultivated for the purpose, and a pound of down produced from forty to fifty plants. Its young shoots are edible like poke, and the flowers produce a honey by compression." *Medical Flora*, Vol. 1 (1828).

12. Audubon's best known plate, that of the wild turkey cock, displays the bird against the background of a cane brake.

13. Rafinesque's *A Life of Travels* (1836) reveals that they were personally acquainted.

14. *Palo de vaca* in Brazil; *Brosimum galactodendron.*

15. "Tooth-pick Bishop-weed, *A[mmi] visnaga*, is so called on account of the use made in Spain of the rays or stalks of the main umber. These, after flowering, shrink, and become so hard that they form convenient tooth-picks." John Lindley and Thomas Moore, *The Treasury of Botany* (1866).

12

Greek Independence

The Greek War for Independence began in 1821 with a revolt led by Alexander Ypsilantes. Remembering their own war for independence, many Americans of all levels of society fervently embraced the Greek cause, though their government tried to remain neutral. Public meetings of students to raise money began at Yale University in December of 1823; they were quickly followed by others elsewhere. Naturally, Transylvania, located in "the Athens of the West," had to join the good work, which it did by holding its rally on the anniversary of the Battle of New Orleans. President Horace Holley was one of the speakers on that occasion, but the star performer was Transylvania's own "Constantine of Byzantium." Not to be outdone, the young women of the Lexington Female Academy held their own gathering two days later. Both groups passed pro–Greek resolutions and collected money for the cause. All told, by the end of 1824, private Americans had raised about $40,000 for the Greeks—a sum greater than that generated in England—and the Americans also contributed firearms, swords, and medical supplies.

Rafinesque's insistence in his autobiography that the suburb of Constantinople where he first saw light of day was Grecian "soil and climate" and that his mother, also a native of Constantinople, was "Grecian born" have caused confusion in biographical accounts since. Like other Europeans and Americans of his time, he did not want to be identified with the Muslim Turks of the Ottoman Empire. Until he was naturalized in Philadelphia in 1832, his citizenship followed that of his French father. The Rafinesques had been Protestant Christians for generations.—*Editor*

* * *

Address
Of a Native of Grecian [*sic*],
To the Citizens of the Western States.

A fellow-citizen who drew his first breath and had his cradle in that Grecian land, the ancient seat of the Muses, Philosophy and Liberty, which now enlists the best feelings of American freemen, and who finds himself thrown by human events and his free choice among you, Western Republicans! takes the liberty to address to you a few suggestions in behalf of the modern Grecians, now struggling against the thral[l]dom of Ottoman despotism.

The holy enthusiasm of Liberty and Independence, kindled by America in 1776, is reviving again throughout this continent, in favour of our fellow sufferers in the cause of freedom, in both hemispheres. The oppressed, but brave, noble and worthy Grecians have particularly excited our interest. From the Atlantic shores a bursting voice was heard, which has reached the remote regions of the West; it spoke aloud—*Let us help our Grecian brethren, let us answer their pathetic appeal*—Transylvania University, the central seat of science in the West, has already followed the example of New York, Philadelphia, Boston, New Haven, Andover[1] & Washington, and open[ed] this liberal path in the West, by coming forward to their relief.

Freeborn Republicans of the West! hasten to follow those worthy attempts, if you value as you ought the institutions which have secured to you your actual prosperity. Revive among you by public meetings and discussions, the holy sentiments of freedom and the heavenly flame of sacred Liberty. Let them stand in bold opposition to the baneful doctrines of the Eastern despots, and the result will become not merely beneficial to the Grecian cause, but to yourselves and to your latent posterity.

The knowledge which I possess of the manners, sentiments and wants of the Grecian confederation, enables me to suggest some of the most proper and efficacious means to relieve, aid and assist the new-born Grecian *Liberty*, and will be my excuse for intruding upon your attention, by attempting to direct your generous boons.

The Greeks want money, but they likewise want arms, weapons, guns, powder, lead, and naval stores above all. I would not suggest to you to send mere money, unless indeed it may be a large sum in the shape of a loan, for which they will gladly give you from 8 to 12 per cent. interest, the usual rate in Greece, and pledge you six millions of acres of land, conquered from their invaders; but I would invite you to send them such articles of your superfluous produce, as would be eminently useful to them. For instance:

Timber, spars and masts, for Ship-building.
Gunpowder, from our own manufactories.
Lead and balls, from the mines of Missouri.
Cables, ropes and twine from our hemp factories, wanted by the Grecian navy.
Hemp cloth, from the same, for canvass, bagging and coarse clothing.
Cotton, even the coarsest, from which they manufacture their sails, using none but cotton sails.
Salted beef and pork, rice, beans, flour &c for food.
Some tobacco and whiskey, both prized by the seamen.
Some small steam engines and models of torpedoes, which might by their novelty and destructive powers become the most potent enemies of the ignorant Turks.

Such are, Western Americans! the products of your soil and industry, which the Greeks will receive with the most grateful acknowledgments. By sending them instead of money the precious metals will not be drained from the Western States, where they are so much wanted; and the Greeks will receive the full amount of your bounty, without the reduction of exchange.

The best manner of forwarding these supplies, will be by our natural channels, the Ohio and Mississippi. Patriotic citizens will be found, who will charge nothing or very little for freight and storage. From N Orleans let one or more vessels be chartered for Gibraltar, Malta or Leghorn, free ports, where the supplies may be deposited in the hands of the Grecian merchants or agents, to be taken away or sent when most wanted. It would be perhaps imprudent to convey these supplies direct under our own flag into the ports of Greece, unless we are willing to give up the Turkish trade.[2]

Such are the hints which my acquaintance with Greece, and my love of liberty, have impelled a glowing heart to address to you.

May the Genius of Freedom, who spreads his golden wings over this continent, smile upon your enlightened deliberations and liberal thoughts! guide the associations and committees which may be formed in every Western State, and pour the choicest blessings over this happy land, the home of good, free and generous souls!

The joyful thanks of the Grecian heroes, widows and orphans await you and will be your reward at the reception of your gifts. Despots may frown, but heaven will smile and register the deed.

CONSTANTINE,
of Byzantium.

Lex: *Ky*: January 8, 1824.

☞ Editors friendly to the cause of Liberty, are requested to give circulation to this address in the Western States.[3]

Published in the *Kentucky Reporter*, January 12, 1824.

1. Transylvania was responding to a circular letter sent out by students of the theological seminary at Andover, Massachusetts, asking "fellow Christians" to help the Greeks. The money collected at Transylvania was forwarded to Andover.

2. Boston merchants especially were conducting a lucrative trade in opium through the port of Smyrna.

3. The only reprint found is that in Lexington's own *Western Monitor*, on January 20, 1824.

VI.

Popular Science

13

The Cosmonist

To reach a larger audience than that of his university lecture room, Rafinesque contributed a number of articles to the newspapers printed in Lexington. No doubt, too, he expected these articles to excite the interest of auditors willing to pay to attend his university classes, for which he received no salary. To the weekly *Kentucky Gazette* he contributed a series of fifteen "Cosmonist" papers in 1822 as part of his campaign to make natural history popular. Since each focuses on his own discoveries it is likely that he expected them to enhance his reputation as well. Under rules for priority established by zoologists and botanists long after Rafinesque's death, the occasional technical descriptions appearing in them constitute "valid publication," even though the newspaper was not a scientific journal and its circulation hardly extended beyond Kentucky.

Cosmonist I begins by explaining the meaning of the title. In numerous other places Rafinesque advocated the term Cosmony in place of Natural History—see his "Survey of the Progress and Actual State of Natural Sciences" in this volume, for instance, as well as his review of Maclure's *Geology*—but, like most of the other neologisms of his invention, it never caught on.

The Cosmonist subjects are as varied as their author's own natural history interests: fossil trilobites, birds, ants, turtles, and, of course, plants, including roses. Cosmonist X that anthropomorphizes ants may give us some inkling of how entertaining his lectures could be. According to one of his former students, as quoted in R. E. Call's biography (p. 43), Rafinesque's "lecture on the ants was peculiarly instructive and interesting ... especially when he described them as having lawyers, doctors, generals and privates, and of their having great battles and of the care by physicians and nurses of the wounded."

Cosmonist VIII, "On the Botany of the Western Limestone Region," though brief, is especially noteworthy. Here was an early discussion of phytogeography (the geographical distribution of plants), a specialty in which Rafinesque pioneered, as well as his clear expression of the concept of ecological succession, another Rafinesque insight in advance of his time.

As a follow-up to Cosmonist XV—unknown to bibliographers until 2001—has been added another article on the same subject published two years later in a Cincinnati magazine. Both relate to what has become since a well-known ornamental tree, commonly called the Yellowwood. This tree's generic name is one of about three dozen of those devised by Rafinesque that have never been disputed.—*Editor*

* * *

The Cosmonist—*No. I*

On a large Fossil Trilobite of Kentucky

Beneath our soil, how many beings lay
Conceal'd and buried deep in huge rocks!
They once had life—like us they breath'd and moved,
Like them we all must die, and sink in graves
Of various sorts, where crumbling bones reveal
Our common fate, or past existence tell.

It is my intention to publish a series of Essays on some of the most important and interesting discoveries and observations relating to the various departments of Natural Science which I have made in the western states, and to offer them under a popular form and style, in order to render them acceptable to the majority of readers, and to those who may feel a share of interest in whatever relates to the knowledge of the valuable productions of this region.

The title which I assume is not a new one, although it may be new to many readers: *Cosmony* and Natural History are synonymous names, like *Physics* and Natural Philosophy, *Ethics* and Moral Philosophy; therefore a *Cosmonist* is a Naturalist, or a man who studies the world and all its wonders. It derives from the Greek word *Cosmos*, which means the world.

The number of Discoveries made within a few years in the various branches of natural history are almost beyond belief. Twenty-five thousand new Animals and Plants have lately been collected in Brazil, South America, the shores of the Mediterranean, Western America, Australia, Polynesia, the East Indies, Nepal and Thibet, by intelligent and laborious Naturalists.

In North America nearly 2000 undescribed Plants and Animals have lately been detected; many of which are fossil remains of extinct species. It has been my happy lot to be the first to observe and describe nearly one half of this number. I shall perhaps return to this subject and delineate these progressive additions to our knowledge. But many of these discoveries are merely interesting to the Naturalists and Botanists, while some among them assume a general interest by some peculiarities, such as remarkable size, beauty, usefulness or rarity.

The Fossil Animal which will form the principal object of this Essay, is of that kind, since only one specimen has been found and its size exceeds that of all other consimilar animals. The specimen alluded to has been presented to the museum of Transylvania University by Jesse Bledsoe, Esq. of Paris; it was found in 1821, many feet under ground, in a limestone quarry on Cane Ridge, Bourbon county, in this state, (Kentucky,) and is quite perfect. The Animal is entirely petrified and become flinty, with some pyritaceous spots; it is fixed to the stone by the whole of its lower surface, but the upper surface is free and in beautiful relief. The body is quite extended, 7 inches long and 3½ wide, which is more than double the length of any consimilar Animal of the same tribe.

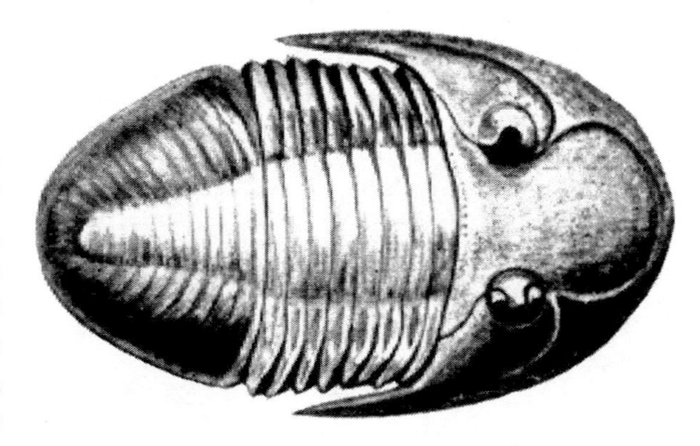

Figure 12. *Isotelus* (DeKay 1824) / *Isoctomesa* (Rafinesque 1822), from Maryland Geological Survey (1906).

When found it was mistaken for a petrified fish, turtle, or frog, while it is totally different from either, having no visible limbs or fins, but the body divided in several segments—This reduces it to the class of Crustacea, where it will belong to the tribe of Trilobites. This name was given in 1811, by [James] Parkinson, in his Organic Remains of a Former World [London, 1804–11], to several fossil Animals, having a trilobated head, two eyes and several segments to the body.[1]

Our *Trilobite* is of a brown colour, quite smooth, with a large semicircular head, having in front a marginated edge, and two large distant lateral eyes. The body is larger than the head, divided in eight equal narrow segments, with an oblique angle on each side. The tail is larger than the body, semicircular and entire.

This fossil Animal must therefore constitute a new genus in the tribe of Trilobites, distinguished by the semicircular head and tail, and the eight equal segments of the body and tail, with a trilobe head.

In a memoir lately written on the Trilobites, (wherein I describe fourteen species, of which seven are American and new,) and sent to Europe for publication in a scientific Journal,[2] I have named this new genus Isoctomesa, which derives from four Greek words, meaning *eight equal segments* in the *middle*, and the species is called *Isoctomesa magna*, or Big Isoctomesa.

As the Trilobites were marine Animals, living in the sea, at the period when our continents were covered by the ocean, and before the formation of the stony strata where they are found Fossil remains of animals attract a great deal of attention among Naturalists since the late labors of Sowerby, Lamarck, Brongniart have shown how interesting they are for the history of our globe.[3]

Petrified Crustacea are scarcely ever perfect, and unique specimens are very valuable,[4] and the *Isoctomesa* belongs to this description. The Trilobites are an extinct tribe of Animals, and no longer present in any sea: The animals now living to which they have any affinity, are those of the Astacian tribe, which have however antenna[e] or horns on the head and appendages to the tail, organs missing in the Trilobites; they had probably articulated feet beneath the body as the Oniscians, but they are concealed in all the specimens.

C. S. Rafinesque.

Published in the *Kentucky Gazette*, 31 January 1822.

1. Actually, the word trilobite (*Trilobiten*, in German) was first used by Johann Ernst Immanuel Walch (1725–1778) in volume 3 of his monograph *Die Naturgeschichte der Versteinerungen* (Nürnberg, 1771).

2. If published, the article has not been found by Rafinesque's bibliographers. He sent many articles to Europe for publication that got lost in transit. See also comment on Cosmonist No. II.

3. Rafinesque returned to the subject of Trilobites a decade later (*Atlantic Journal*, pp. 71–73), where he acknowledged that his *Isoctomesa* has been published by DeKay in 1824 under a different name.

4. Without pricing it, Rafinesque advertised for sale one of two specimens of *Isoctomesa* in his *Enumeration and Account of Some Remarkable Natural Objects* (Philadelphia, 1831), p. 2.

The Cosmonist—*No. II*

[There is no reason to doubt that this appeared in the *Kentucky Gazette*, 7 February 1822, but no copy of that issue has ever been found.

The French paleontologist Alexandre Brongniart (1770–1847) kept cryptic notes about his professional correspondence. One note alluding to publications Rafinesque had sent him says (in translation) "Memoir on the trilobites, cyclorites ... ditoxsus" where the ellipsis sign appears in the original. This might be the Rafinesque memoir alluded to at the end of Cosmonist I. Or it might be a copy of Cosmonist I itself, in which case "cyclorites ... ditoxsus" could refer to a copy of the missing Cosmonist II. The last named organism, later spelled "Ditaxopus," is briefly treated in Rafinesque's *The Good Book*, 1 (January 1840), 66–67—*Editor.*]

The Cosmonist—*No. III*

The Songsters fill the groves, the Swallows fly,
With genial spring from distant climes returning
In happy crowds: the warblers sing their loves,
The fair blue-birds in sprightly mood proclaim
This season of delight, and ev'ry where
The notes of airy tribes, salute the ear.

On the Birds of Kentucky and a new Swallow.

Among all the tribes enlivening animated Nature, there are few, if any, that are so interesting as the Birds; those aerial beings who enjoy the glorious privilege of roaming through the atmosphere, and soaring to the clouds, whence they often may look down with pity upon us confined as we are to creep on the ground. Their lively plumage, and elegant forms charm the eyes, while their melodious voices and varied songs delight our ears.

Ornithology or the branch of Cosmony, which leads us to become thoroughly acquainted with their history and manners, has therefore been cultivated sooner and better than many other branches of natural science. The splendid works and colored figures of Catesby, Edwards, Buffon, Vieillot, and Wilson,[1] have contributed to make known, the greatest number of the beautiful Birds which live in North America.

The magnificent work of Wilson, published in our country, is well known; and although it has remained imperfect by the untimely death of the author, it stands as a monument of genius, science, and taste. It is also a pity that the worthy author was not aware, that another American Ornithology had been undertaken some years before his, (in France, by Vieillot) which has likewise never been completed, where some of his new species were previously described under different names.

The result of Wilson's labors, consist in about 320 figures, belonging to nearly 300 species, among which he has described 56 as new, which might be reduced to less than 50, by comparing them with Vieillot's new species; but increased to about 70, by adding thereto several birds which Wilson did not consider as new, and blended with foreign species, while they are really distinct, and ought to be separated, distinguished and named, as I have done in my manuscript criticism of his work.

Extensive as this number may appear it is less than one half of the real number of our birds. In Ord's Catalogue of the Birds of the United States 573 species are enumerated; but in my Manuscript Catalogue I have ascertained and distinguished above 660 species, among which about 60 species, have been discovered by myself, and described as new. Some of these are already published; but the greatest part are only extant in my manuscript.[2]

Among this number I have already observed and ascertained that upwards of 200 species are found in Kentucky, nearly 40 of which are new for the science of ornithology. These new species belong principally to the Genera or tribes of Warblers, Rails, Hawks, Ducks, Swallows, &c.

Some of our Birds belong even to new Genera, and I published in 1818 in the French Journal of Physics and natural history,[3] the description of a new genus under the name of *Rimamphus citrinus*, to which a single species belongs, which was first discovered in 1808 near Louisville by Mr. Audubon, and mistaken for a Warbler; but it is distinguished from that tribe by its bill open on the sides, and round mandibles. It is besides a silent bird of a pale yellow colour.

There are two [new?] species of Swallows in Kentucky, besides several well known species. One of them[,] the red-head Swallow (*Hirundo phenicephala* in ornithology)[,] was already mentioned in my annals of nature No. 1, spec. 10. It is a rare species; grey above, white beneath, with a scarlet head, the bill and feet black.[4]

The second species I shall now describe and call it the Blue Bank-Swallow. I have given it the scientific name of *Hirundo albifrons*[,] which means the Swallow with a white forehead. It is very remarkable by its unforked tail: almost all the Swallows having a large forked tail, and a few species a large stiff and sharp tail; but in this new Swallow the tail is small and truncate, neither sharp, stiff nor forked; this peculiarity occurs also in a South American species, the Tapera Swallow, (*Hirundo tapera*,) which is however totally different from ours, being black above and white beneath.

Our Blue Bank-Swallow is a small species, about five inches long: it has a black bill and brown feet. Its face or the space surrounding the bill is black, the forehead white, the top of the head blue; the cheeks, throat and upper part of the rump of a reddish ches[t]nut colour, or rufous, the back is blue spotted with white, the belly of a dirty white, the wings brown, with some yellow spots beneath at the base, and the tail is equal unforked, truncate and brown.

This pretty Swallow is found on the banks of the Ohio, where it has only been lately noticed; whether it has lately come there from southern regions or had not been noticed heretofore, may be a matter of doubt, but of little consequence. It appears now to be rather common on some peculiar spots, such as near Newport in Kentucky and Madison in Indiana; it comes late in the Spring[,] builds its nests on the high banks of the river and disappears early. Its nest is singular, in the shape of a reversed bottle, with the opening at the end of the neck; the materials being similar to those employed by the common Swallow. This bird is to be seen preserved with its nest in the museum of Cincinnati. It deserves the further attention of the friends of science.

C. S. Rafinesque.

Published in the *Kentucky Gazette*, 14 February 1822.

1. Mark Catesby, *The Natural History of Carolina, Florida, and the Bahama Islands* (2 vols., London, 1771); George Edwards, *A Natural History of Birds* (4 vols., London, 1743–1751), Edwards also revised Catesby's book; Georges L. Buffon, *Histoire naturelle des oiseaux* (16 vols., Paris, 1770–1781); Louis Jean Pierre Vieillot, *Tableau encyclopédique et méthodique des trois règnes de la nature: Ornithologie* (4 vols., Paris, 1790–1823); Alexander Wilson, *American Ornithology; or, the Natural History of the Birds of the United States* (10 vols., Philadelphia, 1808–1814; Wilson had completed seven volumes and most of the eighth before his death in 1813; subsequently additional material was added by George Ord, and later by Charles Lucian Bonaparte and others).

2. Not found.

3. This bird is described on p. 418 of Rafinesque's "Prodrome de 70 nouveaux Genres d'Animaux découverts dans l'intérieur des États-Unis d'Amérique, durant l'année 1818," *Journal de Physique, de Chemie et d'Histoire Naturelle*, 88 (June 1819), 417–29. Audubon is not mentioned there.

3. This spurious bird was an Audubon prank. The same description is given in Rafinesque's *Annals of Nature*, No. 1 (Lexington, 1820), where it appears as species 16—not 10—in that single number of the publication that ever appeared under this title.

The Cosmonist—*No. IV*

By winter's gales and stormy winds impell'd,
They leave the briny waves, and stray beyond
Their usual haunts, in search of climes unknown.

On the wandering Sea-birds of the Western States.

Extensive tribes of Birds dwell on the Ocean; they have been met one thousand miles from any land; they fly and skip over the waves, swim and dive in search of food, repose and even sleep on the water; they often defy the storms, and come near the shores merely when the need of laying their eggs compels them to seek convenient places and shelters.

The Sea-birds very seldom wander in the continents, and far from their usual element and food, which consists in fishes, sea-animals and sea-weeds. It was therefore with some astonishment that I have observed several of them in Kentucky, Ohio, Indiana, &c. Some appear to follow the meanders of the Mississippi and Ohio, and to ramble at a great distance from the gulf of Mexico, their native place, finding probably an adequate food in the variety of fishes swarming in those noble streams.

Pelicans have been seen and shot on the River Ohio, as far as Louisville, Cincinnati, and Portsmouth, nearly 2000 miles from the gulf of Mexico, by the course of the Rivers, although only one third of that distance in a direct course. Some few individuals have been seen both in summer and autumn; but do not appear to have raised their young in our rivers.

The other sea-birds which I have observed or noticed in the interior of the western states, belong to the genera of Divers, Gulls, Terns, Phalaropes, Grebes, Sea-ducks, &c. They were seen on the Ohio, Kentucky, Licking river[s], &c. or even near Lexington and Harrodsburg.

A Loon was shot near the Kentucky river in the spring of 1821. Several Phalaropes have been shot near Louisville and Henderson. If these birds wandered from the gulf of Mexico, the distance from it in a straight line, was about 600 miles. A Carolina Grebe, (*Podiceps Carolinianus*) was shot at Harrodsburg in March 1821, which came probably from the nearest Atlantic shore of North Carolina, at the distance of 400 miles or more.

These birds must probably be blown from the sea-shore, towards us by some violent storms, and many more in the same predicament may escape our notice. This singular fact in their History deserves however to be recorded.

Among the sea-birds which I have seen in Kentucky, there are two kinds, a Gull and a Tern, which I cannot find described in any book; they might be considered new species. They must probably have wandered here from the distant shores of the Mexican Gulf and Empire, where many unnoticed birds must exist as yet.

The Gull might be called the wandering Gull: I have given to it the ornithological name of *Larus marginatus*, which has a reference to its black-edged wings.

Its total length was one foot; but the dimension of the extended wings reached 28 inches. Bill black, feet of an orange color, with black claws. Head, neck, and belly of a snowy white, back and wings of a pale ash color; but the quills of the wings are white, with a black tip, and the external quill is edged with black, which gives to the whole wing the appearance of having a black edge. The tail is white, and obtuse.

The known species to which it resembles most, is the grey gull, (*Larus canus*) which is found in the United States, and even on our large lakes; but it differs from ours by being much larger, having a yellow bill, greenish feet, several black quills in the wings with white spots &c.

It was shot in January 1821, on a pond near Harrodsburg by Mr. Sutton,[1] and the specimen is in the possession of Doctor Graham[2] of the same town.

The Tern or Sea-Swallow may be called the black-headed Tern; I have given it the scientific name of *Sterna melanops*, which implies the same thing.

This Tern was nine inches long from the tip of the bill to the end of the tail, and the dimension of the extended wings was 21 inches. It was of an ash color above, and white beneath, with the head, neck, and feet blackish. The bill was of a lead color, one inch long, compressed and sharp. The feet had three half-webbed toes, and none behind. The tail was long and forked, white beneath.

This bird is totally different from all the known Terns, and might even perhaps be considered as a new genus, by its long, compressed bill, toes only half-webbed, and want of a hind toe, to which the name of CHLIDONIAS MELANOPS might be applied.

It was shot in June 1821, near Harrodsburg, and was preserved by Dr. Graham, in whose possession I saw it.

C. S. RAFINESQUE

Published in the *Kentucky Gazette*, 21 February 1822.

1. Probably David Sutton, owner of the Harrodsburg Springs spa that aspired to be "the Saratoga of the West."

2. Christopher Columbus Graham, David Sutton's son-in-law. Though 35 years old in 1822, Graham, having failed in business as a silversmith, had just taken his medical degree at Transylvania University.

The Cosmonist—No. V

The fictious Sirens, Salamanders, Sylphs,
No longer sought in vain, in water swim
Near us, and breathe in clumsy shapes or blow
In elegance of forms; but changed all.

On two new Salamanders of Kentucky.

All the fictitious beings which the exuberant imagination of Poets has created, are now removed to the delusive regions of fancy; but the Naturalists have availed themselves of many such poetical names, and given them a kind of relative reality, by adopting them to their nomenclature, and applying them to several beings formerly nameless.

Thus the name of Sylph has been given to a genus of beautiful North-American plants, of the sun-flower tribe. Some very singular animals resembling Eels, but having two feet and external gills, which are found in the U. States and Mexico, have been called Sirens: while all the Lizards without scales nor claws, and living in a tadpole state in their youth, like frogs, form the extensive tribe of Salamanders.

There are at least fifty species of Salamanders in the United States, although hardly ten species had been noticed a few years ago; but several species have lately been discovered by Prof. [Jacob] Green of Princeton and one by Mr. [Jacob] Gilliams of Philadelphia, which I have myself detected and described above 25 new species of this family; of which I have already published many, and the others will be seen in my Monography of the American Salamanders.[1]

Mr. [Joseph] Delafield has lately asserted to have discovered a new American Salamander or Proteus; but it happens to be my *Necturus maculosus*, published in 1820 in my Annals of Nature, species 17. Such mistakes will frequently happen, if former labours are not consulted when we study our animals.

Our Salamanders belong principally to the genera *Triturus*, *Salamandra*, *Meanthes*, *Necturus*, *Sirena*, &c.; but an attentive comparison of our species, has induced me to notice many distinctions among them, which will require the formations of new genera or sections, which among the Salamanders of Kentucky, two new species lately observed must be considered types of two new genera.

I have called the first genus UROTROPIS. It differs from my genus NECTURUS by having the hind feet with five toes and the tail cylindrical with a keel above.

The species upon which this genus is established, was found in the Kentucky river in 1821, and the specimen is now in my possession. It is commonly called Alligator, as are all the large water Salamanders, although they are totally different from the American Crocodile or real Alligator.

My specimen is 21 inches long, of which the tail is one third or 7 inches. The whole color is uniformly of a dull brownish, faintly clouded with darker brown. The head is large, flat above, mouth very large, full of small teeth, upper jaw longer and mucronate: the eyes are above, distant, very small and round. The gills are on the sides of the neck in the shape of rounded holes. The feet are short and thick, with unwebbed toes. The body is oblong, with a flattened back. The tail is pyramidal, the keel is slightly undulated.

I call this animal UROTROPIS MUCRONATA. The generic name applies to the shape of the tail, and the specific name to the shape of the upper jaw.

I have called the second genus EURYCEA LUCIFUGA: the first name is mythological, the specific denomination relates to its habit of avoiding the light.

This animal was discovered near Lexington in 1820, by Mr. [G. F. H.] Crockett, who collected several species, and gave me some of them: it lives in caves and in the dark; but not in water. It differs from the genus *Salamandra* by its large flat orbicular head, jaws with teeth and feet with half-webbed toes. Vulgar name, Ground puppet.

Its length is from 4 to 6 inches, the tail being much longer than the body or five eighths of total length. Color orange, with numberless oblong black dots. Mouth exceedingly large, split to the neck, jaws equally orbicular, the lower with many small unequal teeth, the upper jaw has only some lateral obtuse teeth. Feet slender, the fore feet have four toes and the hind feet five toes, all half-webbed. Tail slender cylindrical. Eyes large elliptical, covered by a black eye-lid. Its manners are similar to those of the land Salamanders, and it feeds on insects.

C. S. RAFINESQUE.

Published in the *Kentucky Gazette*, 28 February 1822.

1. Unknown, either as a publication or manuscript.

The Cosmonist—*No. VI*

[No copy has ever been found of the 7 March 1822 issue of the *Kentucky Gazette*, where this article should have appeared. No hints of its contents are known—*Editor.*]

The Cosmonist—*No. VII*

The soils and stones in many ways are formed,
By water, fire[,] concretion, even dust
And metals, sand, that fall from far above
The clouds, where gases meet and often take
A fluid or solid form by means unknown.

On the Geological Meteoric Formation.

I have ventured to assert[1] that the quantity of atmospheric dust falling on the ground may amount to several inches in the course of a century; but it would be absurd to suppose that it must in consequence have covered the earth with a coat or stratum of more than 50 feet thickness in 4000 years. The greatest proportion of that dust is precipitated by rains, and diluted or carried down the streams with the rain water as soon as it falls: a small proportion alone is mixed with the soil and increases its bulk. It is only in pits, caves, corners and hollows that it may accumulate to a certain extent, and even there compression will speedily reduce it to a smaller bulk.

It would also be absurd to ask whether this dust forms all the rocks and soils on the surface of the earth; but it is reasonable to suppose that it contributes in a certain degree to their increase. Our soil is formed by the decomposition of rocks, the accumulation of vegetable and animal decayed substances, mixed with atmospheric dust.

That it may in some instances form or increase the solid mineral substances and strata, cannot be denied, since this effect takes place before our eyes in cisterns and reservoirs of rain water. The earthy and dusty particles conveyed in them by the water are gradually deposited, forming concretions and stones. This is very evident in the old cisterns of the Eastern continent, which have held water during a long period of time.

Since every thing seems to indicate that there is an extensive and permanent formation and fall of dust in the atmosphere: the permanent succession of this phenomenon must contribute in part to form our soils, our alluvions and some stones, to fill the fissures and hollows of rocks and lavas, preparing them for vegetation; and in former times, when many of our sub-strata were formed, it may have been more abundant, contributing to the formation of some of those strata; but probably seldom exclusively.

This assertion will appear new and strange to some persons slightly acquainted with geological and meteorological phenomena; but not unreasonable to those who have observed nature with some care.

I have ventured to announce in my lectures that another geological formation, THE ATMOSPHERIC or METEORIC FORMATION, must be added to those already known. To this must be referred all those singular geological anomalies which puzzle so much the systematic writers, when they find extraneous stones, gravels, sands, earths, metals and other mineral substances mixed [with] or superincumbent over late or newer formations.

It may perhaps in time be found necessary to ascribe to meteoric formation, those extensive sub-strata and upper strata of sand [or] gravel which cannot properly be deemed alluvial nor volcanic, and those isolated masses and strata of sandstone, grit-stone, &c. the production of which in peculiar spots and in conical hills with regular horizontal strata, does not exactly correspond with our idea on the probable operation of extensive depositions by water.

When our rocks and strata were formed under water by the deposition of the particles floating in water or therein formed, many of their principles may have originated in the briny ocean; but some may have been derived from the atmosphere, and by their fall into the ocean become deposited through its medium.

C. S. RAFINESQUE.

Published in the *Kentucky Gazette*, 14 March 1822.

1. Rafinesque, "Thoughts on Atmospheric Dust," *American Journal of Science*, 1 (1818), 397–400. Reprinted here.

The Cosmonist—*No. VIII*

You blooming fields and you enchanting groves
Where peace and happiness not sought in vain
Reside, from Nature's friendly bard receive
The greeting, tuneful, song, or else his thanks,
Deserv'd by sweet sensations oft conveyed
To his warm soul with purity of thought,
And gratefully accepted, duly nursed.

On the Botany of the Western Limestone Region.

The vegetation of the western states offers several singular peculiarities: among which one of the most remarkable is the paucity of species to be found in the rich soil of the limestone region. For instance[, there are] no more than 600 species of plants that grow spontaneously within 15 miles of Lexington in every direction, while a similar circle of 30 miles diameter round Philadelphia, affords about three times as many or 1800 species.

Many tribes of plants afford but a very small number of species in this region, such as the Ferns, Lichens, Orchideous, Liliaceous, &c.

But to compensate for their small quantity the number of Individuals of the spontaneous growth is very great; some species of grasses, radiacious, and trees, grow in compact social clusters, covering many acres of ground, and with the utmost luxuriance.

This may be ascribed to that region having been covered formerly with an extensive growth of Canes (*Miegia arundinaria*) forming almost a general canebrake under the forests, where but few plants could take a stand.

Another remarkable feature in our Botany, is the casual change of the prevailing plants and trees upon many peculiar spots of grounds. It has been observed by the ancient settlers that the following plants have followed each other in succession in many plants [places?] as the prevailing growth.

The Canes, or *Miegia arundinaria.*
The Butterweed, or *Lupatorium [Eupatorium] urticefolium.*[1]
The Ironweed, or *Vernonia prealta.*
The Nimblewill, or *Panicum dactyton [dactylon].*
The Hardgrass, or *Panicum glaucum.*
The wild C[h]amomile, or *Anthemis cotula*, &c.

There is therefore a kind of natural perennial change of vegetation; when a species has exhausted the soil of a peculiar nutrition which it requires, it gives way to another for a series of years, &c.

The number of plants peculiar to the limestone region is small, I have however detected several new genera and species in it, some of which will be described in my next number.

The other regions or botanical zones scattered through the western states, such as the Alluvial Valleys and river Cliff, the Barrens, and the Hills, are much more fruitful in rare and beautiful plants. The hilly region of the knobs, Cumberland mountains &c. afford a variation nearly similar to that of Virginia but the cliffs, valleys, and barrens of the west offer quite peculiar and interesting features of their own, which deserve the attention of the enquirer into the science of vegetable geography.

The plants of the western country begin to excite some attention among the learned Botanists of Europe, and I am called upon by them to send them all our botanical riches; as being the only zealous and diligent Botanist settled in the Western States, possessing a rich comparative herbarium of nearly fifteen thousand specimens, which I increase every year. My extensive correspondence with the European Botanists is annually enlarged, and they send me large and fine collections of European and foreign plants in anticipation of their demands of ours.

I have lately received 500 specimens of French plants from Prof. Decandotte [i.e. De Candolle] of Ginevra, the best European Botanist, author of the French Flora, the Theory of Botany and the new species Plantarum.

Besides 810 specimens of English plants from Mr. [Adrian] Haworth of London, 300 species of European plants from William Swainson, Esq. of Liverpool, 200 from Prof. [William Jackson] Hooker of Glasgow, several hundred German plants from the Imperial Museum of Vienna; and 748 German plants from Prof. Schultze [i.e. Joseph August Schultes] of Augsburg, Counsellor of the King of Bavaria.

My rich herbarium has therefore been lately increased with about 3000 foreign specimens, and this is I believe the first instance of such valuable vegetable collections being brought to the western country, where I hope that it may become the foundation of the most extensive botanical collection in the united States, if it is not so already.

C. S. RAFINESQUE.

Published in the *Kentucky Gazette*, 4 April 1822.

1. The first two letters of the specific epithet are indistinct; "urticefolium," a Linnaean species, is the only Eupatorium epithet found elsewhere in Rafinesque's publications that fits, but there (*Medical Flora*, I, 180) the common name of this plant is given as "Deerwort Boneset" and it is said to have white flowers. It seems unlikely that "ancient settlers" would call such a plant Butterweed.

The Cosmonist—*No. IX*

Of Nature's secrets, long concealed, I try
To seek, detect, observe, at last reveal
Some wonderful display; or beings new
To many eyes and Science acute survey
Secure with eager hand, and soon contrive
From nullity to raise upon a scale
Of fellowship with those already known.

On two new genera of vernal plants from Kentucky.

ENEMION & STYLYPUS.

The Botanical discoveries which I have made in the immediate neighborhood of Lexington, are not so numerous as might be expected, owing to the peculiar nature of the productions of the limestone region, and my most important discoveries in the vegetation of Kentucky, have been made among the hills, knobs, barrens, cliff and bottoms, at a distance from Lexington.

I calculate to have discovered in 1818, 19, 20 and 21, or within four years, about 25 new genera, and above 100 new species of plants in Kentucky, while among them only 5 new genera and about 15 new species were found near Lexington.

Among the genera totally new are to be noticed two pretty vernal plants, blooming early in spring, and growing in great abundance around this town, and even within its limits in some places; these I now propose to describe in a concise manner.

I. ENEMION.

I have given this name to one of them which has pretty white flowers, and very much the appearance of some Anemones; it was one of their names in ancient Greek.

Its generic characters consist in having a Perigone simple colored, and with five petals. Many hypogenous stamina with club-shaped filaments. From 2 to 5 pistils ovate angular, with a style and a thick truncate stigma. The fruit consists in 2 to 5 capsul[e]s, opening longitudinally inside, each with 2 or 3 seeds.

This genus belongs therefore, to the *Polyandria* of Linneus, and in the natural method to the first class *Ettrogynia* [i.e. *Eltrogynia*], the natural order *Adnantheria*, and the twelfth natural family *Peonidia*, next to the genera *P[a]eon[i]a* and *Caltha*. It has also much affinity with the genus *Thatictrum* [i.e. *Thalictrum*], at least those species with 5 petals, but differ[s] widely by the capsul[e]s not being indetriscent [i.e. indehiscent] and one seeded.

It contains as yet only one species, to which I gave the name of

ENEMION BITERNATUM.

It is a small perennial smooth plant, from 4 to 10 inches high, with fibrous roots mixed with oblong tubercles. The stem is angular somewhat branched with some terminal flowers. The leaves are biternate, the folioles are divided in 3 or 4 lobes or partitions, and veined: the radical leaves have long petiol[e]s while the stem leaves are sepile [i.e. sessile] and with rounded membraneous stipul[e]s. The flowers are white, with 5 el[l]iptical obovate obtuse petals, and somewhat similar in size and appearance, to those of *Anemone thali[c]troides*.

This fine little plant grows in the open woods near Lexington, and also near Harrodsburg, Frankfort, &c. It blossoms in April and May, sometimes a second time in June. I discovered it in 1820.

II. STYLYPUS.

This name means in Greek, having a *pillar beneath*.

This genus has a calyx campanulate persisteus [i.e. persistens], five parted, the partitions reflexed. Five small petals upon the calyx. Stamina many on one row on the calyx. A central cylindrical pillar or gynophorus, bearing several pistils on its top in a round head, they are ovate oblong compressed, gibbous outside with a style geniculated, twisted and smooth. Seeds like the pistils.

It belongs therefore to the *Icosandria polyginia* of Linneus, and in the natural method to the first class *Ettrogynia* [i.e. *Eltrogynia*], the first natural order *Rhodanthia*, and the second natural family *Senticosia*, next to the genera *Geum* and *Dryas*. It differs principally from this last, by the 5 parted calyx and central pillar. Only one species belong[s] to it, which I call

STYLYPUS VERNUS.

It is a spreading an[n]ual plant. The stems are commonly many[,] adscendent[,] simple[,] and pilose. The leaves are pinnate with laciniate stipul[e]s and alternate. The folioles are ovate or obovate laciniate, the last larger and sepile [i.e. sessile]. The flowers are terminal in a decomposed dichotomy with two opposite leaves lanceolate laciniate at the base of it: they are small, green with yellow petals obovate, equal in length to the calyx.

I observed for the first time this plant (out of blossom) in 1818, on the banks of the Ohio near Cincinnati. It is very common near Lexington, and I have also found it near Harrodsburg and Mountsterling. It blossoms in April and May: it grows in woods and meadows.

C. S. RAFINESQUE.

Published in the *Kentucky Gazette*, 11 April 1822.

The Cosmonist—*No. X*

Attend the Bee, if wonders can delight
Your leisure-hours, but th' Ants observe, admire,
Study their arts, their science, policy
Beyond belief: this wonderful display
Of active, social life, a charm affords
That many needful lessons might impart.

On several new Ants of Kentucky and Ohio.

There are few Animals more interesting than the Ants; notwithstanding their diminutive size, they stand very high in the scale of animal intelligence, and offer wonderful signs of the most eminent faculties of the mind. The philosophers who have studied their actions with care and principally [Pierre Samuel] Dupont [de Nemours], [Pierre André] Latreille & [Pierre] Huber, are unanimous in granting them an intellect similar to that of the most intelligent quadrupeds and birds, although they are deprived of a heart and brain! this last organ being merely represented by a very minute nervous gangtrion [i.e. ganglion].

But the observations of these authors, several of which I have repeated, go still further, and might lead us to conclude that the ants are, next to men, the most intelligent beings on our globe; in order to prove this assertion, it will be sufficient, I presume, to detail some of the intellectual and moral actions, sciences and political institutions which have been evidently ascertained to belong to these little animals.

The Ants are social beings, they live in towns and villages, often subterraneous, having sometimes a population of 50 thousand individuals.

They are acquainted with the arts, besides tactics, diplomacy, medicine and surgery! &c.

They know the sciences of geometry, meteorology, natural history and political economy.

They have a language of signs, which is very copious.

They have the civil distinctions and professions of queens, nobility, magistrates, soldiers, nurses and teachers of youth, labourers, hunters, shepherds, masons, pioneers, plasterers, store-keepers, physicians and surgeons, architects and engineers, &c.

They have wars, make treaties of alliance and peace, form confederations, many towns uniting in a great war; they send off colonies, are subject to seditions and revolutions, they cure their sick, wounded or drowned individuals, and bury their dead in peculiar vaults.

All these, and many other similar acquirements, are evident proofs of their high station in the animal creation.

The cattle of the ants form the genus aphis, and are small insects, living in crowds on plants, and affording a peculiar honey; these are raised, nursed and taken care of by the ants for the sake of their honey, on which they feed. They practice agriculture, by sowing and scattering the seeds of many plants. This alone, besides their diplomacy and civil policy, is sufficient to raise them above the most intelligent quadrupeds, such as the beavers, marmots, monkeys, dogs, elephants, &c.

I have always taken a peculiar delight in observing these little intelligent animals. I have already detected twenty-five species of them in the United States, most of which are new, and I have discovered above fifty species of their cattle, or the genus aphis, mostly new.

Among these ants twelve new species have been found by me in the western states, principally in Kentucky and Ohio, which I now propose to notice and describe concisely.

1. Yellow ant, (*formica lutea.*)—It is entirely yellow, body narrow, nearly half an inch long, with short and straight antenna[e], abdomen oblong, slightly tricurpidated [i.e. tricuspidated].

2. Amber ant, (*formica succinea.*)—Of a fine shining and transparent amber color, thorax with three knots, abdomen oboval el[l]iptical, antenna bent, club-shaped at the end, two-fifths of total length. A very small species, one-tenth of an inch in length. It lives in small societies, under stones, in the knob hills of Kentucky.

3. Whitish ant, (*formica albescens.*)—Entirely of a dirty white except the jaws and eyes, which are fulvous; head oblong, thorax with three knots, abdomen oblong acute, with transverse rings; antenna bent, one fourth of total length. In the hills of Kentucky and Ohio.

4. Heart-shaped ant, (*formica cardiogaster.*)—Brownish black, shining, head round, thorax with three knobs, abdomen heart-shaped oval acute; antenna bent, one-third of total length, which is three-sixteenths of an inch. In the knob hills of Kentucky, in small societies.

5. Winter ant, (*formica hyemalis.*)—Blackish, belly and thighs incarnate, head oval, thorax oval oblong simple, abdomen el[l]iptical with 3 rings, slightly pubescent; antenna bent in a right angle; found in logs of wood in winter at Lexington.

6. Black-head ant, (*formica melanocephala.*)—Head round and black, thorax

oval brown simple, abdomen oblong mucronate, fulvous; antenna not bent, one-third of total length. A small species, one eighth of an inch long, found near Lexington.

7. Dicrope ant, (*formica dicropus.*)—Brown, head black and round, thorax with three knots, the second knot heart-shaped, abdomen el[l]iptical ringed, feet brown at the base, fulvous afterwards; antenna bent and short; length one fourth of an inch. Discovered near Lexington in woods, wandering.

8. Biting ant, (*formica morsitans.*)—Shining black, head oboval, thorax simple, abdomen gibbose oblong obtuse, antenna curved, half an inch long. In the barrens. Builds nests four feet high, with many openings; they bite those that disturb them.

9. Gibbose ant, (*formica gibbosa.*)—Blackish, abdomen oblong acute olivacevus [i.e. olivaceous], head round, thorax gibbose simple, antenna bent outwards; less than half an inch. In the mountains Alleghanny [*sic*] and Cumberland.

10. Lobster ant, (*formica astacena.*)—Brown above, fulvous beneath as well as limbs, head oval, thorax with two knots abdomen oboval, antenna bent and long. Found in 1818, near Hendersonville in Kentucky. Length one-third of an inch. It raises a small nest 3 inches high, one foot broad, and with many openings.

11. Spotted ant, (*formica notata.*)—Head black rounded, thorax black oblong simple pedunculated, abdomen oblong obtuse fulvous, with a black spot above, feet fulvous, antenna bent outwards; about half an inch long. In the hills of Pennsylvania, Ohio and Kentucky.

12. Saffron ant, (*formica crocea.*)—Entirely of a saffron or brick color, even the feet and eyes, antenna bent, head oval, abdomen oblong. Length one-fourth of an inch, in the woods of the western parts of Kentucky.

C. S. RAFINESQUE.

Published in the *Kentucky Gazette*, 25 April 1822.

The Cosmonist—*No. XII*

Within their shells the sluggish Turtles live,
They crawl or swim, affording luscious food.

On the Turtles of the United States.

The Turtles, often called by the Indian name of *Tarap[e]n* in the United States, are a peculiar kind of harmless Reptiles, which afford a delicious food, and their shell pretty ornaments.

Linneus only knew eight species of Turtles from the United States, many new species were described by [Johann David] Schuepf [i.e. Schoepf], [Louis Augustin Guillaume] Bose [i.e. Bosc], [William] Bartram, [Charles Alexandre] Lenieur [i.e. Lesueur], &c. and I have myself discovered twelve new species: the total number now known is therefore increased to thirty-five.

Among these there are six species of Sea Turtles and eight Land Turtles. All the other[s] are amphibious, living commonly in the fresh-water, but often creeping on land.

I have called as follows my new species:

1. *Trionyx nasica,*	Long-nose soft-shell Turtle.
2. *Trionyx pusilla,*	Dwarf soft-shell Turtle.
3. *Emyda pugnar,*	Fighting Tarapen
4. *Emyda mordens*	Biting Tarapen.
5. " " *meg[a]lonyx,*	Long claw Tarapen.
6. " " *Striata,*	Striped Tarapen.
7. " " *Semiradiata,*	Radiating Tarapen.
8. " " *nodosa,*	Knobby Tarapen.
9. " " *granularis,*	Granulated Tarapen.
10. *Monoclida Kentukensis,*	Kentucky Box Turtle.
11. *Lepidemy bifida,*	Bifid Tarapen.
12. *Didicla erythrope,*	Red-eye Tarapen.

I wrote in 1819 a Monography of these, and remarks on all the Turtles of the U. States which was sent to the American Journal of Science, but not published: it has been since sent to Europe and published there.[1]

The most important of these Turtles is the Long-nose Great Soft Shell, so common all over the western streams, and which reaches sometimes the weight of fifty pounds. It had been blended with the ferocious Turtle of the Southern States (Testudo ferox) from which it is quite different. I was the first to ascertain that it belongs to the genus *Trionyx* of [Étienne] Geoffroy [Saint-Hilaire], to describe it and draw it correctly.

I will conclude by describing one of my new species No. 10, found in Kentucky, and called the Kentucky Box Turtle. It belongs to the genus Monoclida or Turtle with lower shell shutting like a box. It has some affinity with the Carolina Box Turtle, but is yet very different.

Upper shell 6 inches long, 4 broad, and 2 high: almost elliptical, very convex[,] deeply notched in front, slightly serrated and flattened behind, blackish with some yellow irregular spots, 13 central scales, the first and third with a flat ridge, 25 marginal scales, all the scales slightly striated in concentric waved small wrinkles. Lower shell 5 inches long, 3 broad, elliptical, blackish entire, concave behind, shutting before with 13 smooth scales.

Head black with yellow dots, neck yellow with black dots, whitish beneath, cheeks flat, jaws horny entire, neck without scales, but flat horny warts.

Limbs scaly, tail very short, forefeet yellow with black spots, 5 brown claws, toes united. Hind feet brownish above, pale beneath, four claws, toes soldered.

C. S. RAFINESQUE.

Published in the *Kentucky Gazette*, 23 May 1822.

1. Such an article has not been found by Rafinesque's bibliographers.

The Cosmonist—*No. XIII*

The largest Alligators, dreadful foes
Of finny tribes, to fleetest Lizards are
Nearly related....

On the Lizards of the United States.

The Lizards are a tribe of Reptiles easily known by having feet and scales. The Salamanders differ from them by having no scales. Linneus only knew eight species of Lizards from the United States; but now twenty-eight species are known of which I have discovered eleven. My new species are

Crocodilus ditropurus,	or Double keel Alligator
Tupinambis niger,	Black Lizard.
" *pusillus,*	Dwarf Lizard
" *erythrostoma,*	Red-mouth Lizard.
Ascalabetes plica	Grey Lizard.
Agama velox,	Swift Lizard.
Scincus rufus	Rufous slow Lizard.
" *n[e]gropunctatus,*	Dotted slow Lizard.
" *catesbianus,*	Catesby's slow Lizard.
" *hyacinthinus*	Blue slow Lizard.

Our Alligators have not yet been well studied by real Naturalists. I suspect that many species and varieties exist in the Southern States. The most common species is the *Crocodilus lucius* of Cuvier; there is a sharp snout Alligator in Florida which must be his *Crocodilus acutus*. Linneus has blended above 20 species under his name *Lacerta Crocodilus*. My *Crocodilus ditropurus* is a small species with six longitudinal keels on the back and two on the tail; the snout oval, elliptical, flat, smooth and very obtuse.

A gigantic *Megasaurus* or Great fossil Sea Crocodile has been found embedded in the marl of New Jersey. Dr. Mitchell [i.e. Mitchill] has figured its teeth in his notes to Cuvier's Theory of the earth[1]; many of its bones are deposited in the Museums of New York and Philadelphia. It will be interesting to decide whether and how it differs from the European *Megasaurus* of Maestricht.[2] I suspect that it is different, and ought to form peculiar distinct species, to which I have given the name of *Megasaurus atlanticus*.

I will describe one of my new species, the black Lizard (*Tupirnambis niger*) which is found in Georgia, Florida, Cuba, Kentucky, and probably as far as the Rocky Mountains, if it is the black Lizard barely mentioned by Lewis and Clarke. There are perhaps several black Lizards in the United States, but I have at least ascertained one of them.

Length from 5 to 7 inches, entirely blackish, without spots; head obtuse with broad scales above, back with a longitudinal obtuse keel. Tail two thirds of total length, slightly depressed at the base, but compressed on the remainder, rather slender, obtuse, without keel, the lower scales larger; legs short, toes long.

C. S. RAFINESQVE [*sic*]

Published in the *Kentucky Gazette*, 18 July 1822.

1. Samuel Latham Mitchill, "Observations on the Geology of North America, Illustrated by the Description of Various Organic Remains Found in That Part of the World," pp. 319–431, pls. 6–8, in his edition of Georges Cuvier, *Essay on the Theory of the Earth* (New York, 1818).

2. "Le grand animal inconnu des carrières de Maestricht," as it was known until 1829 when it acquired the name *Mosasaurus hoffmanni*, has been called the world's most famous fossil. This Cretaceous period reptile, dug out of limestone in Holland near Maestricht in 1770, helped to demonstrate that the earth had once been populated by animals now extinct.

The Cosmonist—*No. XIV*

The splendid sweetly fragrant Roses blow,
Adorning our green fields, and hilly groves[.]
In gardens kept, remov'd, they charm the senses,
And all the other flowers, proclaim them Queens.

On the Roses of the United States.

The poets have long ago called the Rose, the Queen of flowers, and I have proved that it ought to stand at the head of the vegetable kingdom, in the correct natural method of Botany, just like man stands at the head of the animal kingdom.[1] This splendid rank is acquired to the Rose by its complex organization, rather than its size and beauty, in which it is excelled by many other plants.

This beautiful genus has lately been much noticed and enlarged in Europe. Linneus knew but 17 species of Roses; bu[t] now above 100 are known. Many works have been published on the English, French, Sweedish [*sic*] and Swiss Roses, and all the European species are ascertained[,] many of which adorn our gardens.

Those indigenous to the United States had been equally neglected, only 12 species are mentioned by Pursh and Nut[t]all, 5 or 6 more have been noticed by other botanists, while I have ascertained already the existence of 34 species and as many varieties of them in the United States, where many more may still exist unknown to me as yet.

I published in 1820 in the 5th volume of the Annals of physical sciences, an introduction to their natural history,[2] wherein 15 new species were introduced and described; they were—

Rosa glandulosa	The Glandular rose.
—— Kentukensis	Kentucky rose.
—— Trifoliata	Three leaved d[itt]o.
—— Enneaphylla	Nine leaved d[itt]o.
—— Elegans	Elegant rose.
—— Globosa	Globular rose.
—— Pusilla	Dwarf rose.
—— Cursor	Running rose.
—— Pratensis	Meadow rose.
—— Acuminata	Sharp leaved rose.
—— Riparia	Bank rose.
—— Flexuosa	Crooked rose.
—— Obovata	Obovate rose.
—— Serrutala	Serrulated rose.
—— Dasistema	Hairy rose.
—— Nivea	Snowy rose.

And I have discovered besides, this year another new species in the knob hills of Lincoln county, which I have called *Rosa viscida* or the Clammy Rose. It is a beautiful species, highly fragrant and deserving to be introduced in our gardens. It forms a shrub two or three feet high, with brown branches, prickles straight, leaves with 5 or 7 folioles, petioles pubescent, folioles obovate serrate, base entire, pale beneath, with the nerves somewhat pubescent. Flowers terminal solitary or geminate, large, of a fine rose colour, peduncles and calyx clammy by pedunculated glands, division of the calyx entire foliose oblong. Fruits globular. It blossoms in May and June, the buds have a fine balsamic smell and the expanded flowers a bal[s]amic rose smell rather strong.

Among my new roses, the following deserve particularly to be cultivated in our gardens.

The glandular rose, which grows in the atlantic States and has very large flowers, sometimes three inches in diameter.

The elegant rose, from the banks of the Hudson, with very fragrant flowers.

The running rose, from the hills of Kentucky, which forms a vine, running to the top of the highest trees or covering a whole wall, when properly supported. It produces as many flowers as the *Rosa multiflora*, but they are pale and inodorous. It has already been introduced in the gardens of Lexington, &c. as well as a variety with double flowers, found on an island in Licking River.

The Bank rose, from the banks of the Potowmack near Cumberland, with purple flowers very fragrant, and sometimes naturally double in the wild States.

The obovate rose, from the Catskill Mountains, with very fragrant large flowers.

The snowy rose, from Illinois, with snowy white flowers, slightly fragrant.

And the globe rose, which grows in the Barrens of Indiana, Illinois, Missouri, &c. forming a large shrub with paniculated flowers.

C. S. Rafinesque.

Published in the *Kentucky Gazette*, 8 August 1822.

1. Though hardly a "proof," he is alluding here to a passage (p. 38) in his *Analyse de la Nature* (Palermo, 1815) where he organizes the animal and vegetable kingdoms as "two scales of living beings. Man will be there at the summit of the animal scale, and the Rose at the summit of the vegetable scale ... for it is useful to study at the outset the most perfect Bodies, and to descend gradually to the knowledge of less perfect beings." (Cain translation.)

2. "Prodrome d'une monographie des rosiers de l'Amérique Septentrionale, contenant la description de quinze nouvelles espèces et vingt variétés," *Annales Générales des Sciences Physiques*, 5 (1820), 210–20.

The Cosmonist—*No. XV*

The shady western groves, may yet conceal,
Tennants unknown, deserving our best care.

On a new and valuable Tree of Kentucky.

I had been told that there was on the cliffs of the Kentucky River, a single tree of an uncommon and nameless kind; the bark of which was yellow and could die [i.e. dye] of that colour: I suspected that it was a forlorn individual of the *Yellow Locust* or *Yellow Bark* of the Mountains of Georgia and Tennessee, which possesses that property and was discovered by Michaux jun. who called it *Virgilia lutea*.

Having made some enquiries for that Tree at the Shakers ferry, it was pointed out to me last September, and I discovered three other trees of the same kind in the bottoms of the Kentucky, two of which were covered with ripe pods. It is not therefore a strag[g]ler from the south; but a native although rare tenant of our river groves.

I ascertained that it was different from the *Virgilia lutea* of Michaux, which has ovate leaves or folioles with yellow flowers, while this has elliptical leaves & is said to have white blossoms: I have not seen however its flowers, which appear in May, but shall not fail to watch them in future. I call it the *Brittle Yellow Locust.*

The genus *Virgilia*, has been separated by the modern botanists from the extensive and anomalous genus *Sophora* of Linneus, on account of its flat pods, and dedicated to the great latin poet Virgil. It contains several species of trees and shrubs growing in Africa, Mexico and the United States. I suspect that the North American species must form a peculiar sub genus, which I propose to call *Cladrastis*, which means brittle branches.

I find in the *Botaniste cultivateur* of Dumont Courset published in 1811, a species of *Sophora* called *Sophura* [*sic*] *Kentukea*, which is said to be a shrub with a blackish bark, and comes very near to my new tree, wherefore I suspect that it is also a species of Virgilia; but its flowers and pods are not described by Dumont. I will call it *Virgilia dumonti.*

All these trees & shrubs have the foliage of the Ash trees, with the flowers like *Laburnum* and the pods like the White Locust. Several species such as the *Virgilia aurea*, and the *lutea* and this new one have a yellow bark, dying of a beautiful orange color.

The North American species may be distinguished from each other as follows:

1. *Virgilia lutea.* (Yellow Virgilia) A tree, folioles oval, acuminate, flowers yellow pendulous.—In Tennessee.
2. *Virgilia fragilia.* (Brittle Virgilia) A tree, folioles elliptical, semi-obtuse, the odd one obovate, obtusely acuminate, flowers white pendulous.—In Kentucky.
3. *Virgilia dumonti.* (Dumonti Virgilia) A Shrub; folioles oval elliptical acute, flowers unknown.—In Kentucky?
4. *Virgilia secundiflora.* (Blue Virgilia) A shrub; flowers blue, all on one side, pods tomentose.—In Mexico.

My new *Virgilia* is a fine tree, of quick growth, rising about 40 feet high. It has many tough horizontal roots, grey and wrinkled outside, yellowish inside. The stem is straight, bark grey as in the blue ash outside, yellowish inside; wood white, brittle, heavy, with a peculiar vapid smell, something like the Fig tree; branches very brittle, breaking abruptly like glass, because they were articulated to the stem in their youth: twigs crooked, brownish outside, yellow inside. Leaves alternate, with 5 or 7 folioles or leaflets, petiol cylindrical, articulated and knobby at the base: folioles smooth, large, light green above, pale or glaucous beneath, alternate or approximated by pairs, shortly petiolated, entire, elliptical, base acute, summit attenuated nearly obtuse, length from 2 to 5 inches, breadth from 1 to 3; the odd leaflet or terminal foliole is very different, being larger obovate or somewhat rhomboidal, very broad, base acute, summit shortly acuminate and yet obtuse, from 3 to 5 inches long and 2½ to 4 broad. The pods hang in pendulous bunches, alternate, pedunculate on an angular branched axis, calyx deciduous, sometimes partly persistent, companulate, hairy, unequally five lobed: pods stipulate, membranaceous, thin, smooth, mucronate, flat, narrow, 3 to 6 inches long, with few seeds, only 3 to 5 beans, which are small, brownish green, elliptical and compressed.

This tree deserves to be cultivated extensively, since its roots will die of a fine and permanent deep yellow color: being scarce it ought to be raised in quantity in our own woods. I have collected a large parcel of its seeds, which I have sent to various parts of the United States and Europe, and I advise those who may find it again in seeds to transmit them to their friends and thus scatter the tree over the country.

C. S. RAFINESQUE.

Published in the *Kentucky Gazette*, 7 November 1822.

On a new tree of Kentucky forming a new genus: CLADRASTIS FRAGRANS.

I discovered in 1822, a new tree on the banks of the river Kentucky, which I noticed in the Kentucky Gazette under the name of *virgilia alba [sic]*; but having seen it only in seed, and not in bloom, I suggested the possibility of its being a peculiar genus. I have last year observed the blossoms of that tree, and found that although decandrous like the *virgilia*, they are not papilionaceous, but irregular pentapetalous like the *cercis*. It must therefore constitute a peculiar genus, which I propose to call *cladrastis*; meaning brittle branches. It bears no peculiar vulgar name, being known by few individuals; the name of yellow locust might be given to it, if it was not already applied to the *virgilia lutea*: it would be preferable to locust-ash, (since it is not an ash-tree, altho' the leaves are more similar to the ash than to the locust,) and also to Fustic, names suggested to me; since it is widely different from the Fustic.

Figure 13. *Cladrastis kentuckensis* (Yellowwood), adapted from *Morton Arboretum Quarterly* (1980).

The tree is remarkable by uniting the characters of several others, it has the stem and wood of the mulberry, the leaves like the ash, and the flowers like the locust or Robinia in color and scent; but in long loose panicles and shaped rather like the *cercis* or red-bud. It grows on the banks of the rivers, Kentucky, Rockcastle, Cumberland, Laurel, Dick, &c. in the alluvions or bottoms, but is not very common. It is a beautiful ornamental tree, rising from 30 to 60 feet, its blossoms which appear between the 15th and 30th of May, have a delightful scent, and more similar to the scent of Orange flowers. Its roots which are yellow, die [*sic*] a citron color, and the wood which is a yellowish white, dies also another yellow shade. This tree is therefore highly valuable and ought to be introduced in our gardens, &c.

DESCRIPTION.

GENUS. *Cladrastis.*

CALIX. Campanulate, curved, base veined, margin 5 lobed.

COROLLA. Five petals unequal, unguiculate, inserted on the middle of the calix; one larger petal obovate, four smaller ones nearly unequal, elliptical, bare equal[l]y cordate or bilobed, end obtuse.

STAMENS. Ten, free, inserted like the petals, filaments unequal filiform, anthers elliptical bilocular.

PISTIL. Germen pedunculate, linear, compressed, hispid:—style curved linear compressed, stigma terminal, acute smooth.

FRUIT. A pedunculate Pod, very flat, membranaceous, with few seeds or beans, small and oblong.

SPECIES. CLADRASTIS FRAGRANS. *Fragrant Cladrastis.*

LEAVES. Genticulate at the base with 7 alternate folioles, shortly petiolate, oval, acute, entire, smooth and pale beneath. *Flowers* in long branching pendulous racemes, scattered irregularly, rachis angular.

C. S. RAFINESQUE

Published in the *Cincinnati Literary Gazette*, 1 (21 February 1824), 60.

14

Lightning

No essay by Rafinesque has received more attention by people who have never read it than this little piece published in 1819.

As one who had spent much time out-of-doors in search of plants and had slept under the open sky during these expeditions as well as during his long voyage down the Ohio River to reach Kentucky, Rafinesque had had ample opportunity to observe the spectacular thunderstorms that sweep over the Ohio Valley from time to time. He must have thought his observations would be of some interest to others who live more sheltered lives.

During his lifetime, while his contemporaries pilloried him for other sins both real and imagined, nobody took exception to his descriptions of the different forms of lightning he had observed. In fact, the editor of the *Cincinnati Almanac,* mindful of his readers' interest in meteorological phenomena, reprinted it verbatim in his 1820 edition. But after Rafinesque's death, when Asa Gray set out to destroy whatever remained of the immigrant botanist's tattered reputation, the Harvard professor pounced on what he called "a paper which Rafinesque many years ago sent to the editor of a well known scientific journal, describing and characterizing, in natural history style, *twelve new species of thunder and lightning!*" Whew! Gray must have thought: *that* ought to put in his place the upstart naturalist accused of "discovering" and naming plants which never existed.

This first of Rafinesque's many contributions to Lexington's *Western Review and Miscellaneous Magazine* surely had not been examined by Gray, because, far from being "well known," the magazine itself soon died for lack of circulation much beyond Lexington's city limits. The magazine emphatically was *not* a scientific journal, either. Whether or not it made any sense for Gray to equate Rafinesque's varieties of lightning flashes with biological species, he apparently couldn't count accurately, because any way you number them Rafinesque had distinguished either more or less than twelve categories. But who would dispute Asa Gray, the Nestor of American botanists?

A generation later, Frederick Brendel, writing on the history of botany in North America, felt impelled to enliven his narrative by repeating the story—on the authority of the unimpeachable Professor Gray, of course. In 1907, Lucien Underwood, writing on much the same topic in the widely read *Popular Science Monthly*, opined that having classified every plant in sight, "when there were no further plants within his reach," Rafinesque "took flight to the clouds and deliberately classified the form of thunder and lightning." Poor T. J. Fitzpatrick, Rafinesque's patient and meticulous bibliographer who *had* read the piece, charged that such writers "only discredit themselves" when they repeat the lightning story; but his disclaimer had little chance against a tale that already had taken on a life of its own. It continues to be repeated in college lecture rooms, it found a place in the standard biography of Asa Gray (1959), and it usually pops up in writing about Rafinesque's contemporaries, as it does in the 1992 biography of Thomas Say.

Although trivial, Rafinesque's essay has been reprinted here in the hope that those who want to discuss the antics of the "mad naturalist" at least will read it before they chortle over it.—*Editor*

* * *

On the Different Lightnings Observed in the Western States, by C. S. Rafinesque, Professor of Botany and Natural History in Transylvania University.

IT is well known, that thunder storms assume in the United States a peculiar degree of violence and offer an intense display of electrical phenomena, hardly inferior to those of the tropical climates, which circumstance has been accounted for by the philosopher Volney, under the supposition of an aerial stream being propelled into the gulf of Mexico by the trade winds of the Atlantic Ocean, through the whole continent of North America (by meeting the obstacle of the lofty Mexican mountains) loaded with an excess of electrical fluid, imbibed during its long and warm course. To this cause our south-west storms, and the south-east storms of the Missouri, may safely be ascribed; but the summer north-west storms of the western country, which are often equally violent, cannot properly belong to the same category. We must account for them by the usual polar streams of both hemispheres, rushing towards the equator during the summer, loaded with an excess of accumulated polar electricity.

But it is not merely during the storms, that a great display of electricity usually takes place. Whenever an excess of that fluid exists any where, it has a tendency to expand itself and seek its equilibrium. This phenomenon happens almost every day and every where; unseen during the day time; but often perceptible in the calm nights, and producing the beautiful sheet lightnings, &c.

It is my intention to give a concise view of the multifarious lightnings, produced by the different modes, directions, and shapes of the electrical light, while flashing and seeking its level. No where have I seen a greater variety of them, than in the western states. I am not aware that the subject has been fully attended to by any one. It has even been doubted by many philosophers, whether any lightnings did actually rise from the earth, notwithstanding many witnesses had asserted the fact, to which I can now add my own actual testimony; and they did not appear to know that a discharge of electrical matter often takes place from the ambient air, or is attracted by it; nor have they ever thought of comparing with electricity the usual light, produced by meteors called fire-globes, aerolites and shooting stars, although no evidence of a disparity can be detected between them.

The following modes of electrical flashes have already fallen under my observation.

1. *Spark Lightning.* Similar to an electrical spark from the usual machine, course straight and very short; unfrequent, happens of course only between two near objects, principally clouds, and is commonly without thunder.

2. *Darting Lightning.* Similar to the foregoing, differing merely by a longer course; but always straight. It is accompanied by thunder, and takes place between the earth and clouds, being seldom horizontal or between two clouds. Its shape is commonly that of a spark followed by a long luminous trace.

3. *Ball Lightning* or real thunderbolt, differs from the foregoing by having the shape of a ball or globe of fire, followed by a large luminous trace and accompanied by violent thunder. This description of electrical discharge is the most dangerous. It commonly falls from clouds and reaches the earth in a straight course, often oblique, making a hole where it reaches it. Thunder-rods are hardly any avail against it. It has been known (in England) to fall on and blow up a powder magazine, where two rods were erected. It is very similar to the aerolites and globes of fire, differing merely in not leaving a stony deposite where it falls. This kind of lightening is unfortunately not uncommon in Kentucky, Ohio, &c.

4. *Comet Lightning.* I have observed this singular kind only once. It fell with a loud clap of thunder in an oblique direction, nearly straight, from a heavy cloud to the earth, and appeared to be similar to a ball-lightening, but it was followed by a long trace similar to the tail of a comet, or broader as it receded from the head. It might perhaps be an aerolite; but it happened in a storm, among many other kinds of lightning.

5. *Sun Lightning.* This happens in calm weather, even without clouds. It assumes at once the appearance of a bright sun, or globe of fire, which suddenly disappears by a diffusion into the ambient air, without having any peculiar direction; when it has, and lasts any time, it would be called a globe of fire.

6. *Star Lightning* or shooting stars. A frequent phenomenon, certainly electrical, although the electrical fluid may be combined with other gasses, as in every other instance of lightning.

7. *Common Lightning.* This is the most frequent, is always accompanied by thunder, and assumes every kind of direction; but shows itself constantly in a zigzag or crooked angular trace. This trace, as in many other lightnings, may be called

Falling or Descending, when it falls from the clouds to the earth, either vertically or obliquely;

Ascending or rising, when it rises from the earth to the clouds or the air, &c.

Horizontal or Lateral, when it goes from one cloud to another, or from a hill to another hill;

Re-ascending, when it appears falling from a cloud and yet reascends to another cloud, without reaching the earth, which happens rarely, yet I have observed it; but I never yet have seen a *re-descending lightning*.

8. *Forked Lightning* differs from the common lightning by dividing itself in its course.

9. *Sheet Lightning* appears at night and in calm weather with or without clouds, and is a diffused electrical discharge, differing from the *Sun Lightning* by having no apparent nucleus, perhaps a concealed one.

10. *Radiant Lightning* happens when the sheet or sun lightning is behind a cloud, wherefore the edge of the cloud alone is illuminated. It may be called *areolar lightning*, when it forms a bright edge to the cloud, and *radiate lightning*, when it forms beams of light round it.

Respecting the primordial directions of lightnings, I conceive that sixteen different ones may be enumerated, although some have not yet been observed by me.

1. From the clouds to the earth or waters, common.
2. " " to the clouds, frequent.
3. " " to the meteors, not yet observed.
4. " " to the air: not unfrequently the sheet and radiant lightnings are discharged in that way.
5. From the earth to the clouds. I have seen the common and the darting lightnings assuming that direction.
6. From the earth to the earth. I have seen an horizontal lightning discharged from one hill to another.
7. From the earth to the meteors, not yet observed.
8. " " to the air, common in volcanic eruptions.
9. From the meteors to the earth, common.
10. " " to the clouds, unfrequent.
11. " " to the meteors, ditto; but seen in some volcanic eruptions.
12. " " to the air, frequent.
13. From the air to the earth, unfrequent.
14. " " to the clouds, unfrequent.
15. " " to the meteors, unfrequent.
16. " " to the air, unfrequent.

The thunder is well known to be merely an appendage to some lightnings, or the noise produced by the refraction of sound in the elastic fluids of the atmosphere. I need not therefore explain any further their connection. The variety of sounds which it produces, can hardly be reduced to any descriptive enumeration: I mean however to attempt it at another time.

Published in *The Western Review and Miscellaneous Magazine*, 1 (August 1819), 60–62.

15

The Milky Way

Because of light pollution almost everywhere, few of us today have ever seen the brilliance and clarity of the cloudless night sky as it appeared to Rafinesque in the three advantageous locations he mentions below in paragraph 3. Ever the naturalist, he examined what he saw in the sky with the same minute attention he gave to plants as tiny as mosses, animals as small as aphids. The same onomastic zeal that caused him to devise nearly seven thousand new plant names motivated him to formulate names for the distinctive parts of the Milky Way he had observed. To do this, he first had to find appropriate generic names for the structures he wished to distinguish. He calls some of the prominent ones "arms" and he freely uses the word "galaxy"; but there is no reason to believe that he understood ours to be a spiral galaxy. His conception of the Milky Way galaxy appears to be somewhat like that of the 18th century English astronomer Thomas Wright of Durham, with the result that Plate XXIII of Wright's book has been inserted here as an appropriate illustration.

Rafinesque may not have known Wright's book when he wrote this essay in Kentucky. He did know it seventeen years later, however, for in Philadelphia he reprinted Wright's *Original Theory and New Hypothesis of the Universe,* embellished with his own preface and explanatory notes. Wright is credited with being the first to conclude on the basis of observation that the Milky Way consists of a three-dimensional arrangement of stars occupying a layer of space, a layer thinner in one direction than the other two. Unlike Wright, Rafinesque was very clear that this layer of space has the form of an irregular disk (paragraphs 20–29) and that our sun is not in the center of the disk (paragraph 19).

Astronomy had progressed to a point by Rafinesque's time that, without instruments, there was little likelihood that he could make any original discoveries himself. He returned to the subject of current astronomical knowledge the following month, in a five-page article he wrote for his own magazine, the *Western Minerva.* But since the magazine itself was suppressed before publication, this article did not see light of day until 1949. And he continued to be an intelligent user of the discoveries others, such as those persons named in paragraph 4. For a little textbook—titled *Celestial Wonders* (Philadelphia, 1838)—that he compiled for his proposed Central University of Illinois, he summarized their discoveries and those of later investigators. In the essay which follows, interest today lies mostly in the astronomers and philosophers he chose to honor, several of whom are very obscure indeed. Needless to say, Rafinesque's proposed names for regions of the galaxy have not been adopted by astronomers.

As originally published, structures such as "the sinus," "the arm," the "elbow," etc. in paragraphs 15 and 16 were printed but once, then a blank of corresponding length was left below to avoid repeating these words. This typographical device may have been necessary because of the printer's limited supply of type. Elsewhere, when a Kentucky publisher printed a Rafinesque poem in French he had to advise its readers to "supply the accents," because his fonts had none. Here the missing words have been inserted between square brackets.—*Editor*

* * *

Enquiries on the Galaxy or Milky-way

1. The contemplation of the Starry Heavens, fills the mind of the enlightened Philosophers with wonder and astonishment; they do not believe with the crowd of vulgar gazers, that Stars are mere specks of fire, drop[p]ing now and then in blazing tracks, and subservient to the paltry use of affording them a glimmering light, in the absence of the luminary of night; but convinced by study, analogy and the feelings of their understanding that the Sun is a Star, and that Stars are Suns, they attempt to enquire into their eminent destinies in the sublime scale of creation, and to detect the laws of their co-ordination.

2. No where do the Stars shine with more brightness than on lofty Mountains or in the middle of the Ocean, in serene nights; the comparative purity of our atmosphere in those situations, allows them to sparkle with increased intensity: it is then that the contemplative soul delights to gaze at their numberless association, and to reflect on the immensity of their distance, their immeasurable size and their other numerous properties, some of which are not even dreamt of by common philosophers; while a few, gifted with perspicuous foresight, daring to rise on the wings of sober and well regulated imagination, and delighting to investigate their unknown, concealed and undetected co-operations, dwell with pleasure and sagacity on the attributes of those mighty and splendid bodies.

3. While benighted on the summit of fiery Mount Etna, while furrowing the surface of the deep Ocean, or while gliding along the gentle stream of the Ohio, my eyes fixed on the celestial expanse of etherial [*sic*] space, I endeavoured to account for the apparent irregular position of the myriads of millions of starry Suns, scattered through it, by unequal velocities in their separate motions, compared with their combined and simultaneous motion.

4. It is well known that every material body suspended in space has a peculiar, simple or compound motion, either rotary or excentric [*sic*], elliptical or circular,

pendular or spiral, &c. and that most of them circulate around common centres: the moons around the planets, these around the suns, &c. Our sun was thought to be provided with a mere rotary and epicydoidal motion on himself; but Piazzi, Laplace and Herschell, have lately ascertained that it has a progressive one besides, which must form part of an orbit.

5. Similar motions have been observed in many stars; but our observations on this subject are of such modern dates that our astronomers have not been able to measure with any degree of precision, the extent and velocity of those motions: future observers will ascertain them now that the respective actual situation of all the large stars (7500) have been accurately fixed by Piazzi[1] in a memoir rewarded by the french institution in 1814.

6. Whatever be the extent and rapidity of those motions, it is evident that they must be commensurate with the size, weight and mass of those huge bodies, of which our sun has been calculated to be one of the smallest; and as various comparatively, as those of the planets of our solar system; whence arise the multifarious appearances of starry aggregations and constellations.

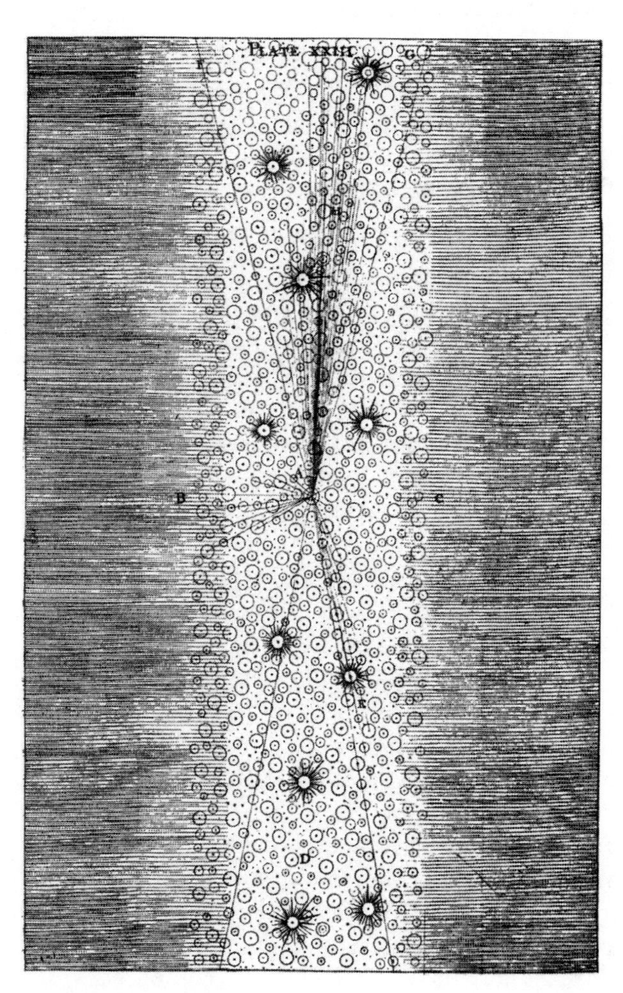

Figure 14. Sight lines demonstrating shape of the Milky Way, from Thomas Wright's *Theory ... of the Universe* (1750).

7. The most conspicuous anomaly in the disposition of visible stars is their peculiar accumulation beyond the largest stars, in an irregular concentric girdle round the etherial canopy, where by their vast number and increased distance they assumed a nebulous light; this second appear[ing] stars being intermediate between the glittering stars and the invisible ones.[2]

8. This belt of Stars, has received the vulgar name of milky way, the astronomers call it the galaxy. These names have a reference to some ancient mythological opinions, hardly worth mentioning. Every mythology, down to the modern Roman mythology which calls it St. James's way, has taken hold of this singular appearance, and connected it with their superstitious opinions.

9. The galaxy has more the appearance of a bright

cloud than of a mass of stars; but the Telescope shows that it consists of numberless multitudes of stars, since Herschell has reckoned 351 thousand of them between two stars of the *Swan*!

10. The irregularity of its shape when compared to a girdle or belt is striking, the breadth being various, the edges wawing [waving?], and there being in the northern hemisphere of our skies two peculiar anomalies, a hole or enclosed unstarry place, between the constellations of *Cepheus* and the *Lizard*, besides a large bifurcation near the *Swan*.

11. The cloudy light of the galaxy, has moreover a different degree of intensity in many parts, which is obviously owing to a difference in the number and distance of the stars included therein.

12. The general shape [of the] galaxy in the northern hemisphere, approximates to that of an irregular bow, whose concavity looks towards the polar star; it is narrow in the middle, and widest at the two extremities, one of which (the western) is divided in[to] two forks.

13. Names ought to be given to the different parts and appearances of such remarkable clusters of stars: they must be dedicated to astronomers and philosophers. I shall now attempt to name some of the most remarkable among those perceptible in the northern hemisphere: the southern hemisphere which has not been quite so well delineated and which I have never seen, will not be attempted, it may receive similar names afterwards.

14. But before affixing those names, it appears necessary to distinguish by appropriate denominations such peculiar appearances and anomalies. I propose to call,

ISTHMUSES, the narrowest parts of the belt,
DILATATIONS, the broadest parts of the same,
GULPHS and BAYS, the hollow sinusses,
CAPES and POINTS, the projecting sinusses,
SINUS, the projecting undulations,
ARMS, the branches surrounding the gulfs, &c.
HOLLOWS, the large enclosed unshining and unstarry places,
ELBOWS, the bents, or incurved undulations,
Clouds, the brightest cloudy spots,
SHEETS, the dullest parts,
VEILS, the transparent parts, hardly perceptible,
SPOTS, the small dull specks, or hollows.

15. I shall now propose my astronomical names for some of those appearances, stating their respective situations.

1. The Isthmus of *Pytheas*,[3] is situated south of the *Goat Star*, in the constellation of the *Driver*.
2. That of *Cepheus*,[4] at the star *Cepheus*, under the polar star.
3. That of *Halley*,[5] south of the star *a* of the *Swan*, at the base of the arm of *Piazzi*.
4. That of *Archimedes*,[6] west of the above, at a small distance.
5. That of *Euler*,[7] near *Ophiucus*.
6. The dilatation of *Newton*,[8] between the *Gemini* and *Orion*.
7. [The dilatation] of *Mairan*,[9] near the *Lizard*.
8. The gulf of *Leibnitz*,[10] between the arms of *Piazzi* and *Herschell*; which is perhaps a Mediterranean or immense hollow.

9. The gulf of *Descartes*,[11] below the dilatations of *Newton*.
10. The bay of *Huyghens*,[12] in the gulf of *Leibnitz*, near the isthmus of *Euler*.
11. [The bay] of *Maskeline* [Maskelyne],[13] outside the isthmus of *Halley*.
12. [The bay] of *Davy*,[14] under the dilatation of *Mairan*.
13. [The bay] of *Franklin*,[15] south of the isthmus of *Pytheas*.
14. [The bay] of *Meton*,[16] south of *Andromeda*.
15. The cape of *Theodosius*,[17] at the end of the arm of *Piazzi*.
16. [The cape] of *Gassendi*,[18] at the end of the arm of *Lacaille*.
17. The point of *Laplace*,[19] above the bay of *Huighens* [*sic*].
18. [The point] of *Hevelius*,[20] outside the isthmus of *Halley*.
19. The point of *Hipparchus*,[21] south of the isthmus of *Pytheas*.
20. The sinus of *Euclid*,[22] north of the arm of *Lacaille*.
21. [The sinus] of *Lalande*,[23] in the gulf of *Leibnitz*, on the arm of *Herschell*.
22. [The sinus] of *Rittenhouse*,[24] below the bay of *Huigens* and point of *Laplace*.
23. The arm of *Lacaille*,[25] stretching out between *Orion* and the *Little Dog*.
24. [The arm] of *Herschell* [Herschel],[26] stretching between the gulf or sea of *Leibnitz*, and the *Lyre*, &c.
25. [The arm] of *Piazzi*,[27] on the opposite side of the gulf of *Leibnitz*.
26. [The arm] of *Kepler*,[28] on the northern side of the hollow of *Galileo*.
27. [The arm] of *Copernic*,[29] on its southern side, between it and the bay of *Davy*.
28. The hollow of *Galileo*,[30] between the two above arms, near *Cepheus*.
29. The elbow of *Olbers*,[31] in the arm of *Herschell*, under the sinus *Lalande*.
30. [The elbow] of *Maraldi*,[32] the isthmus of *Cepheus*.
31. [The elbow] of *Schroeter*,[33] under the point of *Hipparchus*.
32. The cloud of *Flamsteed*,[34] at the base of the arm of *Herschell*.
33. [The cloud] of *Bayer*,[35] in the same arm below the sinus of *Rittenhouse*.
34. [The cloud] of *Casssini*,[36] in the arm of *Copernic*.
35. The sheet of *Biot*,[37] in the arm of *Piazzi*.
36. [The sheet] of *Aratus*,[38] above the gulf of *Descartes*.
37. [The sheet] of *Lemon[n]ier*,[39] between the elbow of *Olbers* and the sinus of *Lalande*.
38. The veil of *Clairanlt* [Clairault],[40] below the cloud of *Bayer*.
39. [The veil] of *Pythagoras*,[41] between *Sirius* and the gulf of *Descartes*.
40. The spot of *Hourcastreme*,[42] in the arm of *Piazzi*.

16. Among the remarkable appearances of the southern hemisphere, I shall only mention a few. I find such a diversity in their delineations on planispheres that I can hardly be certain to which denominations some of them may belong.

1. The arm of *Lambert*,[43] is a continuation of the arm of *Herschell*, being separated however by veils and sheets.
2. That of *Maupertuis*,[44] continuation of that of Piazzi, also separated by veils.
3. That of *Brahe*,[45] stretches beyond the last at great length.
4. The gulf of *Derham*[46] lies between the two last arms.
5. [The gulf] of *Bailly*[47] divides the arms of *Lambert* and *Manpertius* [*sic*].
6. The gulf of *Delille*,[48] lays opposite the gulf of *Derham*.
7. The hollow of *Humboldt*[49] is nearly between them.
8. The isthmus of *Confucius*,[50] below that hollow.
9. The veil of *Hermes*,[51] below the last.
10. The cape of *Condamine*,[52] at the end of the arm of *Brahe*.
11. [The cape] of *Ulloa*,[53] below the veil of *Hermes*.
12. The bay of *Fontenelle*,[54] under that cape.
13. [The bay] of *Ptolemy*,[55] above the gulf of *Delille*, in the arm of *Lambert*.
14. [The bay] of *Dalembert*,[56] above the last.
15. [The bay] of *Plato*,[57] above this last bay, &c.

17. When both hemispheres of the galaxy, shall have been accurately drawn, a

complete map and more enlarged designation of their appearances, may be easily delineated.

18. Since it surrounds as a belt the cluster of stars to which our sun belongs, it must have some peculiar connection with it. I think that they all form a single cluster. Herschell has reached with his huge telescopes, to discern stars of the 1342d. magnitude! they are not the last certainly, and the whole is connected with those of the galaxy.

19. Our sun is not quite in the middle of this cluster: but it is very far from the edges, since it is surrounded by stars on all sides. Thinly on each sides of the galaxy; but thickly in the direction of that belt: whence it follows that the shape of our cluster must be an *irregular disk*, compressed laterally on each side of the vertical direction of the belt, which forms the edge of the disk.

20. This shape accounts for the respective disposition of the stars of our cluster, since the center of a disk of starry particles, must be diaphanous, while the edges, seen from this center must assume a cloudy appearance, nearly opaque, similar to that of our galaxy.

21. The nebulosities and nebular cluster or clouds, scattered through the firmament, appear to be peculiar clusters of various nature; but unconnected with our own cluster. Some appear to be similar to our own, and even of the same discoidal shape. I do not mean to dwell on them at present, but shall proceed to state my ideas on our solar cluster.

22. In etherial moving bodies, the spindle shape is the result of the slowest circulating motion, sphericity of a moderate motion, and the discoidal shape of the accelerated or swiftest circulation, wherefore our cluster must move with *rapid velocity* on itself, or perhaps on an unknown central body, imperceptible because not luminous; and together in another forward progressive course.

23. As the planets and planetary comets, perform their revolutions in unequal times and at unequal distances from the sun and each other, such or somewhat similar must be the various respective revolutions of the stars of our cluster.

24. The only star, the elements of whose revolution and motion have been calculated with probable accuracy, is *Arcturus*; Lalande has found that it moves in longitude at the rate of about 3' 2" per century, that its annual motion is 3,428,000 leagues in one year, and that if it moves in a circle, its circumference must measure 22 millions of millions of leagues, requiring 700,000 years to perform it.

25. Notwithstanding the difference in size and density, this apparent velocity will appear small when compared with the annual motion of our earth, which moves at the rate of 198 millions of leagues in one year; but other astronomers have ascribed to *Arcturus* an annual motion of 80 millions of leagues, producing a circular orbit of 56 millions of millions of leagues.

26. Yet such a motion is a *mere fraction of the actual course*, since it is at best, the mere difference between the velocity of Arcturus compared with the unknown velocity of our sun, who carries the earth along while we endeavour to measure the road of its neighbour star! It is therefore rational to suppose that the velocity of Arcturus is at least ten times greater than that of the earth and its orbit commensurate.

27. All the stars have similar rapid motions within the unlimited bounds of space, all different yet in harmony, with their co-ordinate stars; *Arcturus*, the sun and the stars of the discoidal center of our cluster, have probably a shorter circulation to perform, while those of the galaxy and discoidal edges must perform a much enlarged revolution.

28. From those rapid and combined motions, in a cloud of clustered stars, the discoidal shape of the cluster has resulted; the axis of the disk passing through its compressed sides and the circumvolution of the cluster taking place on the vertical plane of the disk, as in a wheel in motion.

29. Thus the results of my present enquiries are, 1st. *that our galaxy forms the outward edge of our cluster of stars*, 2d. *that our cluster has an irregular discoidal shape*,

3d. *that it revolves (like a wheel) on its edges*, 4th. *with the utmost velocity*, and 5th. *that this shape is the result of that motion.*

30. Considering those millions of stars and of clusters, visible and invisible, and the numberless imperceptible planetary bodies revolving around them; all suited to their peculiar scales of beings, by the mighty ARCHITECT OF THE UNIVERSE; let us exclaim, *how wonderful are thy works, supreme Lord of the creation!*

Published in *The Western Review and Miscellaneous Magazine*, 3 (September 1820), 117–24.

1. Giuseppe Piazzi (1746–1826), with whom Rafinesque had been acquainted in Sicily, published *Praecipuarum stellarum inerrantium positiones mediae ineunte saeculo decimonono ex observationibus habitis in specula panormitana ab anno 1792 ad annum 1813* (Palermo, 1813) in which he cataloged the mean position of 7,646 stars, for which the Institut de France did award him a prize.

2. Figure 14. The lines in the diagram are sight lines radiating from the assumed position of a viewer in the center of the Milky Way galaxy. They show that the number of visible stars varies according to the direction the viewer looks through the starry layer.

3. Pytheas (4th century B.C.); Greek navigator who visited the British Isles; made accurate astronomical observations to determine latitudes.

4. Cepheus; in Greek mythology, the father of Andromeda. As Rafinesque has indicated, Cepheus also is the name both of a constellation and of a star.

5. Edmond Halley (1656–1742), Oxford professor, he studied comets and was first to predict their return, but did not study the one that bears his name.

6. Archimedes (*c.* 287–212 B.C.), Greek mathematician; on the power of the lever he said, "Give me a place to stand and I will move the Earth."

7. Leonhard Euler (1707–1783), Swiss mathematician; wrote prize-winning memoirs on planetary motion and on that of the moon.

8. Sir Isaac Newton (1642–1727), English mathematician whose *Principia Mathematica* expounds his theory of gravitation.

9. Jean-Jacques Dortous de Mairan (1678–1771), French astronomer who studied the orbital motion and rotation of the moon.

10. Gottfried Wilhelm Leibniz (1646–1716), German polymath who extended the Cartesian theory of motion to explain all of nature, including the phenomenon of gravity, as emanating from the ether which penetrates all bodies.

11. René Descartes (1596–1650), French rationalist philosopher whose theory of vortices competed with Newton's theory of gravitation to explain movement in the universe.

12. Christiaan Huygens (1629–1695), Dutch physicist who discovered the ring and fourth satellite of Saturn; discovered polarization of light.

13. Nevil Maskelyne (1732–1811), English astronomer-royal; he measured the density of the Earth and contributed to the method of determining longitude by measurement of lunar distances.

14. Sir Humphry Davy (1778–1829), English chemist whose most notable discovery was that alkalis are compounds of oxygen and metallic bases; he made no contributions to astronomy.

15. Benjamin Franklin (1706–1790), American statesman who clarified the distinction between positive and negative electrical charges and, in his famous kite experiment, proved that lightning and electricity are identical.

16. Meton of Athens (fl. *c.* 440 B.C.), Greek philosopher who discovered that the sun and moon occupy the same relative position after about 19 years; this Metonic cycle became the basis for the Greek calendar.

17. Theodosius of Bithynia, Anatolia, (*c.* 160 B.C.–*c.* 90 B.C.), wrote *Sphaerics*, a book on the geometry of the sphere, to provide a mathematical basis for astronomy.

18. Pierre Gassendi (1592–1655), French mathematician; in 1631 he was the first to observe a transit of Mercury across the sun.

19. Pierre Simon, Marquis de Laplace (1749–1827), French astronomer whose 1796 nebular hypothesis depicted the origin of the solar system in the contracting and cooling of a large, flattened, and slowly rotating cloud of incandescent gas.

20. Johannes Hevelius (1611–1687), German astronomer; he catalogued 1,500 stars and discovered 4 comets.

21. Hipparchus (fl. 150 B.C.); Greek astronomer; he discovered that the sun's path is eccentric and determined the length of the solar year.

22. Euclid of Alexandria (d. *c.* 265 B.C.), most famous mathematician of antiquity; his *Elements* is the basis of geometry, which in turn is fundamental to astronomy.

23. Joseph Jérôme Lefrançais de Lalande (1732–1807); at the Collège de France he developed the best planetary tables of his time.

24. David Rittenhouse (1732–1796), professor of astronomy at the University of Pennsylvania, he observed the transit of Venus of 1769 and built orrerys.

25. Nicolas Louis de Lacaille (1713–1762), French astronomer who was the first to measure an arc of the meridian from South Africa.

26. Sir William Herschel (1738–1822) and his sister Caroline (1750–1848); born in Hanover, they practiced astronomy in Britain, where Caroline discovered several comets and nebulae; her brother discovered the planet Uranus, the motions of binary stars, the period of Saturn's rotation and two of its satellites. Rafinesque probably was not aware of the work of the former's nearly as famous astronomer son, Sir John Frederick William Herschel (1792–1871).

27. Giuseppe Piazzi (1746–1826), identified above.

28. Johann Kepler (1571–1630), German astronomer whose studies of planetary orbits were the foundation for Newton's discoveries. For instance, according to Kepler's Third Law, "the square of a planet's periodic time is proportional to the cube of its mean distance from the sun."

29. Nicolas Copernic[us] (1473–1543), Polish/German scholar whose *De revolutionibus* (1543) proved the sun is the center of our planetary system; he was the founder of modern astronomy.

30. Galileo Galilei (1564–1642) Italian astronomer who discovered four of the satellites of Jupiter; his advocacy of the Copernican system caused him to be condemned by the Inquisition.

31. Heinrich Wilhelm Matthäus Olbers (1758–1840); at Bremen he calculated the orbit of comets and devised a method to determine the velocity of "shooting stars."

32. Giacomo Filippo Maraldi (1665–1729); compiled a catalog of stars at the Paris observatory; assisted his uncle, Gian Domenico Casssini.

33. Johann Hieronymus Schroeter (1745–1816); he built an observatory near Bremen where he studied the moon's surface and estimated the height of many of its mountains.

34. John Flamsteed (1646–1719); England's first astronomer-royal, he catalogued the fixed stars and furnished data for Newton's lunar theory.

35. Johann Bayer (1572–1625), German astronomer who began the practice of using Greek letters to code the stars in his star map *Uranometer* (1603).

36. Gian Domenico Casssini (1625–1712); he set up the observatory at Paris; studied the sun's parallax, zodiacal light, and the periods of Mars, Venus, and Jupiter.

37. Jean Baptiste Biot (1774–1862), French physicist and astronomer; in 1804, he undertook with Gay-Lussac the first balloon flight for scientific purposes by means of which they showed that the Earth's magnetic field does not vary with altitude.

38. Aratus (3rd century B.C.), author of *Phaenomena*, an astronomical poem alluded to by St. Paul in Acts 17:28.

39. Pierre Charles Lemonnier (1715–1799), French astronomer who studied Uranus before it was identified as a planet.

40. Alexis Claude Clairault 1713–1765), Parisian astronomer; in 1736 he took part in an expedition to Lapland to measure the length of the degree of longitude and published on his return *Théorie de la figure de la terre.*

41. Pythagoras (6th century B.C.), Greek philosopher who taught that numbers are the ultimate reality. Though he believed the Earth is a sphere at the center of the universe, he was one of the first to realize that Venus as an evening star is the same planet as Venus as a morning star.

42. Pierre Hourcastremé (1742–1832), French barrister whose principal work, the four-volume *Les aventures de Messire Anselme, chevalier des lois*, despite its title, deals with astronomy among many other subjects.

43. Johann Heinrich Lambert (1728–1777), German mathematician who published a treatise on comets, but his most important works were on optics.

44. Pierre Louis Moreau de Maupertuis (1698–1759), French mathematician and biologist; head of the French Academy's 1736 Lapland expedition to measure a degree of longitude, thus confirming Newton's theory on the exact shape of the earth.

45. Tycho Brahe (1546–1601), Danish astronomer who refined most then-known celestial measurements; his theory of the cosmos attempted to mediate between the Ptolemaic and Copernican systems.

46. William Derham (1657–1735), Anglican clergyman who published meteorological records and on natural history and astronomy; Newton used his measurement of the velocity of sound as the most accurate available.

47. Jean Sylvain Bailly (1736–1793), French astronomer who studied comets and the satellites of Jupiter; he published on the history of science, especially astronomy.

48. Jacques Delille (1738–1813), the "French Virgil," alluded to by Rafinesque in the notes to his own poem "Les Rives de l'Ohio." Delille probably was honored here because of his poem on the

"Trois Règnes de la Nature," but it also is tempting to suppose that he was named in error for Joseph Nicolas Delisle (1688–1768), who devised a widely used method for observing the transits of Mercury and Venus.

49. Friedrich Heinrich Alexander, Baron von Humboldt (1769–1859), German naturalist and explorer; his multi-volume *Kosmos* (1845–1862) attempted to give a comprehensive account of the entire physical universe.

50. Confucius (551–479 B.C.), Chinese sage; the ancient book most prized by him, the *Yi-king* or *Book of Changes*, gives an enigmatic theory of the universe.

51. Hermes, messenger of the Greek gods. It is possible, of course, that Rafinesque had in mind Hermes Trismegistus ("thrice great Hermes"), the Egyptian god of wisdom.

52. Charles Marie de La Condamine (1701–1774), French geographer who tried to determine the length of a degree of the meridian near the equator in a 1735 expedition to Peru.

53. Antonio de Ulloa (1716–1795), Spanish mathematician who participated in the 1735 expedition to Peru; established the observatory at Cadiz.

54. Bernard le Bouyer de Fontenelle (1657–1757), French author who published *Entretiens sur la pluralité des mondes* (1686) and also wrote on the history of mathematics and the philosophy of mathematics and science.

55. Ptolemy (*c.* A.D. 90–168), Alexandrian cosmologist whose *Almagest* gives in detail his mathematical theory of the motions of the Sun, Moon, and planets, the geocentric theory that prevailed for 1400 years.

56. Jean le Rond d'Alembert (1717–1783), though best known for his collaboration with Diderot on the great *Encyclopédie*, this French mathematician and philosopher wrote on many mathematical subjects and in his *Traité de dynamique* (1743) elaborated the principles of mechanics.

57. Plato (427–347 B.C.), Athenian philosopher; among his many works, it is his *Timaeus* that sets out the lineaments of Plato's cosmology.

16

Phenology

Early in the nineteenth century, charting America's climate was no less important than mapping the nation's geography. As early as 1766, Thomas Jefferson had begun making notations on changing weather conditions in his "Garden Book" as a way to gauge the effect of climate in the vicinity of Monticello on the growth of crops there. Among the kinds of observations he thought useful were dates of "the flowering of plants, ripening of their fruit and coming to table of the products of the garden, arrival of birds, insects, etc." Hence, he applauded the effort of Jacob Bigelow and others to collect this kind of information on a national basis.

With his wide ranging interests in all natural phenomena, Rafinesque was one of the naturalists excited by this collective research effort. In Kentucky, in a letter of 21 December 1819 (Filson Historical Society), he advised Charles Wilkins Short to "keep a kind of Journal of the progress of vegetation" at Hopkinsville, as he was planning to do at Lexington. He went on to remark that he had a copy of such a journal from the vicinity of Pittsburgh, and another from the banks of the Wabash in Indiana. Rafinesque never accomplished the "comparative Vernal calendarium of the Western States" he planned to assemble from these observations, but he did publish the phenological data he collected for Lexington during the spring of 1820 in six successive numbers of the *Western Review and Miscellaneous Magazine*.

Rafinesque's floral calendar chosen for reprinting here, however, is an earlier one published in the first number of Benjamin Silliman's *American Journal of Science*, and covering observations made in the vicinity of Philadelphia in the spring of 1816. This is the peculiar year sometimes called "Eighteen hundred and froze to death" at the time, or, later, "the year without a summer." In New England, there were snowstorms that year well into the month of June and the continuing cold, combined with less than normal rainfall, produced crop failures throughout the Northeast. The abnormal cold penetrated as far south as Philadelphia but did not linger as long there. Rafinesque's observations show how the natural world appeared to an observer who was unaware at the time of the unusual season he was experiencing.

Even if Rafinesque had the opportunity to read proofs—which is by no means certain—he had little talent for correcting typographical errors. Since he also practiced several orthographic peculiarities of his own, one cannot always be sure which are mistakes in his scientific nomenclature and which are misprints. Hence, deviations from standard spelling of scientific names are here amended by alternatives inserted between square brackets.—*Editor*

* * *

A Journal of the Progress of Vegetation Near Philadelphia, between the 20th of February and the 20th of May, 1816, with Occasional Zoological Remarks.

The importance of observations on the annual progress of vegetation is obvious, and, as connected with agriculture, gardening, &c., eminently useful. Comparative observations acquire a particular degree of interest, when made by skil[l]ful observers, at the same time, but at different places. Dr. [Jacob] Bigelow, of Boston, issued a circular, proposing that such contemporaneous observations should be made in the spring of 1817; and I wish that his request may have been attended to, when the collection of those observations may afford valuable materials for an American calendar of flora.[1] The blossoming of plants is easily watched, but their foliation and budding ought not to be neglected. Having been prevented, by various causes,[2] from keeping an exact record of the progress of vegetation near New-York in 1817, I submit an accurate journal which I had kept the year before, at Philadelphia, in which I hope that some interesting facts may be noticed. Dr. Benjamin [Smith] Barton has published a sketch of a calendar of flora for Philadelphia, in his Fragments on the Natural History of Pennsylvania;[3] by comparing it with mine, many material differences may be traced, which evince a gradual change of temperature, although the spring of 1816 was remarkably cold and late. The greater quantity of species observed by me may, besides render this journal a sort of vernal flora of the neighbourhood of Philadelphia; and many species found by me are not to be met in the *Flora Philadelphi[c]a* [*Prodromus* (1817)] of Dr. William [P. C.] Barton.

February 20. The *Hyacinthus orientalis* begins to show its flowers, and on the
February 24. In full blossom, as well as *Convallaria majalis*, in rooms.
February 25. The grass begins to look greenish in some parts.
February 26. Seen the first larva of insect in a pond.
February 27. The *Motacilla sialis*, or bluebird, is heard for the first time.

February 28. The first shad (*Clupea sapidissima*) is taken in the Delaware, while on the same day, the first smelt (*Salmo eperlanoides* [i.e. *eperlanus*]) was taken in the Raritan, at New-Brunswick.

March 1. The *Tulipa gesneriana*, and *Hesperis matronalis*, are in blossom at the windows: the suckers (*genus Catostomus*) appear in the fish-market.

March 2. The catkins of the *Alnus serrulatus* [*serrulata*] begin to swell.

March 3. Those of *Salix Caprea* begin to appear.

March 4. The grass looks green by patches in the country.

March 5. The leaves of *Veronica officinalis, Plantago virginiana* [*virginica*], *Saxifraga virginica*, &c. are quite unfolded.

March 6. The new leaves of *Kalmia latifolia* begin to appear.

March 7. The spathas of *Spathyema f[o]etida*, or *Pothos f[o]etida*, begin to appear in blossom.

March 8. The *Alnus serrulatus* [*serrulata*] is in full blossom.

March 10. Found several mosses and ferns in blossom; these last were covered with capsules or old fructification: they were *Asplenium ebeneum, Aspidium marginale, Asp.acrostichoides, Polypodiun medium*, N. Sp. &c.

March 11. Seen the first spider, in the country, brown, oblong, walking. A fall of snow at night.

March 12. Seen in blossom, at the windows, *Narcissus tazzetta* [*tazetta*], *N. janguilla* [*Jonquilla*], and several saffrons, genus *Crocus*, &c.

March 14. The grass looks quite green; the *Draba verna* ?[4] is in blossom in the State-House garden, the *Viburnum tinus, Primula acaulis*, &c. in the rooms, &c. The following fish are at market: white perch, (*Perca mucronata*, Raf.) yellow perch, (*Polyprion fasciatum*, Raf.) mamoose sturgeon, (*Accipenser* [*Acipenser*] *marginatur*, Raf.) elk-oldwives, (*Sparus crythrops*, Raf.) &c.

March 15. The *Populus fastigiata*, Lombardy poplar, begins to show its catkins.

March 17. The big-eye herring (*Clupea megalops*) begin to be seen at the fish-market.

March 18. Many plants begin to grow and show their leaves.

March 19. A fall of snow. The first shad (*Clupea sapidissima*) appear in New-York: they are now common here.

March 20. *Crocus aureus* in blossom in gardens; likewise *Iris persica*, &c.

March 21. *Betula lenta* begin to show the catkins.

March 22. *Galanthus nivalis*, and *Lamiun amplexicaule*, are in blossom in gardens at Cambden [*sic*].

March 24. *Populus fastigiata*, and *Salix caprea*, are in full bloom.—The gooseberry bushes shoot their leaves.

March 25. *Populus angulata* in blossom at Cambden.

March 26. *Salix babylonica* begins to blossom and shoot the leaves. *Viburnum prunifolium* is budding.

March 27. *Draba verna ?* is in seed already in Cambden: the *Rhododendron maximum* begins to shoot in gardens.

March 28. *Juniperus virginiana* is in bloom. *Saxifraga virginica* begins to show its flowers. *Laurus benzoin*, and *Cornus florida*, are budding.

April 1. In the morning, a large flight of wild geese went over the city northwards, making a great noise. In the afternoon there was a thunder storm from the southwest.

April 2. The frogs begin to croak. Found in blossom near Cambden, *Arabis rotundifolia*, Raf., *A. lyrata, Saxifraga virginica, Draba verna* ?[,] *Betula lenta*, &c. *Pinus inops* is budding.

April 3. Seen the first swallow. Found in blossom on the Schuylkill, *Fumaria cucullaria, Anemone thalictroides, Saxifraga virginica*, many ferns and mosses.

April 4. The fresh-water turtle (*Testudo picta*) begins to show itself.

April 7. Found in blossom to-day, *Hepatica triloba*, *Laurus benzoin*, *Sanguinaria canadensis*, *Spathyema f[o]etida*, *Acer rubrum*, &c. The first bee is seen.

April 10. In blossom at the Woodlands,[5] *Viola blanda*, *Luzula filamentosa*, Raf., *Gnaphalium ? plantageneum* [*plantagineum*], &c.

April 12. In blossom at Cambden, *Viola lanceolata*, and *Houstonia c[a]erulea.*

April 14. The apricot trees begin to blossom in gardens. *Acer negundo* is in bloom at Gray's Ferry.

April 15. Seen the first butterfly—it was small and grey. Found in blossom, near Cambden, *Phlox subulata*, *Arabis parviflora*, Raf., and *Vaccinium ligustrinum.*

April 18. Seen in blossom, *Epig[a]ea repens*, *Carex acuta*, and *Taraxacum densieonis.* In gardens, the peach and cherry trees are in bloom. Observed many insects. The *Camellia*, the *Magnolia chinensis*, &c. are seen in the hot-house of the Woodlands.

April 20. The first snake is seen. *Coluber trivittata*, Raf. Also a beautiful large butterfly,[6] red and black. The *Salix vitellina*, and *Capsella bursa* (*Thlaspi bursa-pastoris*) are in blossom.

April 21. Found in blossom, near Gray's Ferry, *Narcissus pseudo-narcissus*, and *Sedum ternatum*, both naturalized. Likewise the *Populus tremuloides*, and *Mespelus [Mespilus] canadensis*. The leaves of *Podophyllum pettatum* [*peltatum*] are fully expanded.

April 23. Seen in full bloom in gardens, the pear-tree, plum-tree, *Riber* [*Ribes*] *grossularia*, and *R. rubrum.*

April 24. Found in blossom along the Schuylkill, *Aguilegia canadensis*, *Hyacinthus botryoides*, *Ranunculus fascicularis*, *Viola papilionacea*. *V. decumbens*, Raf., *Houstonia c[a]erulea*, *Cerastium pumilum*, Raf.

April 25. Found in blossom near Cambden, *Viola pedata*, *V. lanceolata*, *V. ovata*, Raf., *V. primulifolia*, *Arabis parviflora*, Raf., *Cerastium pumilum*, Raf., *Carex acuta*, *Meopilus* [*Mespilus*] *botryapium*, *Laurus sassafras*, *Cercis canadensis*, *Potentilla simplex*, *Andromeda racemosa.*

April 28. Seen in blossom in gardens, *Calycanthus floridus*, *Syringa persica*, *Phlox pilosa*, &c. The leaves of *Liriodendron tulipifera*, *Æsculus hippocastanum*, *Populus fastigiata*, *P. angulata*, are unfolded.

April 30. In blossom on the Schuylkill, *Obolaria virginiana* [*virginica*], *Anemone trifolia*, *Hydrastis canadensis*, &c.

May 1. In blossom in the Neck, *Cerastium vulgatum ?*[,] *Veronica serpyllifolia*, *V. arvensis*, *Ranunculus bulbosus*, *Viola cucultata* [*cucullata*].

May 3. Found above the Falls of the Schuylkill, *Viola striata*, *V. concolor*, *V. primulifolia*, *V. blanda*, *Fumaria aurea*, *F. cucullaria*, *Cha[e]rophyllum procumbens*, *Uvularia sessitifolia* [*sessilifolia*], *U. perfoliata*, *Cercis canadensis*, *Arabis falcata*, *Stellaria pubera*, *Erigeron pulchellum* [*pulchellus*], *Orchis spectabilis*, *Hydrastis canadensis*, *Dentaria diphylla*, *Azalea nudiflora*, &c.

May 4. Found on the Vissahikon, *Arabis bulbosa*, *Panax trifolium*, *Viola pectata* [*pedata*], *V. rotundifolia*, *Cardamine pennsylvanica* [*pensylvanica*], *Krigia virginica*, and several grasses.

May 7. Found in blossom over the Schuylkill, *Laurus sassafras*, *Viburnum prunifolium*, *Aronia arbutifolia*, *A. melanocarpa*, *Fragaria virginica* [*virginiana*], *Cerastium nutans*, Raf., *Convallaria majalis*, naturalized, and several species of the genus *Vaccinium.*

May 10. Found below the Falls of the Schuylkill, *Floerkea uliginosa*, *Viburnum acerifolium*, *Oxalis violacea*, *Cerastium tenuifolium*, *Glechoma hederacea*, &c.: and the following above the Falls—*Trillium cernuum*, *Viola pubescens*, *V. pensylvanica*, *Hydrophyllum virginicum*, *Polemonium reptans*, *Senecio aureus*, *Saxifraga pennsylvanica* [pensylvania], *Staphylea trifoliata*, *Obolaria virginica*, *Caltha palustris*, *Ranunculus abortivus*, &c.

May 11. Seen the first bat.

May 12. Near Haddonfield, *Bartsia coccinea*, *Helonias bullata*, *Trifolium repens*, &c.

May 15. Found between Cambden and Haddonfield, *Trifolium pratense*, *Silene virginica*, *Antirrhinum canadense*, *Lithospermum tenellum*, Raf., *Festuca tenella*, *Seleranthus* [*Scleranthus*] *annuus*, *Oxalis biflora*, Raf., *Poa rubra*, *Vaccinium corymbosum*, *Viola palmata*, *V. parvifolia*, Raf., *Rubus flagellaris*, &c. Also in blossom, *Quercus rubra*, *Q. obtusiloba*, *Q. alba*, &c.

May 20. Found near Burlington, *Plantago virginica*, *Euphorbia ipecacuan[h]a*, *Comptonia asplenifolia*, *Myosotis lappula*, *Senecio obovatus*, *Scirpus acicularis*, *Lithospermum trinervum*, Raf., *L. tenellum*, Raf., &c.; besides several *Carex*.

Published in *The American Journal of Science*, 1 (1818), 77–82.

1. Under the title "Facts Serving to Show the Comparative Forwardness of the Spring Season in Different Parts of the United States" Bigelow published the results of his inquiry in the *Memoirs of the American Academy of Arts and Sciences*, 4 (1818), 77–85, where he acknowledged that it was the Rev. Henry E. Muhlenberg who had suggested to him "that if a series of Calendars of vegetation should be kept for the same year in different parts of the United States, and the whole published collectively, the result would be valuable, by affording an actual view of the comparative forwardness of the season in the various latitudes and situations of the country."

2. Speaking of this period in his autobiography, *A Life of Travels*, Rafinesque said (p. 52) that he had "the intention to settle awhile in New York in trade; but several unfortunate events in 1817 compelled me to change my plans. A perfid Sicilian whom I thought my friend was chiefly the cause of these new losses; the bankruptcy of a [commercial] house in New York, law suits and other troubles, combined to thwart my industry."

3. Benjamin Smith Barton, *A Discourse on Some of the Principal Desiderata in Natural History ... Read before the Philadelphia Linnean Society ...* (Philadelphia, 1807).

4. Here the question mark does not indicate the writer's doubt, but rather his egregiousness. Landing on these shores in April of 1802, "the first plant I picked up," quite possibly also in the vicinity of Independence Hall, was "then called *Draba verna*," he wrote in *A Life of Travels* (p. 14). Believing that few if any New World plants were identical with those of the Old World, he promptly named this humble member of the mustard family *Draba Americana*; but "American Botanists would not believe me." They haven't to this day, and continue to regard the plant as a widely distributed naturalized weed from Europe.

5. The estate of William Hamilton (1747–1813), which later became Woodland Cemetery in West Philadelphia.

6. It is noteworthy that although Rafinesque devised scientific names for many different insects and at least one spider, he never named a butterfly.

VII.

Phytogeography

17

Kentucky Botany

As a widely traveled field botanist, Rafinesque had readily at hand enough information to make a significant contribution to phytogeography. In his earliest letters to fellow collectors such as Henry Muhlenberg, he urged that they note down the physical location where they gathered each specimen. He himself summarized his own records in articles such as the following. The floristic regions he described here may be compared with the description he published 18 months later as Cosmonist VIII. In order to bring together in one place all his writing on botanical geography, an excerpt from one of his letters to Charles Wilkins Short is included here. Along with ten other Rafinesque-Short letters, the letter from which this excerpt is taken was published in the *Filson Club History Quarterly*, 12 (October 1938), but that text is so full of errors that here the letter has been transcribed from the manuscript, which is owned by the Filson Historical Society in Louisville. Finally, Kentucky's flora is considered in wider context in Rafinesque's 1836 essay on the "Flora of North America: Botanical Geography and Localities" also included in this book.—*Editor*

* * *

Botany of Kentucky.

On its principal features, by C. S. RAFINESQUE, *Professor of Botany and Natural History in Transylvania University.*

THE state of Kentucky being situated in the centre of the western country, has a *flora* similar to the generality of the western states and participating in their peculiar features, while it offers in itself a complete specimen of the western botany.

The peculiarities of this botany consist principally in the total want of the maritime and mountain regions, which form such remarkable sections in the local floras of the Atlantic states, and abound with plants peculiar to themselves. Another striking feature in the vegetation of Kentucky and the western states is the propensity which many plants and trees exhibit of growing in a social state, to the almost total exclusion of every other. There are may plants which grow crowded together, all over the United States, such for instance as the grasses, ferns, the *Comptonia*, the *Hudsonia*, &c. but they allow many other plants to grow with them; while, in the western country, many extensive spaces of ground are covered with one or a few crowded species, to the exclusion of many others, which are found in their company elsewhere. The plants which may be quoted as a striking instance of this singular fact are not few, among which I shall select the following:

Vernonia pr[a]ealta,	Iron Weed,
Baptisia c[a]erulea,	Blue Wild Indigo,
Cacalia reniformis,	Kidney Weed,
Hedeoma pulegioides,	Penny-royal,
Chenopodium anthelminticum,	Worm Weed,
Elephantopus scaber,	Elephant's Foot,
Gillenia stipulacea,	Indian Physic,
Miegia arundinaria,	Cane, &c. &c.

I consider the state of Kentucky as divided into four natural sections, or botanical regions, which are all distinguished by some peculiarities in their vegetation. They are

1. The Fluviatile Region. This includes all the valleys, and bottoms of the large rivers, such as the Ohio, Mississippi, Tennessee, Cumberland, Kentucky, &c. with their tributary streams. The bottoms of the valleys are formed of an alluvial soil, or the washing from the hills. They are level and often overflowed: while the sides of the valleys are steep, craggy, and composed of limestone, sandstone, or slat[e]y rocks. The following are some of the trees and plants peculiar to this region and giving a decided character to its vegetation:

Platanus occidentalis,	Sycamore or Button wood,
Hesperis pinnatifida,	Ohio Wall Flower,
Jeffersonia binata,	Twin leaf,
Capraria multifida,	Sand Ragweed,
Solanum Carolinianum,	Sand Briar,
Eupatorium c[o]elestinum,	Sky weed,
Polanisia graveolens,	Stinking weed,
Heliotropium Indicum,	Heliotrope,
Catalpium cordata,	Catalpa tree,
Populus angulata,	Cotton tree,
Porcelia triloba,	Papaw tree,
Synandra grandiflora,	Cow mint,
Nelumbium pentapetalum,	Swamp lily,
Pancratium Liriosme,	Lily,
Houstonia fruticosa,	Rock weed,
Prunus pendula,	Cliff plumb, &c. &c.

These two last are new species from the cliffs of the Kentucky river.

2. The Central Region. It is formed by the limestone tract included between the valley of the Ohio and the hilly ridges or knobs. The ground is slightly broken, very fertile and mostly under cultivation. This section is remarkably poor in the number of botanical species growing spontaneously; I conceive that its flora hardly contains 500 species, including trees, shrubs, and naturalized plants! There are

hardly any species peculiar to it; but the following ones, rare elsewhere, are here very common:

Eupatorium urtic[a]efolium,	White nettle,
Pavia muricata,	Prickly Buck-eye,
Isanthus ceruleus,	Blue Penny-royal,
Polymnia uvedalia,	Scented Sun flower,
Phlox glaberrima,	Pink, &c. &c.

It is also highly singular that in this region, the woods are open as parks, without shrubs and with very few plants, except grass or some social weeds.

3. THE HILLY REGION. It contains the hills and ridges which divide the waters of the Kentucky, Green, Licking, Cumberland and Sandy rivers, &c. being spurs from the Cumberland mountains. Those hills are often called knobs, although they have not always the knobby or rounded appearance. The rocks are limestone, or sandstone, or slate. The vegetation approximates exceedingly to that of Virginia and Pennsylvania.On the Cumberland mountain and the highest ridges, I am told that there is a similarity with the Alleghany regions, and that the *Kalmia latifolia*, Common Laurel, and the *Gaultheria procumbens*, Mountain Tea, grow there; but having not yet visited them, I am unable to ascertain whether they ought to form another distinct region, which might be called the mountain region. The hilly region is rich in plants; I shall mention a few of those peculiar to it in Kentucky:

Iris cristata,	Crested Tris or Flag,
Stylosanthes elatior,	Yellow Pea-clover,
Orchis ciliaris,	Yellow-bunch,
Juniperus Virginiana,	Red Cedar,
Vaccinium album,	Wild Currant,
Pinus rigida,	Pitch Pine,
Lechea minor,	Pin week [*sic*],
Rudbeckia fulgida,	Rough Wort,
Gerardia glabrata,	Yellow Wort,
Asarum Virginicum,	Heart-leaf, &c. &c.

4. THE BARREN REGION, or rather the open region. This has an extensive range in Kentucky, particularly in the western and southern parts of the state. The numerous *barrens* and *licks* compose it, lying scattered and irregularly among the central and hilly regions. The *barrens* are tracts of ground destitute of trees, or with few scattered small ones; but thickly covered with a luxuriant growth of plants; while the licks are almost destitute of them, and those that grow in their immediate neighbourhood are all small, which is owing to their poor, slat[e]y or argillaceous soil. Their vegetation is however similar to that of the *barrens*. Both have a growth of plants very similar to the vegetation of the *prairies* of Ohio, Indiana, and Illinois, and more different from that of the Atlantic states, than the three foregoing regions. The plants peculiar to them are very numerous; I shall mention only a few, among the most remarkable and singular.

Solidago rigida,	Stiff Golden-rod,
Polygala polygama,	Nimble weed,
Rudbeckia purpurea,	Purple Sun-flower,
Ruellia oblongifolia,	Rough Bell,
Andropogon arenaceum [*arvenaceus*],	Barren Oats,
———— *nutans,*	" "

Petalostemon candidum,	Nimble Clover,
———— *purpureum,*	" "
Silphium therebinthaceum,	Turpentine weed,
Silene catesb[a]ei,	Scarlet Pink,
Gentiana amarelloides,	Yellow Gentian,
Buchnera Americana,	Black Wort, &c. &c.

From the above a faint, but correct idea may be formed of the display and peculiarities of the wide range of vegetation in Kentucky and throughout the western states, wherein the same peculiar divisions or regions may be traced.

The vulgar names of the plants above mentioned are such as I found used in some parts of Kentucky; but they cannot claim to be generally understood even in this state, many being merely local or personal. The botanical names are alone to be relied on, being universal and not liable to mislead.[1]

Published in *The Western Review and Miscellaneous Magazine*, 1 (September 1819), 92–95.

1. A note titled "Errata" appeared on p. 128 of the same issue of the magazine stating that "In the article on the Botany of Kentucky owing to the absence of the writer from town, some important typographical errors in the technical names occurred, which the botanical reader will correct as follows." The corrections that are then listed have now been incorporated in the text. For the currently accepted Latin names of the plants as well as their common names now in use, see Ronald L. Stuckey and James S. Pringle, "Common names of Vascular Plants Reported by C. S. Rafinesque in an 1819 Descriptive Outline of Four Vegetation Regions of Kentucky," *Transactions of the Kentucky Academy of Science*, 58 (1997), 9–19.

[Excerpts from a Reply of 25 October 1834 to a Query from Charles Wilkins Short]

Instead of the localities of some few plants I am going to give you ~~one~~ <a general> acc^{t} of the best Botanical localities of Kentucky to my knowl[edge.]

Kentucky is divided into 4 Botanical Regions as I stated as early as 1819 in my Paper in the Western Review—1 Mountains & sandstone hills. 2 Barrens or glades. 3 Limestone Bassin & 4 Alluvial tracts. Each has peculiar rare plants—

I Region. Mts—The Cumberland Mts are the nucleus of this Region which is a table land in Tennessee, & sends a spur N. divid^{g} Virg & Ky, with many spurs & hills W. throughout Kentucky—

1. At the falls of Cumberland (3 falls) the waters have brought many rare plants & it is one of the best locality—I was there in the fall & found *Magnolia macrophylla*, 2 *Kuhnias*, a New Leguminous in seed (not yet properly ascertained for want of fl.) & a crowd of rare plants, *Gentiana heterophylla*[,] *Prenanthes*, sev Sp.[,] my N. G. *Discomela* 4 Sp—many asters &c. I recom[m]end you to visit this locality in the spring & you will no doubt discover some new or rare plants—Nay every good locality ought to be visited in May, July & September to obtain the Vernal, Estival & Autumnal Plts[.]
2. Cumberland Gap & hills near it—Ditto—It would be worthwhile for you to follow the Cumberland Mts. either E or West, or thro' the Valley to the head of Cumberland R, where I meant to go but was prevented by the Rains. A Botanist will be amply rewarded by going thro' those Mts along the Valley of Sandy R. to their end near the Ohio—Yet if he could go South in Tennessee he would be still more successful, because more plants are found as you go S.
3. Rockcastle R to the mouth in Cumberland R. Many rare plants[.] I found a new *Chrysanthemum*, *Cladorhiza*, *Trichost. angustif*[.] &c.

4. Knob hills being spurs, ridges & table lands spread out from the Cumberland Mts—These are filled with fine localities particularly at their contact with the Limestone Regions. It is there I found most of my new plants—*Gerardia levigata* (fol. lanc[.),] *Spermacoce, Eclipta*[,] *Houstonia*, &c &c. I recom[m]end you to visit Harmanlick, Buttonlick[,] Cedar lick, & the Gaps on the Great Virginia road & Southern road—Besides the Olympian Springs & hills near them—The 2 Knob licks are on their margin; there, near Govr. Shelby is my *Pachysandra erecta* or *cespitosa* which I mistook for *P. procumbens* of the same, the name must be changed as it does not apply—

5. Further West is the table land of Green R. called also Knob hills on their S. side & Muldra[u]gh hills on the N. side & reaching to near the mouth of Green R. This region blends with the next as the Barrens cover these hills. The good localities are numerous. I recom[m]end the C[h]ameleon Spring, Mammoth Cave, the hills E of Hardin, those of Panther Creek & near the mouth of Salt R &c. The fine plants found there are too numerous to mention, they would fill a page. *Aconitum, Lepachys, Helichroa, Gentiana, Buchnera* &c.

[6. *Omitted by Rafinesque*]

II Region[.] Barrens—You know them well, I recom[m]end above all those between the Cumberland, Tennessee & Mississip[p]i R, as filled with rare Southern plants, and if you could stretch into Tennessee so much the better. Altho' I consider Missouri, Illinois, Kentucky, Tennessee & Arkanzas, all around the confluens of Ohio &c as a similar Botanical Region. There I found many new plants, above all some new Ombelliferous, not yet published, because I was doubtful about them. But having been seen by Nuttall & Torrey, they are now well ~~ascert~~ ascertained, being 5 N Sp of *Polytenia, Oxypolis, Leptocaulis* &c. You ought to go there in May, July & September—(My N G *Orimaria*[.)]

7. Barrens of your former neighbourhood, but you ought to stretch S. in them to Tennessee in 3 seasons.—

8. Bowling[G]reen & Russel[l]ville—the Barrens S of them are full of fine plants of many G—near Elkton is found *Salvia urticifolia* or a var. my *S. piperita* which is smaller than in Carol[ina]. & burns the mouth like Cayenne pepper. My *Oxalis repens* &c.

9. Barrens near Glasgow &c—Rich in fine plants. Here I found my N. G. *Cauloma, Sisyrinchium niveum, Delphinium virgatum* R. or *virescens* Nut[t]al[l]? Many *Tradescantia, Elephantopus* &c.

10. Barrens of Little Barren R. near Elk lick—The finest of all localities, I found there in May 3 N. G. *Therolepta, Vernasolis* comp. pl.[,] many new *Dodecatheons* &c. Visit this place in 3 Seasons & go up Easterly along the knobs if you can—I want again all my N. Sp & G as I have but a set left. I recom[m]end you this place above all. It has 2 localities, one is Elk lick for hill plants, the other is ab^{t} the wouldbe town of Monroe & S. & E of it—Go there in May & Sept^{r}.—

[11. *Omitted by Rafinesque*]

12. Barrens S. of Hendersonville &c.

III Region. Limestone Country. This you know extends from Maysville to Louisville by Lexington &c. The best places are the Rocks & Cliffs of the R. Kentucky, Licking & Salt.

[13. *Omitted by Rafinesque*]

14. Neighbourhood of Blue Licks—Many fine plants, my *Lobelia nivea*, the *Hypericum dolabriforme* Mx.

15. Cliffs below Frankford—There my *Eupatorium rupestre*, *Lobadium* N G formerly Rhus—

16. Vicinity of Cynthiana, 17[.] The Ridgeroad—18[.] Bigbone lick[,] 19[.] Shelbyville & Floyd Creek—All good localities—

20[.] Boon[e's] Creek near Lexington—here is *Dirca palustris*.

21. Dick R & the Cliffs of Kentucky near it—Here many new Sp *Houstonia rupestris*, *Oxalis rupestris*, *Tradescantia pumila*, *Solidago sphacelata*, *Sedum pulchellum* which is a new S. G of mine[,] *Aetyson* having unequal cal. pet. stem. styles & capsules—Caps 2 sp.

22. Harrodsburg & vicinity as far as Cliffs [of] Ky [River]—many rare pl. My *Cubelium* (Viola concolor auct)[.] My *Cladrastis* N G. 1825 the Virgilia of Mx—not of Cape good hope!

IV Region. Alluvions of the Ohio, Kentucky, Licking, Sandy[,] Salt, Green, Cumberland R & other Streams—Rich in peculiar plants[,] many of Louisiana & Carolina. Several N. Sp. of mine *Hypericum riparium* (Green R)[,] *Commelina*[.] N Sp *Heliotrope indicum* of our Botanists, is a N. G. *Tiaridium* of Lehman, who has descr 4 Sp. but not ours—I call it *T. riparium*.

I have forgotten some other rich localities—Mt Vernon, Crab Orchard & Estil[l] Springs in the Knobs—But I must conclude altho' I have not stated one half of them—I was 8 years exploring them <1818 to 1826>—You will have enough to do to visit them all in 3 Seasons, besides finding other new localities as good—You may publish this Sketch of Botanical localities if you cho[o]se, or give me Credit for it whenever you may publish your own experience—

18

North American Botany

In 1813, when he wrote his *Chloris Aetnensis* in Sicily, Rafinesque acknowledged the relationship between elevation and plant distribution by writing four florulas, each listing the plants of its region. Starting with the Florula Pedemontana—that is, the vegetation of the Piedmont—these rise to the Florula Nemorosa, that of the wooded region; then the Florula Excelsa, that of the heights; to finally the Florula Arenosa, dealing with the scanty vegetation of the lava fields at the summit of the volcano.

Thereafter, during his active years as a field botanist, he remained much interested in various aspects of phytogeography. In his 1818 review (reprinted in this book) of Elliott's *Botany of South-Carolina and Georgia* he included a brief digression on the subject, and he devoted two essays to the botanical geography of Kentucky while he was living in that state. They are also reprinted here. His most mature thoughts on the subject, the "results of 24 years of observations and researches," as he wrote in the dedication of the book where they appeared, were developed in the following essay. It ends with a number of provocative questions.—*Editor*

* * *

Flora of North America. Botanical Geography and Localities.

Botanical Geography has lately been much attended to since Wil[l]denow, Decandole and Humboldt have written upon it.[1] Dr. Pickering alone has specially written upon that of North America; and although I do not admit of all his conclusions, nor think his map quite correct, yet he has opened the way.[2]

The Earth is divided into botanical regions, where a peculiar growth of trees and plants are found; these regions although sometimes well defined in Islands and Physical regions, must necessarily blend in large continents near their limits.

Wil[l]denow supposed that groups of mountains were the nucleus of these regions, and that the floras expanded around; others think that mountains often divide the botanical as well as physical regions. In North America both seem to be partly the case.

Decandole had only three botanical regions in North America, north of Mexico, the Atlantic or Ap[p]alachian extending to Florida and Missouri, 2d the Origonic or the Origon mountains and plains of the West. 3. the Boreal common to boreal Asia and Europe. Pickering has proved that following the level of the land, the Boreal or Canadian extends South over the Alleghany mountains, while the Mexican region extends North into Texas and Arkansas.

[Amos] Eaton has supposed that our Atlantic region was divided in two by the Potomac, the Northern, and the Southern that winds round the mountains to the far West including all the Western States.

These are of course exclusive of the three great regions of the Southern parts, Mexico, Central America and the Antilles.

I have rectified these views since 1832 by increasing our regions to seven; to which I have given the names of Boreal, Canadian, Alleghanian, Floridian, Louisianan, Texan and Oregonian: each of these is perfectly distinct and distinguished both by physical features and peculiar Genera of plants.

1. *Boreal Region*, including the Polar region, Greenland, Iceland, Labrador, Hudson Bay and New Siberia. This wide region, is very similar to the Boreal parts of the Old Continent, Lapland and North Siberia, forming perhaps only one wide circle around the Arctic Pole. It is the poorest of all the American Floras, with very few trees and shrubs, chiefly evergreen, and with the lower classes of plants preponderating, such as Mosses, Lichens, Algas, &c. but few Fungi. The floral season is very short, hardly three months from June to August.

2. *Canadian Region*. This forms a broad belt across the Continent including Nova Scotia, New England, Canada, the countries around the Lakes, and the vast lacustral plains of the West. It has spurs in the northern Alleghanies, the Saranac, Taconick, and Kiskanom mountains. It is distinguished by the prevailing Firs, Willows, and Birches, the Genera *Linnea, Diervilla, Parnassia, Rubus, Ribes, Coptis, Nemopanthes, Comarum, Caltha*, &c. and an abundance of Mosses, Lichens and Fungi, not however exceeding one half of the whole number. The floral season of five months, from May to September.

3. *Alleghanian Region*. This has for nucleus the Alleghany mountains of Pennsylvania, Maryland, Virginia, Kentucky, &c. called Ap[p]alachian south of Potomac and Wasioto or Cumberland to the West: this region winds all around East and West into the hilly or broken country. It is distinguished by the abundance of trees, oaks, radiate plants, fungi, grasses, leguminose, hypericines, with the prevailing genera *Hicoria, Kalmia, Trillium, Azalea, Vitis, Rhododendron, Hydrangea, Heuchera, Lactuca, Solidago, Rosa*, &c. the Mosses and Lichens are yet abundant, but now form only a small proportion of the whole, while the Fungi are become about one third of the whole. It is remarkable that here the trees and shrubs although yet of the same Genera as in Europe are nearly all unlike in species.—Floral season from April to October.

4. *Floridian Region*. This begins in the North by a narrow belt in the marshy and sandy Islands or Shores, pine woods and swamps of Long Island and New Jersey, widening gradually in the plains of Virginia, Carolina, Georgia, ascending the hills and south Ap[p]alachian mountains which run west into Alabama, occupying the whole of Florida, Alabama, &c. It blends in South Florida with the Antillian

flora of Bahama and Cuba, in the mountains with the Alleghanian flora, and west with the Louisianan. It is distinguished by Magnolias and Pines, Palms and Yuc[c]as: the peculiar Genera *Clethra, Fothergil[l]a, Passiflora, Chionanthus, Lantana, Marshallia, Xerophylum, Pin[c]kneya, Halesia, Leiophyl[l]um, Hudsonia,* &c. with many species of *Ilex, Ludwigia, Rhexia, Viburnum, Lobelia,* &c. Here the Mosses, Lichens and Fungi greatly decrease, while the grasses, rushes, ferns, aroides, lilies and akin greatly increase. It is the richest of our Botanical regions in Species, and Vegetation is unceasing, there are flowers in every month, even in January.

5. *Louisianan or Missourian Region.* The actual state of Louisiana hardly belongs to this; but the ancient Louisiana is the nucleus of it, extending into the plains of Tennessee, Kentucky, Illinois, Arkansas, Missouri, and up into the vast plains of the Upper Missouri. The cradle of it appears to be the Ozark mountains. It is distinguished by the Pecan-tree, the Bowtree, the Amorphas, the *Planera, Cladrastis,* peculiar Pavias and Poplars; with abundance of Cacti and Ferns; but scanty Mosses, Lichens and Fungi: with the peculiar Genera *Miegia, Collinsia, Bellis, Leucospora,* &c. Floral season from March to November.

6. *Texan Region.* This extends over the wide plains between Louisiana and Mexico, and probably over New Mexico: the nucleus may be the Taos mountains; botanical spurs extend into Arkansas and the Colorado Valley. It is very little known as yet, but the productions are intermediate between Mexican and Louisianan, with abundance of Cactides and Canebrakes. The floral season lasts nearly the whole year as in Florida.

7. *Origon Region.* This extends over the Origon mountains [i.e. Rocky Mountains] and the hilly country west of it. We do not know it thoroughly as yet, but what we know of it, is very peculiar. We may hope that [Thomas] Nuttall will describe it botanically. It has perhaps several divisions, the mountains, sea shores, and new Albion or new California to the South. It bears the same botanical relation to Siberia than [that?] the Alleghanies do to Europe. It is distinguished by peculiar trees. Firs, Oaks, Maples, the singular Garrya tree, many species of *Ribes, Lupinus, Pentostemon, Cactus, Mimulus,* &c. with peculiar Genera *Calochortus, Eutoca, Lasthenia, Collomia, Aegochloa, Lewisia, Clarkia,* &c.—The Labiate, Hypericines, Grasses, Fungi and Mosses appear to be scanty. Liliacea abound, but the Orchides are very few. The floral season extends from March to November in New Albion, but is much shorter in the mountains and to the North.

Besides the above great Regions of this Continent, there are several other Local Regions, more confined in limits, but well distinguished by peculiar vegetation or growth of trees. The principal of these are

1. The swampy pine region, where grow the *Schubertia* and *Cupressus thyoides,* from New Jersey to Louisiana, with many peculiar water plants. These swamps are unlike marshes by being sandy and having a clear yellow water. In the dry places prevail Pines, Hudsonia, &c.

2. The Sandy Shore Regions all along the margin of our Sea from Long Island to Florida; Ilex and Myricas prevail, also shrubs, plants, with palms in the South. The rocky shores of the North are very unlike this.

3. The Alpine Region or Summits of High mountains, where few trees are

found, while many peculiar plants grow only there. In the Alleghanies they partake of the Canadian flora, and in the White mountains of the Boreal flora.

4. The Prairie Regions of the West, with few trees, but a profusion of fine plants, Dodecatheons, Tradescantias, Helichroas, Gentians, Radiate flowers, &c., some peculiar shrubs and hardly any Acotyle plant. There the floral season begins in March or sooner as you go South.

5. The Limestone Region of Florida, with a profusion of fine plants, Ludwigias, Rhexias, &c. and many peculiar shrubs.

6. The Limestone Region of the Ohio, forming a bassin in Ohio, Indiana and Kentucky: it has a very scanty flora, few shrubs in the woods, no Kalmias nor Vacciniums, but among trees many Asiminas and Pavias, with abundance of social grasses or congregated plants.

7. The Ap[p]alachian and Wasioto Region, or of the hills and mountains from Kentucky to Georgia: this has a distinct flora from the Alleghanies and Florida shores, many Rhododendrons, Azaleas, Magnolias, and a crowd of fine plants; many yet new in the Unaka or Iron mountains, the northern nucleus of it, as well as in the peninsula of Florida, Alabama, &c.

Besides these great localities I will add several smaller localities of great botanical interest by the numerous new plants which they have afforded me. Every botanist knows some similar place; but those which I may boast to have discovered or first well explored deserve to be commemorated. They are

1. The neighborhood of Quaker Bridge and Cedar Bridge in the centre of the Pinelands of New Jersey.—2. The neighborhood of Mullica Hill in New Jersey.—3. The sea Islands of New Jersey—4. The source of the river Delaware in New York at Utsiantha Lake in Oquage mountains—The Kiskanom or Catskill mountains of New York.—6. The Dismal Swamp of Delaware—7. Sherman Valley in the Alleghanies—8. The Cotocton mountains of Maryland and Virginia cut through by the Potomac.—9. Valley of Loyalhannah in west Pennsylvania—10. Falls of the river Potomac—11. Falls of the river Cumberland in Wasioto hills of east Kentucky—12. The serpentine rocks of Chester and Maryland—13. The Wiconisco, Tuscarora and Central mountains of the Alleghanies—14. The summit of the Alleghanies in Maryland.—15. The Cacapon mountains of Virginia.—16. The prairies of Bigbarren river in Kentucky—17. The Wasioto Hills and mountains of Kentucky, or Knob Hills, with their knoblicks—18. The banks of the Wabash, and glades near them.—19. The neck of land between the mouths of Cumberland and Tennessee rivers, with the glades of South Kentucky and Tennessee.—The shores of Lake Erie near Sandusky.

I hardly need add the far famed fall of Niagara, the head and falls of the Hudson, the Taconick and Mattawan mountains, and in fact every ridge of the Alleghanies. They are all interesting botanical spots to visit; but in order to detect all their plants, you must visit them at least three times, in the Spring, Summer and Fall, or every month from May to October, and even some plants of short floral duration may then escape you. How can we then hope to know all our productions, except gradually and by repeated explorations. I have never been able to meet the *Hamiltonia*, nor *Centunculus*, nor *Parnassia* in full bloom, and many rare plants were only found once by me during 24 years of exploration.

It is a mistake to imagine that all our plants are sylvan or nemorose, because forests abound in our Continent. The plants growing underneath the shade of trees are not even the major number, and we must look for many in meadows, glades, fields, swamps, sea shores, banks of rivers, brakes, (these are peculiar places in woods or glades where ferns, canes or grasses prevail,) salt licks, rocky hills and cliffs, mountains tops, near springs and brooks.

The distribution of the plants in these localities, and the distribution of peculiar tribes over the whole continent, would form an interesting theme, which although partly conjectural and never quite exact, might lead to philosophical reflections and deductions. Some problems remain to be solved on the subject. Why are our North American trees chiefly of European and Siberian Genera, but unsimilar in species?

Why are our shrubs still more dissimilar both in genera and species?

Why are our grasses and cyperacea so often similar in species, while the other monocotyle plants are all different?

Why are the minute and simple Acotyle plants, such as Algas, Lichens and Fungi, nearly identic in the two Continents?

Why are Mosses and Hepatica so often similar, while the ferns are less so, and offer many new forms?

Why are many similar forms offered by the tribes of Ranunculacea, Labiate, Juncides, Rosacea, Polygonides, Lysimachides, Veronicides, Borragines, Bicornes, Caryophyles, Aroentacea? &c.

Why are so many dissimilar genera and species found among the Violacea, Irides, Asphodelides, Orchides, Gentianides, Apocynea, Corymbifera, Hypericea, Malvacea, &c. of both continents?

Why are some families quite lacking in N. America? Such as Rutacea, Guttifera, Protacea, Epacrides, &c.

Why do the tropical genera so seldom extend into Florida?

Why is the vegetation of Boreal America totally unlike that of Austral America?

Why have we so few Crucifera, Umbellifera and Narcisssides, compared to Europe and Africa?

Why are the two shores of North America, east and west, so unlike to each other in vegetation?

These queries and others of a similar nature may exercise the ingenuity of speculative Botanists, or amuse their idle hours; but they are facts and as such deserve our notice.

Another interesting study is that of our naturalized plants. We have so many that they appear to invade the fields and drive out the native plants in some instances; but it is by no means certain but that some deemed naturalized, were not really native. Such at least must have been the case with *Verbascum thapsus*, *Hypericum perforatum*, *Daucus carola*, *Anthemis cotula*, *Origanum vulgare*, &c. I gave a long list of the naturalized plants in my dissertation[3] of 1808, some have since disappeared, while others have appeared instead out of gardens. But few American plants have become spontaneous in Europe, *Phytolac[c]a decandra* and *C[a]enotis canadensis* are mentioned as such; but we have received several hundreds, besides some few from the Antilles and inland. Decandole has properly stated that naturalized plants even when not spontaneous, but extensively cultivated in the open air ought to be added to every general flora, and Eaton has followed that advice with us.

It appears that even previous to the discovery of America by Columbus, our Indian tribes had received or imported from abroad or the South, several trees and plants. I have at least evidently ascertained historically that this was the fact with the Maize, *Mayzea cerealis* (Zea mays of L) [,] the Orange tree, the Chicasa plumb *Prunus angustifolia*, the Catalpa tree, the Peach tree, and the holy plant *Nelumbium luteum*! besides several kinds of Beans, Tobacco, Potatoes, Squashes, Pumpkins and Melons that were cultivated by our native tribes before Columbus or at least the invasion of Soto between 1540 and 1543, and subsequent French and English colonies.

The number of Genera and Species of each botanical region, and their proportion of Trees, Shrubs, Plants, Herbs, Grasses, Ferns, Mosses, Lichens, Fungi, &c. is not only variable in each, but difficult to fix, owing to the great number that are common to several blending on the limits. In general the Trees, Shrubs, Ferns and Grasses, increase in number from North to South, while the Acotyle plants diminish; but Fungi are most numerous in the Alleghanian region, by [Lewis D. von] Schweinitz['s] account of them they nearly equal the Phenogamous plants.

However all the calculations heretofore made were erroneous, depending on the peculiar knowledge of the writer and his limited views of Genera. [Johann Reinhold] Forster said they were only 1200 in 1771, [Amos] Eaton in 1833 said exactly 5267! [John] Torrey about 8000! I say 15000 and am nearer the truth; at least 10,000 are Vascular plants in the whole of North America, north of Mexico, with about 5000 cellular plants, whereof nearly 300 are trees or palms. Our largest Flora, the Floridian[,] contains 6000 species at least, while the smallest, the Boreal, has hardly 1000 species, and no more exist in the Limestone region of Kentucky. The number of Genera will depend on the reformed view and correction of them.

Published in *New Flora of North America* (Philadelphia, 1836), I, 23–32.

1. Carl Ludwig Willdenow, *Grundriss der Kräuterkunde zu Vorlesungen* (Berlin, 1792); Augustin Pyramus de Candolle, "Mémoire sur la géographie des plantes de France, considerée dans ses rapports avec la hauteur absolue," *Mémoires de Physique et de Chimie de la Société d'Arcueil*, 3 (1817), 262–322, and *Essai Élémentaire de Géographie Botanique* (Strasbourg, 1820); Alexander von Humboldt and Aimé J. A. Bonpland, *Essai sur la Géographie des Plantes: Accompagné d'un Tableau Physique des Régions Équinoxiales* (Paris, 1805).

2. Charles Pickering, "On the Geographical Distribution of Plants," *Transactions of the American Philosophical Society*, n.s. 3 (1830), 274–84.

3. Unless Rafinesque was alluding to a document unknown to his bibliographers or 1808 is a misprint or, most likely, his memory was faulty, the article in question was his "Essay on the Exotic Plants, Mostly European, Which Have Been Naturalized, and Now Grow Spontaneously in the Middle States of North America," *Medical Repository*, 2 (February–April 1811), 330–45. This essay was dispatched from Palermo in the form of a letter dated 1 April 1810.

VIII.

Natural Science

19

Natural Science in America

There are several other accounts of the development of American natural science that were written before the middle of the nineteenth century. Rafinesque's reprinted here is comparable with most and superior to several of those published by such of his contemporaries as Benjamin Smith Barton (1807), Lewis Caleb Beck (1842), Amos Binney (1845), Walter Channing (1821), De Witt Clinton (1815), William Darlington (1849), James E. DeKay (1826), Sereno E. Dwight (1826), George William Featherstonhaugh (1832), Asa Gray (1841), William Jackson Hooker (1825), David Hosack (1811), Samuel Latham Mitchill (1814), Samuel George Morton (1841), Charles Wilkins Short (1836), and John Torrey (1843). Several of these focused on a single discipline, most commonly botany because botanists have always taken a lively interest in their own history. Rafinesque essayed to cut a wider swath, however; and, having been in the United States only about twenty-eight months since his return from Europe, he must have had to rely on others for some of his information. Not all of it was accurate.

A promised second part of this article apparently never appeared, though when he listed this essay among the "tracts already published" in his *Proposals to Publish by Subscription* (1821) a collection of his works, he said it was "with a supplement." Such a supplement may have been written but has never been found.

As was common with him (and many of his contemporaries, for that matter), Rafinesque was content to list persons by last name only. Meaningful as these may have been for the original readers of his essay, they often are enigmatic today. Here, it has been possible to fill in between square brackets most of the given names, sometimes by searching the publications of his contemporaries listed above.—*Editor*

* * *

Survey of the Progress and Actual State of Natural Sciences in the United States of America, from the Beginning of This Century to the Present Time

Introduction.

MERCHANTS feel an interest in trade, poets in literature, painters in pictures, every one in the objects connected with his pursuits and labours: it is therefore very natural, that those who have devoted a share of their attention to the noble pursuits of science, should likewise feel a desire to take an occasional survey of the progress, situation, and prospects of the various branches of science, which they may have undertaken to cultivate, as well to ascertain their positive advances as their relative improvements.

Among sciences, those connected with the natural and material objects of the universe, claim of course a conspicuous rank, since they relate to every thing which we perceive, or which falls under the observations of our senses. Even the numberless arts which human ingenuity has devised, for the purpose of imitating or modifying those objects, ought to class with them; but custom separates them, while it acknowledges their intimate connexion, and absolute dependence. Natural Sciences are therefore limited to three great branches: COSMONY, or Natural History, which enables us to distinguish, describe, value, and employ the natural objects and bodies: PHYSICS, or Natural Philosophy, which teaches us their functions, laws, and phenomena: CHEMISTRY, or Natural Analysis, which decomposes and recomposes them, reaching the elements of nature. They are divided into many collateral branches, such as Astronomy, Geonomy, Botany, Zoology, Optics, Statics, &c. which are again subdivided into numberless minor branches.

In the last century these sciences were yet in their infancy in the United States, as was every thing else; but nevertheless, that first period of their cultivation was adorned by the following eminent or worthy writers, Winthrop, Franklin, Jefferson, Rittenhouse, Clayton, Bartram, Walter, Barton, Muhlenberg, Priestley, Drs. Mitchill, Colden, Garden, Marshall, Carver, Belknap, Cutler, &c. and among the visitors or travellers, Catesby, John Mitchill, Kalm, Bosc, Castiglione, Vieillot, Palissot-Beauvais, Volney, Mason, Mackenzie, Frazer, Dupratz, Charlevoix, Michaux, Schoepf, &c.—some of whom belong to both centuries, and will be noticed again hereafter.[1]

Since 1800, a great impulse has been given to some branches of these sciences; many societies have been established for the purpose of fostering their study; museums have been formed in many cities; professorships established to teach every branch; and, at present, a great number of young and able observers or writers begin to appear every where, who bid fair to reflect honour on themselves and their country. To encourage the disposition which is manifesting itself is the design of this review. The record of the labours of their predecessors, whilst it is a grateful tribute for past services, will tend to excited the emulation of the rising generation, and may serve to enlarge the ideas of European writers, in reference to our general and national character.

All those who pursue the noble path of natural knowledge are united by a friendly bond; although strangers, although distant, as soon as they become known to each other, either personally or by fame, they are friends: it is our object, if practicable, to strengthen those ties, not merely among ourselves, but between American and European writers.

Let no national rivalry interfere—it ought to be unknown among men of enlightened and enlarged minds: and let no mean jealousy arise among ourselves—it can never be fostered by the generous and the wise. But above all let us disregard those snarlers and sneerers, whose profound ignorance prevents them from conceiving the scope and use of our pursuits, and without allowing ourselves to deviate from the honourable paths of knowledge and improvement, let us steadily persevere in observing, collecting, and imparting, useful facts and truths—in improving ourselves and mankind.

We shall divide this subject into two parts: first, collective improvements and labours—second, individual labours and discoveries, concluding by some remarks on what remains to be done.

PART I. COLLECTIVE IMPROVEMENTS AND LABOURS.

At the beginning of this century there were only three learned societies in the United States, which included natural sciences within their range; and even they did not assume their study as the base of their labours.

These were the [American] Philosophical Society,[2] at Philadelphia, founded in 1744; the American Academy of Arts and Sciences, at Boston, founded in 1780; and the Connecticut Academy of Arts and Sciences, founded in 1799. Some other societies had directed their pursuits towards some of the auxiliary branches; such as the Agricultural Societies of Charleston [i.e. South Carolina Society for Promoting and Improving Agriculture, 1785], and Massachusetts, the Society for the Promotion of Useful Arts and Agriculture [1798], &c. of Albany, and several Medical or Chemical Societies in different cities.

While these societies are mentioned, it will not be improper to notice their labours in this century. The Philosophical Society of Philadelphia has published three volumes of transactions in 1802, [180]4 and [180]6, containing many valuable communications; but it has promulg[at]ed nothing since, and the Magellanic premium,[3] of which it has the disposal, has not been awarded for many years. The Academy of Arts and Sciences of Boston has published two volumes of transactions, 2d. and 3d. The Society of Albany has published, occasionally, some tracts connected with its views. The Connecticut Academy has published a volume of transactions in 1810. The other societies have not published any thing to our knowledge; but they have probably been instrumental in imparting knowledge to their members, and nourishing a taste for their pursuits.

Since 1800 the following learned societies have been established, mostly, as will be perceived, for the cultivation of natural sciences.

The Linnean Society of Philadelphia,[4] founded in 1804: whose first president was Dr. Benj. [Smith] Barton, and whose actual president is Dr. W. P. C. Barton. It has not been very active, and had even become nearly extinct; but has lately been revived. None of its labours have been published except an address of the first president.

The Linnean Society of Boston,[5] founded in 18**. Its actual president is Judge [John] Davis [1761–1847]: it has not published any transactions.

The Columbian Chemical Society of Philadelphia, founded in 1811, which has published a volume of memoirs in 1813. [Its first president was James Cutbush, 1788–1823.]

The Literary and Philosophical Society of New-York, founded in 1814. President

Dr. Dewitt Clinton. It has been very active, has published in 1815 a first volume of transactions highly valuable, and is preparing a second for the press.

The Literary and Philosophical Society of Charleston,[6] founded towards 1814. President, Stephen Elliott, who has published his Introductory Discourse.

The Academy of Natural Sciences of Philadelphia,[7] established in 1815. President, Mr. Patterson: the members meet weekly, and instruct each other by lectures; an example worthy of imitation. It has formed a museum; and since May, 1817, has begun to issue a monthly sheet, under the name of Journal of the Academy of Natural Sciences of Philadelphia, and after the plan of the *bulletins of the Philomatic Society of Paris,* which is the first of its kind in the United States, and will materially contribute to spread natural knowledge.

The Cabinet of Sciences,[8] of Philadelphia, established in 1815. President, Dr. Clymer. Nearly on the plan of the foregoing society; but not so active, it has undertaken to form a botanical garden.

The Columbian Institute, of Washington, was established in 1816: the president is Dr. Cutbush;[9] it has for its object to encourage the cultivation of sciences and arts; but as it meets only twice a year, it will not be able to become of much avail.

The Lyceum of Natural History, of New-York, was formed in 1817: President, Dr. [Samuel Latham] Mitchill—it holds weekly meetings. Within a few months, this society by the activity of its members, has begun a museum, and an herbarium: it has appointed lecturers on all the branches of Natural History,[10] and travelling committees, and proposes to publish its transactions in a short time.

Besides the above, another learned society, under the name of School of Arts and Literature, has been established at Cincinnati,[11] in Ohio, towards 1814; but we are unacquainted with its officers or labours: it deserves attention, however, as the first instance of such an institution in the Western States.

Several other minor societies, for auxiliary branches of natural sciences, have likewise been established at different periods; such as, the *Agricultural Society of Philadelphia,*[12] of which Judge [Richard] Peters [1744–1828] is the worthy president, and which has been very active, having published two volumes of important papers: the *Historical Society of New-York,* which has lately assumed the subject of natural history, and formed a museum, &c.[13]—besides some new Medical Societies, to whose lot it falls to elucidate the natural history of man; and three Botanical Societies in Utica, Philadelphia, and Boston, lately established.

The collective labours of these societies have been surpassed by the personal labours of their members, and other individuals, which we shall notice at length in the second part: but we mean to give here an account of the gradual means employed by them.

Only two small museums of natural history existed in the United States in 1800, in Philadelphia and Boston. These establishments, which increase the taste for natural beings, or even create it, when the simple survey of nature cannot inspire it, have become numerous and splendid of late; some of them begin to equal the best European museums; among which, those of [Charles Willson] Peale in Philadelphia, and [John] Scudder in New-York, deserve particular notice for elegance of taste and abundance of objects. There are also public museums and menageries, or exhibitions of living animals, in the following cities: Boston, Salem, Baltimore, Charleston, Norfolk, Lexington, New-Haven, &c. They have all been collected by individual exertions, and the liberal patronage of the public has generally well rewarded them; in some instances legislative or municipal patronage has been extended to them, by the grant of suitable rooms, &c.

Private collections are increasing every day in number and value; almost every University and College has a small museum, or a collection of minerals, shells, &c.: many gentlemen and ladies begin to delight in procuring collections, which has a general tendency to increase the taste for rational and innocent amusements. Among those private collections, the following deserve notice, as the most rich and valuable. The mineralogical collections of Dr. [Archibald] Bruce in New-York, of Col. [George] Gibbs, in the museum of the Historical Society of New-York, and of Yale College at New-Haven, &c. The conchological collections of John G. Bogert, Esq. of New-York, and of the Academy of Natural Sciences, in Philadelphia, &c. The entomological collections of Mr. [Thomas] Say in Philadelphia, of Mr. [John] Torrey in New-York, &c. The general collections of Dr. [Samuel Latham] Mitchill in the University of New-York, of the Academy of Natural Sciences of Philadelphia, &c.

Herbariums, or collections of specimens of plants, have been made by many, but they are generally confined to American plants; the most valuable are those of the late Rev. Dr. Henry Muhlenberg, in the possession of his son, Dr. [Frederick Augustus] Muhlenberg of Lancaster, of Mr. [Stephen] Elliot[t] of Charleston, of Mr. [Zaccheus] Collins of Philadelphia, of Dr. [John] Eddy of New-York, of Dr. [Jacob] Bigelow of Boston, of Dr. W. P. C. Barton of Philadelphia, of Mr. [John] Torrey of New-York, of Mr. Rafinesque of New-York, &c.

Botanical gardens are connected with botany, medicine, agriculture, horticulture, and becoming useful appendages thereto, when properly directed; but no such public gardens have been endowed as yet in the United States, upon the liberal European system. Mr. [John] Bartram's private botanical garden was perhaps the only one in existence at the beginning of this century; since which period many similar private gardens have sprung, such as Mr. W[illiam]. Hamilton's at the Woodlands, near Philadelphia, Dr. [David] Hosack's at Elgin, near New-York, several in the vicinity of Boston, and one in Charleston, &c. The garden of Elgin has lately been purchased by the legislature of New-York, and given to the University; but it is much to be regretted, that it has meanwhile been neglected, and almost destroyed, because no able director was appointed. Several new botanical gardens are in completion, by subscription, in Philadelphia, New-York, and elsewhere; but unless they are liberally endowed, they will not become of permanent utility. The botanical garden at Cambridge, forms however a partial exception, and is an useful appendage of that University.

Gardens on a more moderate scale, but not less useful, are common near Charleston, Alexandria, Baltimore, Philadelphia, New-York, Boston, &c. where useful and ornamental plants, trees, and seeds are raised for sale: those of Mr. Macmahon [*sic*][14] near Germantown, and of Mr. Prince[15] at Flushing, &c. may be quoted as examples. The establishments more closely connected with agriculture, such as nurseries, seed-stores, &c. have also increased in proportion, among which the nursery of fruit trees of Mr. Cox[e],[16] near Burlington in New-Jersey, has ranked among the most valuable.

Agriculture, the base of our real wealth, is of course attended to with unceasing care, and a few worthy individuals, such as Chancellor [Robert R.] Livingston,

Dr. [James] Mease, Judge Peters, John Lowell, Esq.[17] &c. have been endeavouring to study it and teach it as a science; but their attempts have generally failed, because the great mass of farmers conceive they know enough! Enlightened proprietors and farmers, are not however, [un]willing to admit of improvements, and to allow their practice to be directed by a wise theory.

Horticulture, both practical and ornamental, is likewise become fashionable among our wealthy citizens. The cultivation of our native ornamental plants and shrubs is spreading everywhere, and exotics are not neglected; green-houses are quite common, and some hot-houses are to be found in the vicinity of every city.

Extensive public libraries, on a liberal plan, had been established last century; they have gradually increased their stock of books, where valuable materials for the study of natural sciences are to be met; those of Philadelphia, Baltimore, New-York, and Boston, hold the first rank. The libraries of Hospitals, Colleges, &c. have likewise been materially and usefully increased; the libraries of Dr. Benjamin [Smith] Barton, and Dr. [David] Hosack, have been purchase by the hospitals of Philadelphia and New-York, and are exceedingly rich in rare books of natural history.

Most towns, and even many villages, have established circulating or subscription libraries on improved plans; they convey useful publications into every corner of the Union. Athen[a]eums have been formed in Boston, Philadelphia, Lexington, and elsewhere, whose object is to collect useful works, and where the literary journals of America and Europe are regularly received; as they are likewise in the Literary Rooms of Messrs. Eastburn and Co. of New-York.[18] The Athen[a]eum of Boston contains one of the most extensive and valuable libraries in the United States. Reading-rooms and athen[a]eums, on a minor scale, are not uncommon throughout the Union.

Public instruction has kept pace with our rapid increase of population, which is evinced by the general increase of students, and the endowment of many new colleges and academies, particularly in the western and southern States, among which may be mentioned those of Lexington in Kentucky, of Millegeville in Georgia, of Columbia in South Carolina, &c. In all the colleges of the United States, which amount to more than forty, natural philosophy is taught; in some of them chemistry; in a few natural history.

In the Universities, all those branches have professors, often men of ability; but they are generally annexed to the schools of medicine. In the [Harvard] University of Cambridge, however, a distinct course of lectures on Natural Sciences, is delivered by professors in all the branches of those sciences. In the University of Pennsylvania, since the death of Dr. Benjamin [Smith] Barton, a faculty of natural sciences has been established last year: this is the first instance of the kind in the United States. The following professorships were appointed and filled; of natural philosophy, of botany, of natural history, particularly zoology, of comparative anatomy, of mineralogy and chemistry applied to the arts;—those of the institutes of chemistry and materia medica, being left united with the medical faculty.

It is to be regretted that professors are sometimes appointed who have yet to learn what they are to teach: instruction will flourish with more rapidity when they shall be selected, in all instances, among the most worthy and learned candidates.

Public lectures on the most popular branches of natural sciences have been given by many private lecturers, e.g. by Mr. Correa[19] in Philadelphia, on botany, and the natural method; by Mr. Witlow [*sic*],[20] on demonstrations of botany, in New-York, Philadelphia, Albany, New-Haven, Boston, &c.; by Mr. Hare,[21] on chemistry, in Philadelphia; by Dr. Bruce,[22] on mineralogy in New-York, &c.

Natural knowledge has been gradually diffusing itself by all these means, as well as by the individual exertions of the observers of nature, their writings and publications; among which, periodical ones are not to be reckoned the least useful. Even newspapers and literary journals have often been the vehicles of much useful knowledge on the physical and geological geography of our country, the natural history of our shores, meteors, &c.: and even those daily papers which appear to be the most hostile to knowledge and science, cannot help to convey, occasionally, valuable facts belonging to, or connected with, natural sciences. The periodical works dedicated to literature, such as the Port Folio, the Analectic Magazine, the Portico, &c. have not forgotten to bestow a share of their attention on sciences. But it is in the periodical medical journals, (and scientific publications,) that the greatest share of natural knowledge has been introduced, as if we would imitate the early periods of European science, in connecting natural with medical sciences. The Medical Repository of New-York conducted principally by Dr. [Samuel Latham] Mitchill, and alternately by Dr. [Edward] Miller, Dr. [Samuel] Akerly, and Dr. [Felix] Pascalis, and which began in 1797, has lasted with success ever since, includes the greatest mass of facts and knowledge on the natural history of our country, and on physical and chemical improvements. Dr. Benjamin [Smith] Barton's Medical and Physical Journal, which lasted only from 1804 to 1808, ranks next, and contains many valuable tracts on natural history. The other works of a similar nature, which may deserve our notice, are Mease's Archives,[23] Cooper's Emporium,[24] the Medical Journals of Philadelphia, Baltimore, and Boston, the Medical and Philosophical Register[25] of New-York, the Eclectic Repertory,[26] &c.

Periodical works, exclusively dedicated to natural science generally, or to peculiar branches of it, have not yet been numerous, owing, perhaps, to a deficiency of public patronage, although it would appear that the numbers of individuals feeling an interest in such studies, might afford a sufficient encouragement. Whenever a regular and general work of that nature shall be countenanced it may become of permanent utility. Dr. Bruce's Mineralogical Journal, which began to appear in 1810, but of which only a few numbers have been published, at irregular intervals, was perhaps the first ever attempted in the United States; and it contains much valuable matter on mineralogy and geology: it is contemplated to be continued occasionally. The Monthly Journal of the Academy of Natural Sciences, of Philadelphia, begun this year, is the next; it assumes zoology and botany principally, and its concise shape will not diminish its utility.[27] The Annals of Nature,[28] which were to begin this year, have been postponed for a few years, and the Annals of the Lyceum of Natural History are soon to be undertaken on a suitable plan.

Notwithstanding the collective utility of the above works, they are liable to one objection, if they contribute to spread and diffuse knowledge, yet they scatter it too much; as it is difficult to become acquainted with, or to possess the whole collections; so that if it were possible to embody, in some suitable shape, the most interesting, or new matter, which they contain, the acquisition of such knowledge would be greatly facilitated hereafter. Many tracts and pamphlets are often lost or forgotten, which might by this means be rescued from oblivion.

Although a particular notice of the works of each author will belong to the second part of this Essay, it may be proper to indicate here which were the principal works on natural sciences, published during the period we allude to, or some of those which we conceive to have a claim on our gratitude, as having aided to enlarge the sphere of our knowledge.

Those that deserve the first rank are Wilson's Ornithology of the United States, which, for brilliancy of style, elegance, accuracy and novelty, can boldly be compared with Buffon's natural history of birds.—Muhlenberg's botanical works now in the press.—Elliot[t]'s Botany of the Southern States, which has just begun to be published.—Mitchill's Fishes of New-York, an original work, on a subject entirely new.—Lewis's and Clarke's travels on the Missouri and to the North-West Coast of America, which are replete with new facts and discoveries, &c.[29]

The following claim also our attention, although of minor importance: Cleaveland's Mineralogy, Barton's elements of Botany, Clinton's Discourse, Drayton's view of South-Carolina, Drake's view of Cincinnati, Williamson on the climate of America, Mease's Geological survey of the United States, Ellicot[t]'s astronomical and meteorological observations, Morse's geography of the United States, &c.[30]

Many valuable works have been published in Europe, which have a reference to our country, and are grounded on observations made in it; they belong, therefore, (in part at least,) to our scientific attainments. Among these the following deserve our notice: Volney's view of the climate and soil of the United States, Michaux's Flora Boreali-Americana, Pursh's Flora of North-America, Vieillot's Histoire naturelle des Oiseaux de l'Amerique Septentrionale, (a work anterior to Wilson's, but unknown to him,) Robin's voyages à la Louisiane, Michaux Junior's Trees of North-America, &c.[31]

A small proportion only of the valuable works published in Europe on the natural sciences are reprinted, or even imported into the United States; but some popular works are occasionally or periodically republished, which help us materially to improve ourselves; among which we deem the following worthy of notice; Davy's philosophy of chemistry and agricultural chemistry, Rees's Encyclopedia, the Edinburgh Encyclopedia, (which contains the latest improvements in natural sciences,) Chaptal's chemistry, St. Pierre's Studies of Nature, Volney's view of the United States, Aikin's Mineralogy, some scientific journals, the travels of Salt, Barrows [*sic*], Humboldt, &c. A few original translations have even been undertaken, such as Molina's history of Chili, Richard's Botanical Dictionary, &c.[32]

The printing of useful works has generally increased in this century, not however in proportion with the publication of books of a less permanent value. It is much to be wished that our publishers, without divesting themselves totally of their private views and purposes, would encourage and patronize works of original nature, or an useful tendency, in preference to those of a lighter cast.

The States which have conspicuously distinguished themselves in the cultivation and promotion of sciences, or the number of eminent characters they have produced, are New-York, Pennsylvania, Massachusetts, and South Carolina. Those which appear to have been the least conspicuous, are New-Jersey and Vermont, the small States of Delaware and Rhode-Island, and the new States of Tennessee, Louisiana, Indiana, and Mississippi. Let us hope that a noble emulation may arise between them, destitute of any little jealousy, each striving to excel the other in liberality, energy, and accumulation of knowledge.

We shall proceed to give a survey of the collective labours of each class of society, and each class of scientific writers, wishing thereby to inspire them likewise with a spirit of emulation, that each may endeavour to do most, and become preeminent.

The class of Physicians has stood till now, foremost in point of numbers and qualifications; their liberal education and extensive instruction fits them for study and scientific pursuits. They fill generally the chairs in the universities; and some of our most eminent writers belong to this class; such as Dr. [Samuel Latham] Mitchill, Dr. [Benjamin Smith] Barton, Dr. [Benjamin] Rush, Dr. [David] Hosack, Dr. [James] Mease, &c. They are continually increasing in number and respectability; and when it is considered, that nearly eight hundred students of medicine annually attend the lectures in the Universities of Philadelphia, New-York, Boston, Baltimore, &c. and more than one hundred receive annually their degrees, it may easily be conceived that their body is not likely to diminish.

We shall probably be unable to notice, in our account of individual labours, all the members of the medical class who have communicated to the public, through the journals, or otherwise, partial labours, connected with natural sciences, but we avail ourselves of this opportunity to mention the names of some of those, who have added to our stock of knowledge.

Some facts connected with the natural history of man have been partly elucidated by Dr. [John Collins] Warren of Boston; Drs. [Felix] Pascalis, [John W.] Francis, and [David] Hosack, of New-York, Dr. [John] Davidge of Baltimore, Dr. [Caspar] Wistar, and Dr. [Philip Syng] Physick of Philadelphia, Dr. [Benjamin] Rush, Dr. [John] Brickell, Rev. Dr. S[amuel]. S[tanhope]. Smith, &c.

Mineral Springs have been analyzed by Dr. [Valentine] Seaman, Dr. [John Fothergill] Waterhouse, Dr. [William] Meade, Dr. [John] Rouelle, Dr. [Ashel] Green, Dr. [John Honeywood] Steele [i.e. Steel], &c.

Many important parts of materia medica have been illustrated, particularly those derived from the vegetable kingdom, by Dr. [James] Macbride, Dr. [John] Lining, Dr. [Nathaniel] Chapman, Dr. [Samuel] Akerly, Dr. [James] Glen, Dr. [Jacob] Bigelow, &c. and of course Dr. B. Barton and Dr. S. L. Mitchill, who have paid attention to this and every other part of medicine connected with nature.

Among the inaugural dissertations or theses, published annually to obtain the degrees of M.D. several are on the properties of our native plants, which have, thereby, been often thoroughly investigated. A collection of those theses, or rather an epitome of their contents would be a valuable addition to our knowledge of practical botany. It may not be improper to notice a few: on *Phytolac[c]a decandra*, by Dr. [Benjamin] Shultz, on *Fucus edulis*, by Dr. [Augustus Rupert] Griffin [i.e Griffen], on *Rhus glabrum* and *vernix*, by Dr. [Thomas] Horsefield [i.e. Horsfield], on *Arbutus uvaursi*, by Dr. J. J. Mitchill [i.e. John S. Mitchell], on *Magnolia glauca*, on *Eupatorium perfoliatum*, on *Liriodendron tulipiferum* [i.e. *tulipifera*], on *Cornus florida*, on *Pyrola maculata*, on *Asclepias decumbens*, &c. &c.

Fewer parts of animal materia-medica have been investigated; we can only remember the memoirs of Dr. Isaac Chapman, on some new American species of officinal Meloes, and Dr. [T. W.] Waterhouse, Junr. on some new species of American officinal Leeches or Hirudos, &c.

Some other medical gentlemen have paid attention to the economical uses of organized bodies, and their natural history, such as Dr. [Edward] Bancroft on vegetable dyes, Dr. [Adam] Seybert, on fixing the dyes afforded by plants, Dr. [William] Barnwell, and Dr. [John] Fothergill on the power of habit in plants and animals, &c.: but we are yet in want of a general work on our native dyes, and all the economical uses of our native plants.

The names and labours of many other worthy physicians may be seen on perusal of the Medical Repository, and other medical journals, which we unwillingly omit, from a fear of swelling this essay beyond our original intention.

After the medical faculty, the next class is that of our enlightened clergymen: many of whom do not disdain to enlighten the minds as well as the souls of their fellow-citizens. Their attainments and leisure enable them to devote much learning and time to useful pursuits whenever they are so inclined. Their influence is great over the bulk of the nation, and their examples might find many imitators. They generally fill the literary professorships in the colleges, and on them devolves therefore the instruction of youth. Among the worthy members of the clergy who have studied or taught to advantage, the natural sciences, we shall mention the Rev. Dr. Henry Muhlenberg, the Rev. Drs. [Manasseh] Cutler, Collins [i.e. Nicholas Collin], [Timothy] Dwight, S. S. Smith, [Jacob] Vanvleck [i.e. Van Vleck], [Frederick G.] Schoeffer [i.e. Schaeffer], [Daniel] Steinhover [i.e. Steinhauer], [Christian F. H.] Dencke, [Frederick V.] Melsheimer, &c.

Our enterprising merchants have it in their power to do much in favor of science, and to forward or increase our knowledge of foreign countries and productions by employing enlightened supercargoes and captains, or by directing them to bring home useful and rare productions, with which they may chance to meet. Let us consider that we scarcely know yet one third part of the fishes and animals that swim in the sea, that the whole productions of the east and west shores of Africa, the whole west shore of America from Cape Horn to Behring's strait, and nearly the whole of the eastern shore from Cape Florida to the islands of Falkland, besides Australia, Polynesia, Borneo, &c. are nearly unknown; even their plants, reptiles, and shells! What a vast field of inquiry for whoever has the least share of self-pride or good will! Our mercantile gentlemen and travellers may therefore add greatly to our general knowledge, and raise our national character. They have begun to do it, and we could name many merchants and mariners who have added to our collections and museums; and not a few who have increased the stock of our knowledge: but our catalogue would be too incomplete to do justice to this class of our fellow-citizens.

Gentlemen of the navy, and consuls have the same means in their power abroad, and officers of the army on our frontiers. I shall mention with pleasure the names of Com. [Stephen] Decatur, Capt. [David] Porter, &c. of the navy; Mr. [Obadiah] Rich, consul at Alicant[e]; Mr. Warren [i.e. David Bailie Warden], consul at Paris; and Capts. [Meriwether] Lewis and Clarke [i.e. William Clark], and Major [Zebulon] Pike, &c. of the army; as having partly been engaged in enlarging the sphere of our knowledge.

The profession of the law would appear the least likely to afford men of science, yet we feel proud to notice among its members some of our most eminent and useful citizens, such as Chancellor [Robert R.] Livingston, Judges [Richard] Peters, [William] Cooper, [John] Davis, [Augustus Brevoort] Woodward, &c.

Among our wealthy citizens, planters, proprietors, &c. we notice with pleasure the names of Thomas Jefferson, De Witt Clinton, [Captain-]General [David] Humphreys, Col. Geo. Gibbs, Messrs. [William] Dunbar, Z. Collins, R. Haines, [John E.] Leconte [i.e. LeConte], W. Hamilton, [Nicholas] Herbemont, &c. Such gentlemen possessing wealth and leisure, employ both to advantage and honour by patronising those who are deficient in either, or by attending carefully to the pleasing studies which natural objects afford.

Some other classes or professions have also produced men of talents and zeal: we shall notice among the professors, Messrs. [William Dandridge] Peck, [John] Griscom, [Robert] Patterson, [Parker] Cleaveland, &c.; among the engineers and surveyors, Messrs. [Andrew] Ellicot[t], [Robert] Fulton, [Simeon] Dewitt, [Alden] Partridge, [Benjamin] Latrobe, [Loammi] Baldwin, &c.; among the geographers and travellers, Messrs. [John] Mellish [i.e. Melish], [Horatio Gates] Spafford, [Jedidiah] Morse, [William] Darby, [Thaddeus Mason] Harris, [Henry M.] Brackenridge, [Henry Bellenden] Ker, [Thomas] Hutchins, &c.; and among the historians, [Jeremy] Belknap, author of the history of New Hampshire; [James] Sullivan of Maine, [Samuel] Williams of Vermont, [David] Ramsay of South Carolina, [Hugh] Williamson of North Carolina, [William] Smith of New-York, &c.; some of which were published at the close of the last century. They have all added something to the knowledge of our country.

Many ladies begin to show a taste for useful pursuits; they attend botanical and chemical lectures; but none have, as yet, distinguished themselves. The only one that may deserve mention, is Mrs. [Anna Rosina] Gambold, who sent plants to the Rev. Dr. Muhlenberg from the Cherokee country; others are satisfied by feeling a lively interest in the pursuits and success of their relatives.

Writers, or eminent professors, on natural sciences, may be divided into seven classes; 1. astronomers, 2. philosophers, 3. chemists, 4. geognosists,[33] 5. mineralogists, 6, botanists, 7. zoologists. We shall notice under each class a few of those amongst us, whose reputation or real worth entitles them to be known.

Our best astronomers are, or have been, during this century, Dr. [Samuel] Webber, Messrs. [Andrew] Ellicot[t], [Simeon] Dewitt, [Samuel] Williams, [Nathaniel] Bowditch, [Caleb] Gannet, [John] Winthrop, [Stephen Row] Bradley, [William] Lambert, [John] Farrar, &c. &c.

Our philosophers are Messrs. [Thomas] Jefferson, [DeWitt] Clinton, [Benjamin] Vaughan, [William] Bentley, [John] Winthrop, [Robert] Patterson, [Hugh] Williamson, [John] Griscom, [John] Wood, Dupont [i.e. Eleuthère Irénée Du Pont], [William W.] Woodward, Rafinesque, &c. Drs. [Benjamin] Rush, [James] Cutbush, [Samuel Latham] Mitchill, [David] Ramsay, [Joseph] Priestley, &c. &c.

Our chemists, Drs. [William James] Macneven, [Aaron] Dexter, [Benjamin] Silliman, S. L. and T. Mitchill, [John Redman] Coxe, [James] Cutbush, [Adam] Seybert, [Joseph] Priestley, [John] Gorham, &c. Messrs. [Thomas] Cooper, [Robert] Hare, [John] Griscom, &c. &c.

Our geognosists, Drs.[Samuel Latham] Mitchill, [Daniel] Drake, [Parker] Cleaveland, [James] Mease, &c. Messrs. [Constantin François Chasseboeuf, Comte de] Volney, Rafinesque, [James] Cutbush, [George] Wood, [Meriwether] Lewis, [William] Dunbar, [Jeremiah] Day, [William] Maclure, [Horace] Hayden, &c. &c.

Our mineralogists, Drs. S. L. Mitchill, [Parker] Cleaveland, [Archibald] Bruce,

[Adam] Seybert, &c. Messrs. Steinhover [i.e. Daniel Steinhauer], [George] Gibbs, [John G.] Bogert, [Zaccheus] Collins, [Solomon W.] Conrad, &c.

Our botanists, Messrs. Michaux, father [André] and son [François], [Frederick] Pursh, [William Dandridge] Peck, Rafinesque, [Zaccheus] Collins, Leconte [i.e. John Eatton LeConte], [Stephen] Elliot[t], [William] Bartram, [John] Bradbury, [Thomas] Nuttall, [John] Torrey, [Claude C.] Robin, [Joseph] Correa [da Serra], [Obadiah] Rich, &c. Drs. [Henry] Muhlenberg, Benjamin [Smith] Barton, W. P. C. Barton, [Jacob] Bigelow, [Francis] Boot[t], [Caspar Wistar] Eddy, [John] Brickell, Waterhouse, father [Benjamin] and son [John F.], [Manasseh] Cutler, &c. &c.

Our zoologists are, or have been the following, during this century, which we divide into general zoologists, such as, Drs. Benj. [Smith] Barton, S. L. Mitchill, S. Akerly, S[amuel G]. Mott, [John F.] Waterhouse, Jun. &c. Messrs. Bartram, Rafinesque, [Charles Alexandre] Lesueur, [Louis A. G.] Bosc, [John E.] Leconte [i.e. LeConte], [Thomas] Say, [Charles Willson] Peale, [George] Ord, &c.; and partial zoologists, who have only studied one branch of the science, such as ornithology, ichthyology, entomology, conchology, or zootomy,[34] which are Messrs. [Alexander] Wilson, (o) [John] Torrey, (e) [Friedrick V.] Melsheimer, (e) [James] Clemen[t]s, (z) [John G.] Bogert, (c) [D'Jurco] Knevels, (c) Dr. [Caspar] Wistar, (z) &c.

They cannot boast to have made so many discoveries as their fellow-observers in England, France, Germany, &c. particularly in the former branches; but yet they have somewhat increased the general stock of science, and have materially added to the physical and natural knowledge of our country, and North-America in general.

Our astronomers and philosophers have observed, with accuracy, the different celestial phenomena visible in our part of the globe, such as comets, eclipses, solar clouds or spots, &c.; longitudes have been taken or verified, new magnetic properties ascertained, several ingenious theories offered, and scientific principles taught with care.

Our chemists and mineralogists have discovered many substances heretofore not detected in North-America, and even some new substances; they have verified the European discoveries, and in a few instances anticipated them in some measure; mineral waters, metallic substances, and fossil bodies, have been analyzed; some improvements in nomenclature, apparatus and experiments have been introduced, and experimental chemistry has been eagerly taught to all the classes of society.

Our geognosists have been very successful in the study of our atmosphere, waters, and solid earth. Many meteors have been observed and described, such as par[h]elias, aurora-borealis, meteoric stones, unusual lights, shooting stars, globes of fire, &c.; new theories of tides have been proposed, the Atlantic currents have again been examined, the floating islands of ice discovered in their southern course, their influence proved; the theory of our winds completely investigated, and their influence on our climate ascertained; our mountains have been thoroughly explored, their heights measured, and their structure explained; the Missouri has been navigated to its source, five thousand miles from the sea, and many other rivers accurately surveyed; a sort of tide has been detected in our great lakes, and the beds of our ancient lakes have been perceived. The limits of our different soils have been fixed, the ancient state of some districts properly inquired into: many organic remains have been found all over the alluvial and secondary stratas; those of the huge Mastodon or Mammoth were brought to light nearly entire, Elephants, Rhinoceros, Megasaurus, Sharks, (equally bulky,) &c. have been dug from their graves; numberless fossil shells, and polyps, have been met every where; beds of coal are found from the shores of the Atlantic to the foot of the rocky mountains; many mines and native metals have been discovered; and every part of geonomy relating to the United States more or less illustrated. Yet much remains to be done in order to acquire a complete knowledge of our part of the earth, or even to bring that knowledge to a level with the geognosy of Europe.

Our botanists have succeeded in enumerating nearly five thousand species of plants, (one half of which were new,) within our territory; the eastern productions have been thoroughly examined, and probably three-fourths of the species actually existing, within the limits of the Atlantic States, are now described and named; one half of those living in the western States, and one-fourth of those inhabiting our territories and immense western region. Two general Floras have been published. The economical and medical properties of many of our trees, shrubs, and native vegetables, have been likewise attentively investigated; their geography and natural history have been carefully attended to, their physiology and disorders partly inquired into, and some parts of their botanical pathology have been brought into notice. Our trees have nearly all been ascertained, and the greatest proportion of our shrubs: the study of phenogamous plants has been well attended to, and that of our criptogamous plants attempted in many instances. The elements of botany have been taught with success on the Linnean principles, with a few improvements occasionally; but not with all those lately introduced in Europe. Many parts of botany, such as etymology, biography, bibliography, anatomy, and the knowledge of exotic species, have very seldom been attempted. The cultivation of useful and ornamental trees and plants, in fields or gardens, has met with much attention and success.

Those philosophers and naturalists who have taken up the subject of man, and animated beings, have been enabled to add much to our previous, but scanty knowledge of the American aborigines and animals: their exertions have been rewarded by luminous discoveries. Many new nations, and tribes of the American race have been visited; and it has been ascertained that the Malay breed has widely contributed to the population our continent, in addition to the Atlants, Tartars, Samojeds, Scandinavians, Europeans, and Africans: the natural and civil history of those nations has begun to be elucidated: and the physiology and medical history of the human species has been greatly enlarged, and its unity demonstrated. More than eighty new species of quadrupeds have been detected within our possessions; nearly as many new species of birds; about the same number of reptiles; nearly one hundred and sixty new species of fishes from our seas, lakes, and rivers; about five hundred new species of insects; fifty of crustaceous, one hundred and eighty of living or fossil shells and molluscha, besides many new species of worms, polyps, &c.; but some of them have not yet been described, and no general enumeration of our animals has been attempted. The manners and life, faculties and history of many species have been ably illustrated, particularly among the birds, quadrupeds, and fishes. General zoology and zootomy have begun to be taught in the universities; but, with the exception of medicine, the other auxiliary branches of zoology, have not yet attracted our attention; and entomology, polypology, and zoochrony, as well as exotic zoology, have been scarcely noticed, or are much neglected: merely one half of our animals have been described as yet.

Such have been our labours within the short period of seventeen years: from this outline, what has been done may be seen, and how much remains to be done may be conceived.

C. S. R.

Published in the *American Monthly Magazine and Critical Review*, 2 (December 1817), 81–89.

1. Very likely writing from memory and with little or no editorial assistance, Rafinesque misspelled a number of these names. Here the persons are listed in the same order as his, with spellings corrected, dates added, and a brief tag for each to indicate why they were included in Rafinesque's list of contributors to natural science.

John Winthrop (1714–1779), taught natural philosophy at Harvard; observed sunspots and the transits of Mercury and of Venus. Benjamin Franklin (1706–1790), whose experiments led to the concept of electricity as a single "fluid" and positive and negative charges; studied the Gulf Stream. Thomas Jefferson (1743–1826), contributed observations on natural history in his *Notes on the State of Virginia.* David Rittenhouse (1732–1796), designed and constructed scientific instruments, including astronomical ones. John Clayton (1694–1774), collected, identified, and forwarded specimens of plants for the *Flora Virginica* of Gronovius. John Bartram (1699–1777), collected plants to cultivate in the botanical garden he established. Thomas Walter (c. 1740–1789, collected South Carolina plants for John Fraser's *Flora Caroliniana.* Benjamin Smith Barton (1766–1815), University of Pennsylvania professor who published on most areas of natural history. Gotthilf Henry Muhlenberg (1753–1815), many of whose botanical discoveries were published by Willdenow. Joseph Priestley (1733–1804), studied electricity, optics, and discovered oxygen. Samuel Latham Mitchill (1764–1831), published on ichthylogy and geology; established an early scientific journal. Cadwallader Colden (1688–1776), sent botanical observations to Linnaeus, studied optics, and criticized Newton's theory of gravitation. Alexander Garden (1730–1791), sent natural history specimens from South Carolina to Linnaeus. Humphry Marshall (1722–1801), published first book in America on trees and shrubs. Jonathan Carver (1710–1780), whose *Travels* in the Great Lakes region describes both plants and animals. Jeremy Belknap (1744–1798), the third volume of whose *History of Vermont* contains natural history observations. Manasseh Cutler (1742–1823), repeated Franklin's electical experiments and published on New England botany.

Each of the travelers had something to say about North America's flora and fauna and, sometimes, geology, usually in a narrative of his travels. Mark Catesby (c. 1679–1749), *A Natural History of Carolina, Florida, and the Bahama Island*, 3 vols. (1731–1748). John Mitchell (1711–1768), *Map of the British and French Dominions in North America* (1755). Pehr Kalm (1716–1779), *En Resa til Norra Amerika*, 3 vols. (1753–1761). Louis A. G. Bosc (1759–1828), whose observations on American natural history mostly found publication in the works of Lacépède, Latreille, and others. Luigi Castiglioni (1757–1832), *Viaggio negli Stati Uniti dell'America Settentrionale* (1790). Louis Jean Pierre Vieillot (1748–1831), *Histoire naturelle des oiseaux de l'Amrique Septentrionale* (1807).

Ambroise M. F. J. Palisot de Beauvois (1752–1820), *Catalogue raisonné du museum, de Mr. C.W. Peale* (1796) and *Insectes recueillis en Afrique et en Amerique* (1805). Constantin François Chasseboeuf, Comte de Volney (1757–1820), *Tableau du climat et du sol des États-Unis d'Amérique*, 2 vols. (1803). Francis Masson (1741–1805), botanical collector for Sir Joseph Banks; his North American collections (1798–1805) were published by others. Alexander Mackenzie (1764–1820), *Voyages from Montreal ... through the Continent of North America* (1801). Simon Fraser (1776–1862), did not publish though he kept a journal of his exploration in western Canada seeking a river route to the Pacific. Antoine Simon Le Page Du Pratz (1689–1775), *Histoire de la Louisiane*, 3 vols. (1758). Pierre François Xavier de Charlevoix (1682–1761), *Histoire de la Nouvelle France* (1744). André Michaux (1746–1802), *Histoire des chenes de l'Amerique* (1801) and *Flora Boreali-Americana*, 2 vols. (1803). Johann David Schöpf (1752–1800), *Reise durch einige der mittlern und südlichen vereinigten Nordamerikanischen Staaten nach Ost-Florida und den Bahama-Inseln*, 2 vols. (1788).

2. The American Philosophical Society was in fact founded a year earlier, in 1743.
3. The Magellanic Premium dates from 1785, when John Hyacinthe de Magellan of London gave

the American Philosophical Society 200 guineas for a prize to encourage discoveries in navigation and natural philosophy.

4. According to *The Picture of Philadelphia* (1811), pp. 303–305, by James Mease, who was present in Philadelphia at the time as Rafinesque was not, the American Botanical Society, founded in 1806, broadened its mandate to include all of natural history and on March 4, 1807, renamed itself the Philadelphia Linnean Society. Dr. Benjamin Smith Barton was elected president.

5. The correct name is the Linnean Society of New England. It was founded in 1814.

6. Rather, it was incorporated in 1814.

7. Living in New York when he wrote this essay and not yet a member of the Academy of Natural Sciences of Philadelphia, Rafinesque erred in several respects about the ANSP. It had been organized in 1812 by a small group of naturalists meeting at the apothecary of John Speakman, and Gerard Troost was elected president. It was in 1815 that the academy acquired the use of its own rented building; and the academy was chartered by the state in 1816. William Maclure, whose wealth contributed to the rapid growth of the academy and made possible the launching of its Journal in 1817, was elected president that year, following Troost. Maclure was reelected annually thereafter—even when he was living in Mexico and far removed from Philadelphia—until his death in 1840. The Journal, however, soon faltered and could not be issued on a regular basis until 1821.

8. Clymer is an old Philadelphia family, George Clymer (1739–1813) having been a signer of the Declaration of Independence and a long-time trustee of the University of Pennsylvania; from 1805 until his death, he was vice president of the agricultural society, but no other reference to his being president of the Cabinet of Sciences has been found, nor is there any reason to believe he was a physician. If the Cabinet was founded in 1815, as Rafinesque says, perhaps its president was another Clymer. In 1816 Joseph Cloud, Joseph Kersey, Thomas Allibone, and John F. Waterhouse addressed the university trustees on behalf of the Cabinet of Sciences to seek their assistance in establishing the proposed botanical garden. Nothing came of the proposal.

9. Navy surgeon Edward Cutbush, brother of James the president of the Philadelphia chemical society of similar name. The Columbian Institute was chartered by Congress in 1818.

10. As published in the June 1817 number of the *American Monthly Magazine and Critical Review* report on "Transactions of Learned Societies," these lecturers were: Samuel Latham Mitchill (Ichthyology, Plaxology, Apalogy, Geology), C. W. Eddy (Botany), F. C. [i.e. G] Schaeffer (Mineralogy), John LeConte (Mastodology, Erpetology, Glossology), B. P. Kissam (Ornithology), James Clements (Zootomy), P. S. Townsend (Oryctology), John Torrey (Entomology), D'Jurco Knevels (Conchology), and C. S. Rafinesque (Helminthology, Polypology, Atmology, Hydrology, Taxodomy). A member of the committee making these assignments, Rafinesque also had invented names for some of the strange-sounding "specialties," among them Plaxology (Crustaceans), Apalogy (Mollusks), Zootomy (Comparative Anatomy), Atmology (Atmosphere), and Taxodomy (Classification).

11. Actually founded in 1813, by Daniel Drake.

12. The Philadelphia Society for Promoting Agriculture was founded in 1785.

13. Founded in 1804, the New-York Historical Society set up a "cabinet of minerals" in 1816 and began offering lectures in other branches of the natural sciences the following year. After several years of rivalry between its president DeWitt Clinton and the president of the Lyceum of Natural History Samuel Latham Mitchill, the society transferred its scientific collections in 1829 to the Lyceum, which in turn prospered and eventually became the New York Academy of Sciences.

14. Bernard McMahon (d. 1816), Irish immigrant, also conducted in Philadelphia a seed store which was a meeting place for botanists.

15. William Prince (1766–1842) whose Linnaean Botanic Garden and Nurseries on Long Island was a successful commercial venture and a source for the introduction to the United States of many foreign plants, including those of Australia.

16. William Coxe (1762–1831), known as the father of American pomology.

17. John Lowell (1769–1840), a founder of the Massachusetts Agricultural Society; he was a frequent contributor to the *New England Farmer*.

18. James Eastburn & Co. "at the Literary Rooms, Broadway, corner of Pine Street," publishers of Sir Walter Scott's novels, medical texts, and early numbers of Silliman's *American Journal of Science*.

19. José Francisco Correa da Serra (1751–1823), Portuguese ambassador to the United States.

20. Charles Whitlaw (*c.* 1776–1829), a Scotsman who had worked at the Edinburgh Botanic Garden before he toured Canada and the United States to lecture.

21. Robert Hare (1781–1858), noted for his invention of the oxyhydrogen blowpipe that made possible higher temperatures for chemical reactions.

22. Archibald Bruce (1777–1818), at various times a professor of mineralogy and materia medica. His *American Mineralogical Journal*, mentioned later by Rafinesque, was the first American periodical devoted exclusively to a science, but it lasted only through one volume.

23. *Archives of Useful Knowledge*, edited by James Mease in Philadelphia, went through three volumes (1810–1813).

24. *The Emporium of Arts & Sciences*, edited in Philadelphia by John Redman Coxe and Thomas Cooper, went through five volumes (1812–1814).

25. In New York City David Hosack and John W. Francis edited the *American Medical and Philosophical Register; or, Annals of Medicine, Natural History, Agriculture and the Arts* through four volumes (1810–1814).

26. Ten volumes of the *Eclectic Repertory and Analytic Review, Medical and Philosophical* were published in Philadelphia (1810–1820).

27. The *Journal* could not keep to its proposed monthly schedule during the early years, and was suspended entirely for two years, 1819–20. Nevertheless, by the time it was replaced by other Academy publications in 1918 it had published 24 volumes.

28. Rafinesque's own periodical, *The Annals of Nature*, was proposed several times, and finally resulted in a single tiny 16-page pamphlet published in Lexington in 1820.

29. Alexander Wilson, *American Ornithology*, 9 vols. (1808–14); Gotthilf Henry Muhlenberg, *Descriptio uberior graminum et plantarum calamariarum Americae Septentrionalis indigenarum et cicurum* (1817); Stephen Elliott, *A Sketch of the Botany of South-Carolina and Georgia*, 13 fascicles (1816–24) or 2 vols. (1821–24); Samuel Latham Mitchill, "The Fishes of New York, Described and Arranged," *Transactions of the Literary and Philosophical Society of New-York*, 1 (1815), 355–92; Paul Allen (ed.), *History of the Expedition under the Command of Captains Lewis and Clark*, 2 vols. (1814).

30. Parker Cleaveland, *Elementary Treatise on Mineralogy and Geology* (1816); Benjamin Smith Barton, *Elements of Botany* (1803); DeWitt Clinton, "An Introductory Discourse," *Transactions of the Literary and Philosophical Society of New-York*, 1 (1815), 21–184; John Drayton, *A View of South-Carolina, as Respects Her Natural and Civil Concerns* (1802); Daniel Drake, *Natural and Statistical View, or Picture of Cincinnati and the Miami Country* (1815); Hugh Williamson, *Observations on the Climate in Different Parts of America* (1811); James Mease, *Geological Account of the United States* (1807); Andrew Ellicott, sixteen articles on astronomical observations published in the *Transactions of the American Philosophical Society*, 3–6 (1793–1809); Jedidiah Morse, *The American Geography* (1789).

31. André Michaux, *Flora Boreali-Americana*, 2 vols. (1803); Frederick Pursh, *Flora Americae Septentrionalis*, 2 vols. (1814); Louis Jean Pierre Vieillot, *Histoire naturelle des oiseaux de l'Amérique Septentrionale, contenant un grand nombre d'espèces décrites ou figurées pour la première fois* (1807); C. C. Robin, *Voyages dans l'intérieur de la Louisiane, de la Floride occidentale, et dans les isles de la Martinique et de Saint-Domingue, pendant les années 1802, 1803, 1804, 1805 et 1806*, 3 vols. (1807); François André Michaux, *Histoire des arbres forestiers de l'Amerique Septentrionale*, 3 vols. (1810–13).

32. Humphry Davy, *A Discourse, Introductory to a Course of Lectures on Chemistry* (1802) and *Elements of Agricultural Chemistry* (1813); Ephraim Chambers, *Cyclopedia, or an Universal Dictionary of Art and Sciences* (1728), revised and enlarged by Abraham Rees, 5 vols. (1778–88); "A Society of Gentlemen in Scotland, printed in Edinburgh," *The Encyclopedia Britannica*, 3 vols. (1771); Jean Antoine Claude Chaptal, *Elémens de chimie*, 3 vols. (1790); Jacques Henri Bernadin de Saint-Pierre, *Études de la Nature*, 3 vols. (1784); Constantin François Chasseboeuf, Comte de Volney, *Tableau du climat et du sol des États-Unis d'Amérique*, 2 vols. (1803); Arthur Aikin, *A Manual of Mineralogy* (1814); Henry Salt, *A Voyage to Abyssinia and Travels into the Interior of that Country* (1814); John Barrow, *An Account of Travels into the Interior of Southern Africa*, 2 vols. (1801–1804) and *Travels in China* (1804); Alexander von Humboldt and Aimé Bonpland, *Voyage de Humboldt et Bonpland*, 23 vols. (1805–1833); Giovanni Ignazio Molina, *The Geographical, Natural, and Civil History of Chili* (trans. published in Middletown, Conn., 1808); Louis Claude Richard, *A Botanical Dictionary* (trans. by Amos Eaton and published New-Haven, Conn., 1816).

33. In the terminology Rafinesque first developed in *Analyse de la Nature* (1815), Geonomy is to the earth what Astronomy is to the heavens. Geognosy, one of two branches of Geonomy, "concerns the terrestrial globe in general" (p. 14) and contains the sub-branches Atmology (science of the atmosphere), Hydrology (science of waters), and Geology (science of the solid earth).

34. By analogy, Zootomy is to animals what Geonomy is to the earth. Zoochrony, mentioned later as an "auxiliary branch of zoology," is nowhere defined in Rafinesque's writing.

20

Review of Elliott's Botany

Yale graduate Stephen Elliott (1771–1830) was as versatile as Rafinesque himself. Though known today mostly for the book under review here, he served several terms in the South Carolina legislature, where he wrote laws creating the public school system and the state bank. From 1812 until his death he served as president of the bank, an office that caused him to shift his residence to Charleston from his plantation near Beaufort, where he had gathered most of the material for his book on botany. Interested in all branches of natural science, he was also one of the founders in Charleston of the Literary and Philosophical Society of South Carolina as well as the South Carolina Academy of Fine Arts. He contributed numerous articles on various subjects to the *Southern Review*. Posterity has generally agreed with Rafinesque's high opinion of the book reviewed here.—*Editor*

* * *

A Sketch of the Botany of South-Carolina and Georgia. By Stephen Elliott, Esq. &c. &c. Charleston, 1817. 5 Numbers, 8vo, each of 100 pages, with some plates; to be continued.

UNDER the above unassuming title, one of the most learned and elaborate works, ever published in the United States, on Natural Sciences, is making its appearance: being at the same time the first botanical work, written in our country, in which, original, accurate and complete descriptions of our indigenous plants, are given in our vernacular language and on scientific principles. The modesty of its author can only be equalled by his talents; and the multiplicity of his discoveries and researches, by the happy manner in which he conveys to us the knowledge of their results. We have not often the opportunity to witness such a worthy association; and we feel proud in this instance to have it in our power to delineate some of its features. We, therefore, avail ourselves of it at an early period, and before the completion of the work, since the parts before us afford a fair specimen of the whole; and we entertain no doubt that the remainder, which is in forwardness, will appear in a state of improvement rather than otherwise.

We have perceived with pleasure some late attempts to convey the botanical knowledge of our plants, in the English language: in Pursh's Flora of North America, and in the translation of the Flora of Louisiana, although the generic and specific characters are given in Latin, the old classical language of Botany, yet the occasional descriptions and observations are in English; while in Bigelow's Florula of Boston, and in the Manual of the Botany of the Northern States, the whole is in that language; but in this last work, short definitions only are given, and in the former, mere short and often imperfect descriptions.[1] The work before us has not only entire and complete English descriptions, but also generic and specific definitions in both languages: uniting, therefore, the advantages derived from both modes. Local Floras may always be written, with great propriety, in the vernacular language of the country for which they are intended; while general Floras, if written in such languages, ought to have the characters of at least all new genera and species, in both languages, Latin and vernacular, as Mr. Elliott has given them; or have a separate Latin synopsis, after the manner of that for Decandolle's valuable French Flora;[2] although the French language is, next to Latin, a classical one in Europe. These additions are required in order that the works may be read by all the botanists and men of science, of different nations, spreading thereby with rapidity individual discoveries. But if the Latin language may be dispensed with in many instances, it is not so with Latin binarian names, which are the real botanical names, common to all nations of European origin: every work neglecting them must be deemed unclassical and unworthy of notice.

Figure 15. Stephen Elliott, from *Garden and Forest* (1894).

The southern states are richer in vegetable productions than the northern, since they approach nearer to the tropical climates, where are the seats of luxuriant vegetation, and they enjoy a lengthened period of warm temperature, fit for the support of vegetable life. We find, accordingly, that they afford a numberless variety of brilliant flowers and conspicuous plants, which have attracted, at all periods, the notice of botanists and gardeners, most of which are peculiar to their climate, and unknown to the northern states, disappearing gradually as they advance toward the pole. There are two principle nucleus [*sic*] in the botany of the Atlantic states, one exists in the chain of Alleghany mountains, from which the plants springing therefrom, extend on each side to the northward, while many are confined to the mountains towards the south: the second is to be traced on the Atlantic shore, and possesses features of the most peculiar character. Its range, wider in the south, becomes narrow towards the north, and in the New England states, it is confined to the margin of the sea-shore. An investigation of this subject would perhaps be interesting, but might lead us into remote discussions. It may, however, be safely inferred, that out of 3000 species, growing in Carolina and Georgia, only 1000 are also found north of Maryland, while the remainder are peculiar to those states, except a very few common to Virginia and Maryland. Many genera are peculiar to the southern region, and unknown north of the Potomac; such as *Zamia, Cham[a]erops, Dionea, Brunnichia, Eriogonum, Boerhavia, Pislia* [i.e. *Pistia*], *Epidendrum, Tillandsia, Thalia, Elytraria, Callicarpa, Stillingia, Bejaria, Gordonia,* &c. and many more.

Notwithstanding the exuberant luxuriancy of vegetation, in Carolina, which appeared to invite the attention of European travellers and settlers at an early period, we find that its vegetable treasures have not begun to be collected and investigated, until long after those of the more northern states; which may partly be accounted for, by the later settlement of the country, and the unhealthy state of the climate. Catesby appears to be the first who, nearly a century ago, began to explore that state for natural productions, and he has figured many trees and shrubs, together with some plants, in his great work[3] on the birds and animals of Carolina, &c.; but the imperfect state of natural sciences in his time, render his unmeaning descriptions, obsolete names, and inaccurate figures, of little use at present, except as historical references, [Alexander] Garden and [William] Bartram visited that country after him; but few of their discoveries were published, and a long period elapsed before Walter, who had resided a long time in Carolina, published, in London, his Flora of that state.[4] His work was in Latin, and in the Linnaean style, containing a vast number of new plants, most of which, were however, so concisely characterised, that they could hardly be distinguished from their congenera; the existence of many was even doubted; but Mr. Elliott has since had the honour to confirm nearly all of Walter's discoveries. Walter had also many new genera which were fully characterised; but for which he had not the ability to frame names! ushering them under the term anonyma. The consequence has been that they have been named by other botanists, who have reaped all the honour, since the name of the author of a new genus, is only affixed to it, when it is introduced into the nomenclature by receiving a botanical name, and a good one. Michaux resided likewise, at different times, in Carolina, and has published his discoveries in his General Flora of the United States.[5] Many other travellers, such as, [John] Fraser, [John] Lyon, [Aloysius] Enslen, [Matthias] Kin, [Thomas] Nuttall, &c. have visited South-Carolina and Georgia, and their discoveries have been partly published by Lamar[c]k, Sims, and Pursh.[6] This last author having never visited those states, is very deficient and inaccurate in the enumeration of southern plants, included in his Flora of North America, which renders still more valuable the additions which Mr. Elliott has been able to make to our knowledge of southern botany. These additions, exclusive of the many restored plants of Walter, amount to more than we could have anticipated, and will certainly claim the best attention of all botanists, not only at home, but in Europe likewise.

Mr. Elliott appears to have received considerable aid from many gentlemen residing in South-Carolina and Georgia: we were not aware that there existed so many zealous botanists and amateurs in those states; we had the intelligence with high gratification; and feel a pleasure in the expectation, that this work is likely to extend the taste for the blooming objects of botanical science; a science which is continually unfolding the secret stores of divine wisdom; which nurses the best sentiments of the heart, and is constantly supplying means to increase our comforts and relieve our wants.

Among these generous contributors, we ought to notice particularly Mr. Laconte,[7] one of our ablest botanists, who has visited all the Atlantic states, and whose labours and discoveries will soon be published in a Botanical Synopsis, upon the construction of which he has been engaged for many years: Dr. Baldwin, who has studied with attention the plants of Georgia: the late Drs. Brickell and Macbride, whose extensive acquirements have thrown much light on many natural subjects: (this latter gentleman particularly, has communicated many valuable notices on the medical properties of some plants:) Lewis de Schweinitz of North-Carolina, and many other gentlemen of South-Carolina and Georgia, such as Messrs. Herbemont, Jackson, Oemler, Pin[c]kney, Moulins, Bennet, Green, Habersham, &c.[8] Mr. Elliott had also kept up a regular correspondence with the late R[ev]. D[r]. Henry Muhlenberg of Lancaster, and has acquired, by a communication of specimens with him, a perfect knowledge of the results of his unpublished labours, many of which appear now, for the first time, in this work, although they have been enumerated in Muhlenberg's Catalogue,[9] but not described.

We have the first five numbers of this work[10] before us, which include, from the class Monandria to the class Decandria, or about one third part of the whole labour, and contain nearly 1000 species, unnoticed by Pursh, and described for the first time in this work. Several new genera are also introduced here for the first time, at which rate the whole work will add about 25 new genera and nearly 400 new species, to the actual knowledge of American botany, rather more than were added by the Flora of Pursh, to which this work is superior in almost every point of view. Among the new species described in these five numbers, 14 had been already named by Muhlenberg in his Catalogue; 8 have been discovered by Dr. Baldwin; 4 by Mr. Laconte [*sic*]; some by Dr. Macbride and Mr. Lyon; while nearly 100 have been discovered, determined, described and named by Mr. Elliott himself. These new species belong to the following genera: Gratiola 3 N. Sp., Lindernia 1, Micranthemum 1, Utricularia 4, Lycopus 2, Salvia 2, Collinsonia 2, Erianthus 2, Xyris 2, Rhynchospora 4, Cyperus 4, Mariscus 1, Scirpus 9, Dichromena 1, Paspalum 3, Panicum 20, Agrostis 3, Poa 6, Aristida 3, Andropogon 5, Aira 2, Uniola 1, Eleusine 1, Houstonia 1, Ludwigia 4, Villarsia 1, Hottonia 1, Phlox 1, Lysimachia 1, Ophiorhiza 1, Sabbattia 2, Viola 1, Asclepias 3, Hydrolea 1, Eryngium 2, Hydrocotyle 2, Ammi 1, Sium 2, Drosera 1, Tillandsia 1, Pontederia 1, Allium 1, Juneus [i.e. Juncus] 3, Rumex 1, Tofidda [i.e. Tofielda] 1, Trillium 2, Rhexia 1, Polygonum 1, Baptisia 1, Cassia 1, Andromeda 1.

Besides the above material addition of new species, we find that many genera contain the descriptions of a great number of species, becoming almost complete monographies of said genera; among those we shall mention the following genera: Panicum, which contains 45 species![,] Gratiola 8, Utricularia 9, Collinsonia 7,

Cyperus 24, Scirpus 31, Paspalum 11, Andropogon 12, Poa 19, Ludwigia 15, Phlox 17, Asclepias 18, Trillium 9, Andromeda 16, &c.

The new genera will deserve our particular attention, since they become the types of the most important collective aggregate of individuals, which derive their name and characteristic features from them. They are scattered in the following order.

Lachnanthes. Mr. Elliott gives this new name to the *Heritiera* of Gmelin and Michaux, or *Dilatris* of Persoon and Pursh, which he proves to be distinct from the last genus, while the former denomination has now changed its object: the *Convitylis* [i.e. Conostylis] of Pursh, or rather *Lophiola* of Bot. Mag. is quite different from it, by the double number of stamina.

Aulaxanthus. Triandria digynia. Flowers in panicles. Calyx 2 valved, 1 flowered; valves equal furrowed. Corolla bivalve, valves nearly equal. A N. G. differing from *Panicum* by the furrowed calyx and absence of an accessory valve. The type of it is the *Phalaris villosa* of Michaux, which Elliot[t] calls *A. ciliatus*, and to which he adds a second species[,] *A. rufus*.

Monocera. Triandria digynia. Flowers lateral. Calyx 3 valved multiflore, valves awned below the summit. Herm. fl. Corolla 2 valved, unequal; the exterior valve awned below the summit. Neut. fl. Corolla 2 valved unawned. This N. G. is intermediate between *Eleusine* and *Chloris*: it is formed upon the *Chloris monostachya* of Lin. but the name is erroneous, there being already a genus of univalve shells called by a similar name by Lamar[c]k, &c. It must, therefore, be changed into *Triatherus*, meaning three bristles, since the calyx or glume has so many: the specific name will be *T. aromaticus*.

Lyonia. Pentandria digynia. Pollen masses 10 smooth pendulous. Stamineal crown 5 leaved, the leaves flat erect. Stigma conical 2 cleft. Corolla 1 petal, campanulate. Follicles smooth. This N. G. is formed upon the *Ceropegia palustris* of Pursh, or *Cynanchum angustifolium* of Muhlenberg. The name happens to be erroneous as the above, upon two evident principles: 1. because it is almost identical in sound with the genus *Allionia*; 2. because a genus was already dedicated to Mr. Lyon, in 1803, by Rafinesque,[11] in the Medical Repository, (also erroneous in name) which he has since rendered exact by calling it *Lyonella* [*nom. nov.*]. This genus might therefore be dedicated to the late worthy Dr. Macbride, and called *Macbridea*: specific name *M. maratima*.

Acerates. Differing from *Asclepias* by having no appendage in the auricles or crown. A similar name has been given previously by Persoon to a different Genus: this, therefore, which ought perhaps to be a mere subgenus or *Asclepias*, must receive the name *Acerotis* [*nom. nov.*], meaning auricles without horns: the *A. viridiflora* of Rafinesque and Pursh may be united to it.

Podostigma. Corpuscle on a pedicel, pollen masses 10, &c. smooth, pendulous. Stamineal crown 5 leaved, leaves compressed. Corolla campanulate, follicles smooth. Formed with the *Asclepias viridis* and *A. pedicellata* of Walter. A good name.

Lepuropetalon. Pentandria trigynia. Calyx 5 parted. Petals 5, resembling scales inserted on the calyx. Capsul free near the summit, 1 celled 3 valved. Next to *Turnera* and scarcely distinct from it, the ovary is probably free altogether and covered by the base of the calyx at its base. Muhlenberg had united this genus with *Pyxidanthera*, which was wrong, since it has scarcely any affinity with it. The name *Lepuropetalon* is rather too long, being in the same predicament with *Symphoricarpos*,

Anapodophyllum, which have been shortened. This might, therefore, be shortened into *Petalepis* [*nom. nov.*], which has the same meaning.

Monotropsis. Schweinitz. Decandria monogynia. Calyx 5 leaved, leaves upright hooded, base unguiculate-gibbose. Corolla monopetal campanulated fleshy quinquefid. Nectary quinquefid. Stamina 10, a pair between each angle of the nectary. Ovary 5 gone, 1 style, stigma 5 valved.—This new genus, which has been discovered in North-Carolina, by Mr. Schweinitz, belongs to the same natural family than [*sic*] the genera *Monotropa* and *Hypopythis* [Rafin.], notwithstanding the monopetalous corolla, since the stamina are not inserted thereon. The name given by the discoverer being objectionable, Mr. Elliott proposes to substitute therefore the name *Schweinitzea* [i.e. -zia], which we trust, will be acceded to. It contains only one species, *S. odorata*, which has the smell of the violet, the habit of *Monotropa*, aggregate flowers of a whitish colour, &c.

Mr. Elliott might have established several other new genera, and he has, in some instances, intimated the propriety of it; but a timidity, too general among the botanists of the strict Linnaean school, has prevented him from executing what he considered advisable. The following axiom ought to become a botanical rule: *All the species differing generically from their supposed congenera, must form separate genera*, since it flows from the evident botanical laws, that, *a genus is a collection of consimilar species*, and that *consimilar objects are to be united, while dissimilar objects are to be divided.* The multiplicity of genera, far from being contrary to the correct principles of the science, as some botanists have wrongly conceived, is conducive to the gradual improvement of it, since it takes place only when new observations of characters prove the necessity of such an increase.

The shape and style of the whole work is strictly Linnaean; but in the synoptical view of genera belonging to each class, they are deprived of their definitions, which is, perhaps, an oversight, but an objectionable one. The characters of the genera are only synoptical, they are given in Latin and English, as well as those of the species: a selection of synonymes follows them, next a complete English description is given of all the species which the author has seen, and they are by far the greatest number. Many valuable observations are added, including their native situations and soils; times of blossoming; vulgar names; medical and economical properties, &c. Among these properties several are entitled to notice, some are new, and many have been communicated by Dr. Macbride, &c. We deem worthy of attention those belonging to the following species,

Salvia lyrata,
Tris [i.e. Iris] versicolor,
Spigelia marilandica,
Convolvulus macrorhizon [i.e. -hizus],
Lobelia inflata,
Gonolobus macrophyllus,
Chenopodium anthelminthicum [i.e. -ticum],
Acorus calamus, &c.

The classification of this work is also Linnaean, without scarcely any variation. We regret exceedingly this general infatuation for the absurd sexual system, which is as yet prevalent in our country; however, it may be considered as an imperfect alphabet, competent for those who are acquainted with its principles and anomalies. No reference to natural affinities is made in this work; but as it is rather a species than a genera plantarum, the deficiency is less remarkable in this instance.

While a servile adherence is shown to the erroneous Linnaean systematical classification, notwithstanding its defects were well known to its author, and probably to Mr. Elliott himself, and ought to claim the serious consideration of all botanical writers, many of whom have been led thereby to reject it altogether, and supersede it by the real natural principles of classification and botanical affinities;—while we must blame such a blind compliance with errors long ago detected, our astonishment increases when we observe, that a deviation from the wise and correct Linnaean rules of nomenclature, is in some instances adopted. Certainly, if our writers will follow the steps of Linnaeus, whether right or wrong, as some philosophers of yore used to follow the principles of Aristotle or Zeno, to the exclusion of any other, and sometimes even against the dictates of common sense, let them at least be consistent in their principles, and tread steadily in the footsteps of their adopted school. But to deviate from its correct principles, while they adhere to those that are evidently erroneous, is certainly absurd. They do not consider that those errors in nomenclature, are generally adopted upon the authority of some eminent botanists, who, convinced of the blunders of the Linnaean sexual system, were often led thereby, and somewhat hastily, to condemn even his admirable principles of nomenclature. We hope, that, in future, our botanists will attend to this dilemma, and for the sake of consistency at least, will either adopt or reject altogether the Linnaean principles; although we advise them by all means, if they would improve the science, to adopt more correct principles, and exercising a careful discrimination, endeavour to reject errors and adopt truths, whether they originate with Linnaeus or any body else.

We notice the following deviations from the Linnaean rules, in the numbers before us

The generic name of *Arundinaria* Michaux, formed from the previous genus *Arundo*, is adopted instead of *Miegia*, which was properly substituted by Persoon.

Spartina, which is derived from *Spartium*, is adopted instead of *Limnetis* or *Trachynotia*; this last appears the best.

Centaurella, derived from *Centaurea*, is adopted instead of *Bartonia*, a former and better name.

Polygonatum, derived from *Polygonum*, instead of *Axillaria* [Rafin.].

Smilacina, derived from *Smilax*, instead of *Sigillaria* [Rafin.] or *Majanthus* [*nom. nov.*].

Onosmodium, derived from *Onosma*, is adopted instead of *Osmodium*.

Catalpa, including *Talpa*,[12] instead of *Catalpium* [Rafin.], &c. besides *Monotropsis*, which, however is proposed to be superseded by the name of *Schweinitzea*.

The absurd name of *Ammyrsine* Pursh, is however rejected for the previous and better name of *Leiophyllum* Persoon: while the posterior name of *Syena* Schreber, is adopted instead of the first name *Mayaca* Aublet; both being equally good, it would appear that the first ought to have claimed the preference.

A variety of specific observations and important synonymes are scattered through the whole; some changes in the nomenclature of species, appear to have been requisite, which are often proper; yet objections might be made to some: we shall notice here a few instances, and add some miscellaneous remarks.

The *Statice caroliniana* of Walter is quite a peculiar species, which we have seen growing as far north as Long-Island, it is here blended with the *St. limonium* of Europe, which is totally different.

The *Salvia verbenaca* of Muhlenberg, &c. is properly introduced as a new species, under the name of *S. claytoni.*

Houstonia cerulea var. minor. is made a N. Sp. *H. patens.*

The genus *Pyxidanthera* is united with *Diapensia* in imitation of Pursh, &c. but it appears to differ essentially by the insertion of the stamina in the sinus of the corolla, &c.

Hottonia palustris of Pursh, is properly made a N. Sp. under the name of *H. inflata.*

The *Convolvulus tenellus* of Lin. and Elliott, is evidently a peculiar genus, having a 4 celled capsul, 2 cleft style, 2 globose stigmas, and a 10 toothed corolla: we propose to call it *Stylisma* [*nom. nov.*], meaning cleft style. The essential distinction between the genera *Ipom[o]ea* and *Convolvulus*, far from residing in the shape of the stigma, which affords quite a secondary character, does consist in the capsul, the *Ipom[o]ea* having a three celled one, and the *Convolvulus*, a two celled one.

Atropa physaloides does not belong to that genus, but to the genus *Nicandra.*

Rhamnus minutiflorus [i.e. -flora] of Mich. Pursh and Elliott, belongs probably to the genus *Cassine.*

Ceanothus perennis of Pursh, adopted by Elliott, is the *C. herbaceous* [*nom. sphalm.* = herbaceus] of Rafinesque, a previous name.

Viola clandestina of Pursh, is totally different from the *V. rotundifolia* of Michaux: we have seen both.

Collinsonia anisata belongs to the genus *Hypogon* of the Florula ludoviciana, having 4 fertile stamina.

Gratiola acuminata Walt. and Ell. must form a new genus, intermediate between *Gratiola* and *Herpestis*, having the corolla of the former and the stamina of the latter: it may be called *Endopogon* [*nom. nov.*], meaning bearded within.

The author of *Asclepias quadrifolia* is Jacquin, unnoticed by Elliott.

The American *Hydrocotyle vulgaris* of Mich. and Pursh, which we had long ago observed to be different from the European plant bearing the same name, is here named *H. interrupta* with Muhlenberg.

The genus *Sarothra* is correctly introduced again; but all the species of the G. *Hypericum*, with a monolocular capsul must be united to it; the character of the genus laying in the capsul, not in the stamina.

The genus *Baptisia* of Ventenat is adopted for all the North American species of the genus *Podalyria.*

The genus *Elliotia* [i.e. Elliottia] of Muhlenberg, is adopted and described, being next to Clethra, &c. &c.

We have gone through this work with the utmost gratification. We feel proud that our country may now boast of such an enlightened and accurate botanist as Mr. Elliott. His labours entitle him to be ranked with some of the best European writers, and having been directed towards one of the least explored quarters of the U. S. they have greatly benefitted the science which he cultivates. This we venture to assert notwithstanding the systematical school which he follows, and the occasional errors and oversights in which he may be detected, but which are scarcely separable from extensive labours. We shall be happy to see the conclusion of this valuable and classical work, which certainly deserves better the name of Flora, than Walter's It shall be our duty to notice further discoveries which it may convey; and we feel inclined to believe that the remainder will improve as he proceeds, and the corrections of errors and omissions may probably be thrown together in the shape of a supplement.

C. S. R.

Published in *The American Monthly Magazine and Critical Review*, 3 (June 1818), 96–101.

1. Frederick Pursh, *Flora Americae Septentrionalis* (2 vols., London, 1814); C. S. Rafinesque, *Florula Ludoviciana* (New York, 1817); Jacob Bigelow, *Florula Bostoniensis* (Boston, 1814); [Amos Eaton], *Manual of Botany for the Northern States* (Albany, 1817). Rafinesque's two-part review of Pursh's book had been published in the preceding January and February issues of the same magazine, and his review of Bigelow's in the March number; his review of Eaton's anonymous book had appeared there in October, 1817. He roundly condemned all three books.

2. Augustin Pyramus de Candolle (1778–1841), Swiss botanist. It is not clear what Rafinesque means by "his valuable French Flora." Candolle's multi-volume *Prodromus* had not yet begun when Rafinesque wrote and, dealing only with succulent plants, his *Plantarum historia succulentum* (Paris, 1798–1837) could hardly be considered a *flora*. By this time, however, Candolle had published a number of monographs on various plant families which had the bilingual characteristics the reviewer is recommending. He may have had in mind Candolle's *Icones plantarum Gallie rariorum* (Paris, 1808), even though it had only 16 pages of text to accompany its 50 plates depicting rare French plants.

3. Mark Catesby, *Natural History of Carolina, Florida, and the Bahama Islands* (2 vols., London, 1731, 1743).

4. Thomas Walter, *Flora Caroliniana* (London, 1788).

5. André Michaux, *Flora Boreali-Americana* (Paris, 1803).

6. The travelers alluded to are John Fraser (1750–1811), English plant collector (he also published Walter's *Flora Caroliniana*); John Lyon (1765–1814), English gardener and plant collector who worked for a time in Philadelphia; Aloysius Enslen (d. *c.* 1810), plant collector for Prince von Liechtenstein; Matthias Kin (d. 1825), German plant collector resident in Philadelphia; Thomas Nuttall (1786–1859), English naturalist. As regards their scattered publications, Rafinesque is referring to J. B. A. P. M. de Lamarck's contributions to the *Botanique* sections of the *Encyclopédie méthodique* (13 vols., Paris, 1783–1817) and to its *Tableau encyclopédique et méthodique des trois règnes de la Nature* (6 vols., Paris, 1791–1823). John Sims (1749–1831) edited Curtis's *Botanical Magazine* in London, and Pursh included many plants in his *Flora Americae Septentrionalis* that had been discovered by others.

7. John Eatton LeConte (1784–1860), with Rafinesque a founder of the New York Lyceum of Natural History. He published twelve papers on botany including one on the plants of New York City, but he never published the "Botanical Synopsis" promised here.

8. John Brickell (1749–1809), a physician resident in Savannah; James MacBride (1784–1817), also a physician, resident in Charleston; Lewis David von Schweinitz (1780–1834), Moravian official, at this time resident in Salem, North Carolina, later in Bethlehem, Pennsylvania. The "other gentlemen of South-Carolina and Georgia" mentioned here have left little record in the annals of natural science. Nicholas Herbemont—according to Rafinesque elsewhere, a botanist who "never published any thing"—lived in Columbia and taught French at South Carolina College; James Jackson was a professor at the University of Georgia; Augustus Gottlieb Oemler was a Savannah pharmacist; Charles Cotesworth Pinckney (1746–1825) made use on his plantation near Charleston of the botany he had studied in France; Moulins, Bennett, and Green have not been identified; probably the last named by Rafinesque was Joseph Habersham (1751–1815), planter, banker, and politician of Savannah.

9. Gotthilf Henry Muhlenberg, *Catalogus plantarum Americae Septentrionalis* (Lancaster, Pa., 1813).

10. When Elliott's two-volume book was completed in 1824 it consisted of 13 parts. Its plant descriptions in parallel columns of Latin and English occupied 1,300 pages.

11. Either Rafinesque's memory was faulty or the compositor misread his handwriting. The Rafinesque genus dedicated to Lyon was published in the *Medical Repository* in 1808, not 1803. However, it is Nuttall's 1818 *Lyonia* that is recognized today.

12. Though he did not bother here to explain his objection, Rafinesque would reject *Talpa* because it is a genus of European mole.

21

Review of Maclure's Geology

This book review has been little noted, perhaps because it was overlooked in T. J. Fitzpatrick's bibliography and was not brought to light until the 1982 revision of that book. At a time when Rafinesque's personal letters still exhibited considerable awkwardness in his use of the English language, the review is so well written that he must have benefited from editorial assistance. Not only does it give the reader a concise account of this, the most important of Maclure's scientific publications—as a good review ought to do—but the essay also opens with an interesting observation on how sciences have originated in answer to our needs, with a comment on the relationships among them, and with Rafinesque's quite correct judgment that American geology was in its infancy when he wrote.

Born in Scotland, William Maclure (1763–1840), in about two decades as a merchant acquired the fortune that enabled him to devote the rest of his life to natural science and philanthropy. He became a naturalized U.S. citizen in 1796, but spent more years of his life in Europe than in America and died in Mexico. While in the United States he helped to finance both the Academy of Natural Sciences in Philadelphia and Robert Owen's communitarian experiment in New Harmony, Indiana. His personal interest in geology led him to undertake a one-man geological survey of the region east of the Mississippi, which he published as a map in 1809. It was an enlarged and corrected version of this, with accompanying text, that Rafinesque is reviewing here.

Geology was never a central interest for Rafinesque. Nevertheless, he states here a spectacularly wrong but bold assumption he taught in his classes at Transylvania: his claim that coal was formed by "ancient submarine volcanoes." And he also gives here a spectacularly perceptive observation, one at least a decade ahead of his time, when he writes that "we must especially collect and describe all the organic remains of our soil, if we ever want to speculate with the smallest degree of probability, on the formation, respective age, and history of our strata." Maclure's geology was lithological, based on the surface collection of rocks; it remained for Amos Eaton to attempt stratigraphic analysis, while Lardner Vanuxem is said to be the first American geologist to favor fossils as an index to the classification of strata—but that was not until 1829.—*Editor*

* * *

Observations on the Geology of the United States of America; with some Remarks on the Effects produced on the Nature and Fertility of Soils, by the Decomposition of the Different Classes of Rocks, and an Application to the Fertility of every State of the Union, in reference to the accompanying Geological Map. With two Plates. By William Maclure. 8vo. pp. 128. Philadelphia. 1817.

SEVERAL years ago Mr. Maclure communicated to the Philosophical Society of Philadelphia, some observations on the geology of the United States; he has now somewhat enlarged and corrected his former memoir,[1] increasing it at the same time with an attempt to apply geology to agriculture, in which he is highly commendable, as we have no doubt that his endeavours will be found practically useful, even by those who do not entertain any high idea of scientific researches. Every science is connected with the wants of mankind; and many sciences are indebted for their origin to those wants, which increase in proportion to civilization and refinement. Agriculture sprung from the inadequacy of nature's spontaneous supplies of food for a large population, and has but lately become a science; medicine sprung from the natural desire of relieving our pains and lengthening our lives; geometry from the necessity of ascertaining the extent and limits of our fields; geography from the importance of knowing the strength and resources of our own country, and the means and dispositions of our neighbours; astronomy from the exigencies of shepherds and navigators; physics from the need of becoming acquainted with the phenomena which surround us, as well to avail ourselves of their co-operation, as to avert some of the dreadful disasters of which they are sometimes the cause; cosmony[2] from the cravings of nature, which instigate us to learn what animals, plants, or minerals may be made subservient to our use, or afford us food, raiment, weapons, tools, &c.

All the divisions of knowledge to which we have given names of arts or sciences, have, therefore, a common origin—our wants! a common object—our uses! a common view—our improvement! These selfish motives are those which govern the majority of mankind; but philosophy refines and elevates them. This common origin and object of the sciences has often led to the belief of their identity, as if they were all concentrated in a universal science. This hypothesis cannot now have many adherents, since the different scientific pursuits have been so well illustrated and distinguished; yet every one must be aware of the intimate connexion which exists between all the sciences. For instance, botany and geometry, which appear so widely distinct, are yet so far connected that botany must borrow part of its language from geometry, and geometry some of its forms from botany.

In a peculiarly improved stage and extended state of the sciences, the necessity of dividing them into minor sciences or branches begins to be felt, and such a division usually takes place shortly afterwards. It is to such a period that we are indebted for the new science of geology, or the knowledge of the solid part of the earth. This science was for a long time blended with natural history, mineralogy, astronomy, cosmogeny, mythology, history, to which it is more or less connected, without properly belonging to either; but it has in recent days been raised to the dignified station of a separate science, and can already number among its votaries such men as Cuvier, Werner, Hutton, Patrin, Lametherie, &c.[3] in Europe, while in the United States many enlightened men do not disdain to cultivate it for the benefit of the present generation and of posterity.

Among the latter Mr. Maclure stands conspicuous for zeal, assiduity, perspicuity, liberality, utility, and an early attention to this important subject. It is not by the size of his work that we must judge of its value; but by its intrinsic merit. We believe that in the small number of pages of his volume, more essential facts and useful truths are disclosed than in many thick volumes of yore. We shall endeavour to collect such of them as our limits will allow, and such that a tolerable idea of the value of his observations may be formed; and the few imperfections which we may have occasion to notice, will but slightly invalidate its real merit.

We agree altogether with our worthy author, when he states the fallacy of the numberless presumptive theories of the earth, which have so often been set up. While we have scarcely studied one-fourth part of the *surface of the earth*, and while the interior of our globe is totally unknown, all speculative theories must be considered as the *novels of geology* rather than its history. How many of them have even been founded upon a few local facts, which are belied by so many different facts elsewhere! Mr. Maclure mentions that those animals whose bones have been found in northern climates, while they (or their congenerous species) are now found only in tropical climates, might have been migratory, as the wild Buffaloe of America is at this time;—he might have added, that most of them being different from the now living species, were probably (as the mammoth of Siberia was to a certainty) covered with a thick fur suitable to the climates they dwelt in. Yet to account for this simple fact, a supposition has been advanced, that the equator was once where the poles are now, and vice versa! If the mutation of the poles could only be supported by this false reasoning, every supposition of the kind would fall to the ground. Fire and water were, till lately, considered as the only

Figure 16. Physiognotrace of William Maclure, courtesy of Working Men's Institute, New Harmony.

agents acting over the earth,—now galvanism is allowed to have also its share; but electricity, magnetism, light, gases, air, frost, compression, and animal and vegetable agency, &c. have certainly also their share; wherefore every theory founded upon a simple or single agent, becomes an erroneous system.

Our author adopts Werner's classification of rocks; but he is not satisfied with his distinctive names of primitive and secondary; he might have added his transition, which denomination is certainly illusive. The fact is, that there are but four formations of rocks and earths, *all of which*, even granite, are stratified; they are the crystallized, the deposited, the volcanic, and the organic formations; the first originates in crystallizations, the second in depositions, the third in emissions, and the last in organic remains; if a fifth formation was to be added, it ought to be the agglomerated formation. The transition formation belongs to all the formations in various instances, and the alluvial to the deposited formations. All these formations often happen to be blended, which destroys altogether the theories of universal separate formations, since suppositions must yield to facts; and strata vary from the thickness of a sheet of paper to the immense thickness of several thousand feet, so far as they have been penetrated or seen.

The uniformity of the formations in the United States, and the regularity of their dispositions, strike every observer who has witnesses the disparity and irregularity which are exhibited in the formations of Europe. Mr. Maclure traces an able parallel between the two continents, and describes next the outlines and limits of the formations, rocks, mountains and strata of our continent, being the result of nearly thirty different excursions across their nucleus, which runs from northeast to southwest. He describes the whole in general results, disdaining minute investigation of insulated rocks and detached masses: yet if there are some of such, which may throw light upon the approximating formations, why should we neglect them altogether? We shall not follow him through his leading remarks, and his divisions; a single glance at his map will convey a better idea of his principles, the results of which are, that nearly all the New-England states, the northern part of New-York, and a broad stripe as far as Georgia, are primitive; that the alluvial formation extends from Long-Island to Louisiana, from the Atlantic to the granite up the Mississippi as far as the mouth of the Ohio; that the limestone, or secondary formation, extends all over the western states, as far as the lakes, including most of New-York, and that it is divided from the primitive by a transition region. A formation of sandstone exists in the primitive, in New-York, Maryland, Connecticut, &c.

Notwithstanding the able researches of our author, we cannot but regard his results, as well as those of Volney,[4] as mere attempts towards the knowledge which he means to convey; we know of several instances in which the limits assigned to some formations are not altogether correct, nor can they ever be completely known, but after a series of long, minute local observations all over the United States: and even then, how are we to know when those limits are absolute or relative? We would advise observers to notice the angle of inclination of the strata at the place of their disappearance, whence a probable calculation may be made of their further depth and extent. A long period must elapse before we can acquire a complete knowledge of the soil we inhabit; we must sink wells and shafts, dig mines and coal-pits to great depths, ere we can assert which is the predominant formation in the strata we tread upon; but we must especially collect and describe all the organic remains of our soil, if we ever want to speculate with the smallest degree of probability, on the formation, respective age, and history of our strata. Mr. Maclure has altogether omitted these accessories or auxiliaries, which have received, with much propriety, the name of *medals of nature*: he says little or nothing of the numberless animal remains, shells, polyps &c. found all over our deposited and agglomerated soils, or alluvial, limestone, sandstone regions. He omits the alluvial found in Ohio and New-England, &c. The regions north of the lakes are a blank in his map; they are probably of primitive or granitic formation. The present great lakes of North-America, and those which have to a certainty existed elsewhere in ancient times, have had more influence on some parts of the soil than he is aware of. He has not mentioned any volcanic soils and rocks in the United States; yet there are certainly some, which he has classed, with the Wernerian school, among transition and secondary: but the trap[rock], wa[c]ke, coal, and clay formations which are found in many parts, are here, as in Europe, evidently of volcanic, or emitted formation. Volcanoes do not always emit fire and lava, nor heap up mountains and craters; they often vomit water and mud, and, when they are covered by water, their smoke and ashes form, under the water, strata of various substances: such have been the ancient submarine volcanoes of Connecticut, New-York, Pennsylvania, Virginia, Alabama, &c.

The second plate of this work contains five transverse sections of the United States: 1. across lake Champlain and the White Hills; 2. from Plymouth to lake Erie; 3. from Egg-Harbour to Pittsburg[h]; 4. from Cape Henry to Abingdon; 5. from Cape Fear to Warm Springs. They give a tolerably good idea of the succession of formations; but we hope, that by leading each formation to the level of the sea, it was not meant to imply that they really reach it, else we should ask how was it known to be so?

We now proceed to the second part of this work, or the practical part thereof, wherein the author relates, with much propriety in the preface, how various are the practical results to be derived from the study of geology; it is by such a study that we are safely guided in our search for coal, salt, gypsum, limestone, sandstone, millstones, grindstones, whetstones, marble, clay, marl, slate, ores, &c. For instance, those who should search for coal in a primitive region, or under granite, would lose their time and money: those who mistake pyrites and mica for ores, find soon their delusion to their cost. It will teach you to pave turnpikes with quartz, which will wear two years, instead of limestone or any soft stone, which will not last three months. When clay contains too much calcarious matter, it cannot make good bricks, and when limestone contains too much argillaceous matter, it cannot make lime.

The theory of the decomposition of rocks is treated with great ability and perspicuity; it is worth while for every enlightened agriculturalist to become acquainted with it: the results are, that the best soils for agricultural purposes are those proceeding from the decomposition of wa[c]ke, limestone, lava, tuffa, &c.[;] that the worst are those resulting from clay, salt, sand, quartz, &c.[;] that alluvial and transition formations partake of such formations as they have been washed from; that vegetable mould is the common manure of nature, that gypsum is the next, marl and clay, of sand, and vice versa, &c.

In the last chapter Mr. Maclure enters at length into an investigation of the probable effects which the decomposition of rocks may have on the nature and fertility of the soils of the different states of North-America, when such soils are in their pristine state, since, when covered with vegetable and animal manure or mould, their fertility lasts as long as such mould remains. In result it appears that Pennsylvania and New-York possess the greatest quantity of good lands among the Atlantic states, while all the western states enjoy an equal fertility, being all situated in the limestone formation. All the alluvial region fronting the ocean appears to possess a peculiar character, the soil being almost every where light, dry and sandy, or swampy; this soil, when mixed with marl, which is generally found under it, forms a good cultivable ground. It is probable that cotton, the staple produce of this region south of the Chesapeake, will, at a future period, be found suitable to the whole region, and cultivable as far north as Long-Island, and on those Hempstead plains, now thought almost unfit for cultivation, as were formerly thought the pine barrens of South-Carolina.

Mr. Maclure indulges sometimes in digressions in which some happy thoughts are discernible: his great division of the states, into states east and west of the Alleghany, is quite natural, and the probable consequences of their respective features are truly delineated. Happily the Atlantic states are divided also naturally in three districts: New-England states, east of the Hudson and lake Champlain; middle states, whose territories extend west of the mountains or natural limit; and southern states, where slavery prevails; while the western states will soon be divided in three natural districts,—north of the Ohio, south of the Ohio, and west of the Mississippi, whose features and interests will also assume their own peculiarities, the presumable result of which will be a happy balance of indivisible interests.

We wish that a hint of Mr. Maclure's might meet the eyes of some of those who direct among us the education of youth. He insinuates that we may reasonably hope that, ere long, some portion of time will be appropriated, in our colleges and universities, to studies of evident utility, and that the knowledge of substances, their properties and their uses, will be permitted, in some degree, to encroach on the study of mere words, or the smattering of dead languages. His hopes begin to be partly realized, and the utility of the study of our soil, our waters, our minerals, our fossils, our plants, our animals, &c. is becoming daily more evident; let us hope that these studies will soon be taught every where, together, at least, with those of a less permanent and general utility. We shall conclude in the words of this author,—"The earth is every day moulding down into a form more capable of producing and increasing vegetable matter, the food of animals, and consequently progressing towards a state of amelioration and accumulation of those materials, of which the moderate and rational enjoyment constitutes great part of our comfort and happiness. On the surface of such an extensive and perpetual progression, let us hope that mankind will not, nay, cannot remain stationary."

These remarks bear evidence that our worthy author is gifted with a philanthropic and philosophical mind. The style and the details of his work bear the stamp of the same modest, unassuming, and plain philosophy, and give the author a title to the highest reward of a good citizen, the gratitude of his countrymen; and should his labours be rewarded with the praise that greeted his predecessor Volney, we doubt not he will feel his anticipations fully realized.

C. S. R.

Published in the *American Monthly Magazine and Critical Review*, 3 (May 1818), 41–44.

1. William Maclure, "Observations on the Geology of the United States, Explanatory of a Geological Map," *Transactions of the American Philosophical Society*, 6 (1809), 411–428. The book under review also was first read 16 May 1817 before the American Philosophical Society, then published by the author the same year, and finally printed in the Society's *Transactions* the following year.

2. "This name derived from *Cosmos*, a [G]reek term for world (and beautiful) was first used by me in 1815 [*Analyse de la Nature*, p. 9]; it must not be blended with *Cosmogony*, that inquires into the origin of the world, nor with *Cosmography*[,] that describes it like *Geography*." Rafinesque then goes on to say that there are "3 great sciences[:] 1, Physics or Natural Philosophy, teaching the laws, functions and phenomena of bodies—2, Chemistry or Natural Analysis, teaching to decompose and recompose the elements of bodies—3, ... Cosmony ... the most important and primary, and that may almost include the whole." And within this third class he includes three great divisions, each of which has several subdivisions: "Astronomy, science of celestial bodies"; "Geonomy, science of terrestrial bodies"; "Somiology, science of living bodies." Rafinesque, "Classification of the Natural Sciences and Objects," *The Good Book*, 1 (1840), 5–12.

3. Georges Cuvier (1769–1832), French paleontologist, wrote *Discours sur les révolutions du globe* (1812); Abraham Gottlob Werner (1750–1817), German "Neptunist" geologist; James Hutton (1726–1797), Scottish "Vulcanist" geologist; Eugène Louis Melchior Patrin (1742–1815), French geologist who wrote *Histoire naturelle des minéraux ... avec la géologie ou histoire de la terre* (1801); Jean Claude de La Métherie (1743–1817), French geologist who published a five-volume *Théorie de la terre* (1795).

4. Rafinesque had in mind the *Tableau du climat et du sol des États-Unis* (1803) that was based on his tour of the United States by Comte C. F. Volney (1757–1820), not that author's better known essay on the philosophy of history, *Les Ruines, ou méditations sur les révolutions des empires* (1791). Volney's *View of the ... United States* was translated by Charles Brockden Brown and published in Philadelphia in 1804.

22

Discovery of Aerosols

It would be another century before the word *aerosol* was coined, but Rafinesque's familiarity with the phenomenon of "dusty particles ... formed in the great chemical laboratory of our atmosphere" earns him another credit for being ahead of his time. "The most remarkable sets of explanations on the formation and behavior of atmospheric aerosols were provided in several articles by Rafinesque," states a recent specialist on the subject. Another eighty years passed before the experiments of the Scottish meteorologist, John Aitken (1839–1919) confirmed Rafinesque's speculations. See Rudolf B. Husar, "Atmospheric Aerosol Science Before 1900," pp. 25–36, *History of Aerosol Science*, ed. O. Preining and E. J. Davis (Vienna, 2000).—*Editor*

* * *

Thoughts on Atmospheric Dust. By C. S. Rafinesque, Esq.

"WHEN we find the ruins of ancient cities buried under ground; when the plough uncovers the front of palaces and the summit of old temples, we are astonished: but we seldom reflect why they are hidden in the earth. A sort of imperceptible dust falls at all times from the atmosphere, and it has covered them during ages."

These are the words of the worthy and eloquent philosopher [Julien Joseph] VIREY, in his article Nature, Vol. XV, p. 373, of the French Dictionary of Natural History. Even before reading them I had observed the same phenomenon, and I have since studied their effects in various places. I could quote one thousand instances of the extensive and multifarious operations of this meteoric dust: but I mean to give the results merely of those that fall daily under notice, and are yet totally neglected; wishing to draw on them the attention of chemists, philosophers, and geologists.

Whenever the sun shines in a dark room, its beams display a crowd of lucid dusty molecules of various shapes, which were before [as] invisible as the air in which they swim, but did exist nevertheless. These form the atmospheric dust; existing every where in the lower strata of our atmosphere. I have observed it on the top of the highest mountains, on Mount Etna, in Sicily, on the Alps, on the Alleghany and Catskill mountains in America, &c. and on the ocean.

It deserves to be considered under many views: which are its invisibility, its shape and size, its formation and origin, its motion, its deposition and accumulation, its composition, its uses, and its properties.

The size of the particles is very unequal, and their shape dissimilar; the greatest portion are exceedingly small, similar to a whitish or grayish spark, without any determinable or perceptible shape; the larger particles are commonly lamellar or flattened, but with an irregular margin, and the largest appear to be lengthened or filiform; the gray colour prevails. Other shapes are now and then perceptible with the microscope.

Among the properties of atmospheric dust are those of being soft, as light as atmospheric air, of reflecting the rays received directly from the sun, of possessing a kind of peculiar electricity, which gives it a tendency to accumulate on some bodies more readily than on some others, and of forming an earthy sediment, which does not become effervescent with acids.

This dust is either constantly or periodically formed, but chemically in the atmosphere like snow, hail, meteoric stones, honey-dew, earthy rains, &c. by the combination of gaseous and elementary particles dissolved in the air. Its analysis has never been attempted by chemists; but the earthy sediment which is the result of its accumulated deposition, proves that it is a compound of earthy particles in a peculiar state of aggregation, and in which alumine appears to preponderate, rather than calcareous or silicious earths or oxides.

Its motion in calm weather, or in a quiet room, is very slow; the particles appear to float in the air in all directions, some rising, some falling, and many swimming horizontally, or forming a variety of curved lines; what is most singular, is that no two particles appear to have exactly the same direction; yet after awhile the greatest proportion fall down obliquely, somewhat in the same manner as a light snow in a calm day. When a current of air is created naturally or artificially in the open air or in a room, you perceive at once an increased velocity in their motion; they move with rapidity in all directions; but when a strong current or wind prevails, they are carried with it in a stream, preserving however, as yet, their irregular up and down motion.

Its formation is sometimes very rapid, and its accumulation very thick in the lower strata of our atmosphere, but the intensity is variable. Whenever rain or snow falls, this dust is precipitated on the ground by it, whence arises the purity of the air after rain and snow; but a small share is still left, or soon after formed. In common weather it deposits itself on the ground by slow degrees, and the same in closed rooms. It forms then the dust of our floors, the mould of our roofs, and ultimately the surface of our soil, unless driven by winds from one place to another.

I have measured its accumulation in a quiet room, and have found it variable from one-fourth of an inch to one inch in the course of one year; but it was then in a pulverulent fleecy state, and might be reduced by compression to one-third of its height, making the average of yearly deposit about one-sixth of an inch. In the open air this quantity must be still more variable, owing to the quantities carried by the winds and waters to the plains, valleys, rivers, the sea, &c. or accumulated in closed places or against walls, houses, &c. I calculate, however, that upon an average, from six to twelve inches are accumulated over the ground in one hundred years, where it mixes with the soil and organic exuviae, to form the common mould.

The uses of this chronic meteor are many and obvious. It serves to create mould over rocks, to increase their decomposition, to add to our cultivable soil, to amalgamate the alluvial and organic deposits, to fertilize sandy and unfruitful tracts in the course of time, to administer to vegetable life, &c. It does not appear that it has any bad influence on men and animals breathing it along with air, unless it should be accumulated in a very intense degree.

At Segesta, in Sicily, are to be seen the ruins of an ancient temple; the steps, which surround it on all sides below the pillars, are built on a rock, on the top of a hill detached from any other higher ground. Yet now all the steps and the base of the pillars are under the ground, which has accumulated from this dust and the decay of plants (not trees) to which it has afforded food. There are from five to eight feet from the rock to the surface of this new soil, which has chemically combined in a variety of hardness. This soil has arisen there in about 2000 years, notwithstanding the washings of rain. I quote this as a remarkable instance of the increase of soil by aerial deposits, among many which have fallen under my personal examination.

It is commonly believed that the dust of our rooms is produced by the fragments of decomposed vestments, beddings, furnitures, &c.; this cause increases it, and produces a different dust, which mixes with the atmospheric dust; but it is very far from producing it.

The dust of the open air is ascribed to that raised from roads and fields, by the pulverization of their surface; but this secondary and visible dust is only a consequence of the first. From whence could arise the dust observed by the means of the sunbeams in a dark corner, in winter, when the ground is frozen, or when it is wet and muddy, or at sea, or on the top of rocky mountains?

It is therefore a matter of fact, worth taking into consideration by geologists, that the air still deposits a quantity of dust, which must have been much greater in former periods. Just the same as the sea deposits still a quantity of earthy and saline particles dissolved in it, and which were superabundant at the period when the rocky strata were formed on its bottom. Water being more compact, deposits rocks. Air, which is less dense, deposits a pulverulent matter!

Published in the *American Journal of Science*, 1 (1819), 397–400.

* * *

Remarks on Atmospheric Dust, in reply to Mr. Rafinesque. To Professor Silliman.

Sir,

BEING a subscriber to your journal, I observe, (Vol. I. No. IV. p. 397,) an article from the ingenious and learned Mr. *Rafinesque, on Atmospheric Dust.* I confess I can hardly agree with that gentleman in several opinions which he there suggests, and if I am wrong in dissenting from him, science can never suffer from a free and liberal investigation into its principles. Many of the facts stated in the article referred to, are doubtless true, but, as I apprehend, attributed to wrong causes. I am not disposed to question that dusty molecules are visible on the highest mountains, and on the ocean, but I think all the phenomena may be accounted for by supposing that they arise from the roads, fields, woods, and other matter on the surface of the earth, disengaged by various causes. A brisk wind will raise it directly from the earth, and waft it to a great distance, it being so exceedingly subtle and tenurious as that the atmosphere will support it even in a perfect calm. Perhaps in our climate there is not a day in the course of fifty years in which there is not a sufficient breeze sometime in the course of the twenty-four hours, to set in motion what we call atmospheric dust. And occasionally immense quantities are raised. In the stillest times, vegetables and trees are constantly depositing decayed matter, and some part of this, before it reaches the earth, doubtless floats away on the air. But this, says Mr. R. "is only a consequence of the *first*," meaning that dust which it is the property of the air to deposit. Yet surely the clouds of dust, which are every where visible in a windy day, and that which is seen in a room when any extraordinary motion is produced, do not proceed from the atmosphere primarily. The only fact which he mentions, as tending in the least to invalidate the commonly received opinion on this subject, is, that dust is seen at sea. Now, whatever is supposed to be the origin of these molecules, certain it is that they are capable of floating a great while in the air, and of being carried to an immense distance. Is it absurd to suppose that the specific air which we once breathed sitting in our libraries, may now be floating 1500 miles off over the Atlantic? If not, the dust with which it was charged *here*, may still accompany it *there*. Besides, the dust which is visible at sea, is visible only when the ship is nearly or quite becalmed; and may it not then arise in a great measure from the deck of the ship?

Mr. R. "calculates that on an average, from six to twelve inches are accumulated over the ground in one hundred years." Taking his lowest estimate, six inches for one hundred years, the medium thickness of the deposit on the surface of the earth in 1800 years, will not be less than nine feet. But Mr. R. goes farther, and supposes that in former times the deposit must have been much more abundant than at present. So that I apprehend we should do his theory ample justice, by saying that the diameter of the earth is now, from this single cause, twenty-seven feet greater than it was at the birth of our Saviour. But if we examine the surface of the earth, we shall find there has been no such change. How happens it that rocks and stones are every where to he met with? Are they made by a fortuitous concurrence of atoms from aerial deposit? Do the minerals, so various in their primitive substances, in their kinds and composition, which are spread all over the surface of the earth, and which are collected and form the cabinets of the curious, do they owe their origin to atmospheric dust? Has the atmosphere the property of depositing one substance here, and another there, so as to make one tract of country clay, another gravel, and another rocks, and all lying in the same vicinity? But without pursuing the subject farther, I think the ideas already suggested are sufficient to show that Mr. R.'s theory, instead of accounting for any facts, is wholly irreconcileable with what we every where observe with respect to the operations of nature.

I am respectfully, your obedient servant.

X. Y. Z.

P. S.—*Sir*—If you think the foregoing remarks may deserve a place in your instructive Journal, please insert them.

Boston, Oct. 1, 1819.

Observation.—I have not the account at hand, and only advert from memory to the astonishing quantity of extremely fine, indeed *impalpable* dust, found not long since in the castle of Edinburgh, in Scotland, on opening an apartment, and a chest containing the Regalia of the ci-devant kingdom. My impression is, that they had been closed ever since the union, viz. two centuries, and that the dust, in a form light as down, was several inches thick. Whatever theory of atmospheric dust be adopted, this fact is very curious, and well worthy of being more accurately stated and preserved.—[Benjamin Silliman,] *Editor.*

Published in the *American Journal of Science,* 2 (1820), 134–36.

* * *

Letter on Atmospheric Dust, addressed to Governor De Witt Clinton, Albany.

DEAR SIR,

I published[1] in 1809 some ideas upon this subject, in the American Journal of Science; an anonymous reply to my remarks, has since appeared in the same Journal, which is calculated to mislead; and as I have not been able to avail myself of the same vehicle,[2] in order to state more fully and explain the motives of my belief in the atmospheric spontaneous production of a great part of the dusty particles floating in the air, I take the liberty to address you some additional remarks on this subject, which should my conjectures prove correct, will form an important link in the economy of nature.

The anonymous writer contends with the generality of authors, that these dusty particles are altogether lifted by the winds and carried every where. I do not deny that winds raise the terrestrial dust, and often carries it to a distance; this happens whenever the ground is dry and the winds blow; but I assert that it is impossible that this terrestrial dust should be raised above the clouds or when the ground being totally wet or frozen *cannot afford any*. Yet as a dust exists in the atmosphere as far as the clouds *at all times,* I venture to believe with Virey, Patrin, Deluc[3] and other philosophers, that there must be another independent formation of dust in the atmosphere, besides the scanty terrestrial supply wafted by the winds.

To prove this assertion I need merely refer you to the observation of a very common meteoric phenomenon, which has seldom been noticed. Look at the clouds, towards sun rise or sun set principally, when the sun is concealed behind them, and an opening happens to take place, through which the sun may shine obliquely; a pyramidal beam will immediately appear, similar to the luminous and dusty beam appearing in a room into which the sun shines obliquely. This common occurrence has received the vulgar name of *Sun-beams*; but it is evident that it is not a mere beam of light, since it is not so bright nor dazzling as the bright sun rays, nor is it an optical reflection of the enlightened atmosphere, since it is brighter and not azure. It must therefore be a beam of atmospheric dust, and its identity with the beams produced by a hole in a screen or a window in a room is evident. If several openings exist among the clouds, many beams will be seen, and this phenomenon is sometimes visible without openings, when many clouds act as screens.

It remains to prove that this phenomenon happens when there can be no terrestrial dust in the air, else it would be contended that this dust rises (like balloons) to the clouds. Choose for your observation a short time after a long and heavy rain or snow, which must have precipitated all the terrestrial dust to the ground, and you will perceive the same *Sun-beams* under similar circumstances. Whence it must follow that this beam of dust must have preexisted above the sphere of the storm, and fallen since from above the clouds, and as it cannot be admitted with plausibility that any great quantity of terrestrial dust can exist permanently above the clouds, so as to be able to form immediately such immense volumes of dusty beams, or rather to fill *all the space* between the ground and the clouds, I think it rational to presume that this atmospheric dust, is continually formed or evolved in the atmosphere, and falls down after the rain to fill the vacuum.

The insight given us by modern chemistry into the gaseous formations of solid substances, will be amply sufficient to account for this spontaneous formation. We know now that sulphurated arsenic and mercury, sulphur, muriate of ammoniae, &c. can be formed by the sublimation of gases, that smoke, soot, manna, volcanic productions, meteorolites, earths and even stones or metals, &c. may be spontaneously combined by a casual meeting or mixture of gaseous emanations. It is not therefore difficult to conceive how dusty particles may be formed in the great chemical laboratory of our atmosphere.

A singular instance of atmospheric formation, has been recorded in the travels of La Perouse.[4] He saw, in a storm, on the east coast of Tartary, the actual formation of a number of slender threads, similar to spider webbs [*sic*]. The numerous instances lately ascertained of earthy rains, containing many oxides, come still nearer to the point; they only differ from the common dust, by their tenuity, colour, locality and composition. They are local phenomena end productions, while the atmospheric dust, is a permanent and universal phenomenon.

It is absurd to suppose that the atmospheric dust ought to have covered the earth with a coat or stratum 27 feet thick in 1800 years, as the anonymous writer wishes to suppose. Even if the average of dust falling in one century should be ascertained to be six inches, it must be remembered that the greatest proportion is precipitated by rains, diluted, and carried down the streams with the rain water; a small proportion alone is mixed with the soil and *increases its bulk*. It is only in hollows, caves, corners, pits, &c. that it may accumulate to a certain extent, and compression will greatly reduce it.

It is also absurd to ask whether this dust forms all the rocks and soils on the surface of the earth. But it is reasonable to suppose that it contributes to a certain degree to their increase. Our soil is formed by the decomposition of rocks, the accumulation of vegetable and animal decayed substances mixed with this atmospheric dust.

That it may in some instances form or increase substances and stony strata or conglomerations, cannot be denyed, since this effect takes place under our eyes in cisterns, and reservoirs of rain water. The earthy and dusty particles conveyed into them by the water are gradually deposited, forming concretions and stones. This is very evident in the old cisterns of the East which have held rain water during a long period of time.

Every thing therefore seem[s] to indicate that there is an extensive and permanent formation (and fall) of dust in the atmosphere; that it contributes to form our soils, our alluvions and some stones; to fill the fissures and hollows of rocks and lavas, preparing them for vegetation; and that in former times, when many of our substrata were formed, it may have been more abundant, contributing to the formation of some of those strata.

This may appear paradoxical to some persons slightly acquainted with geological and meteorological phenomena; but not unreasonable to those who observe with care. I have ventured to announce in my lectures that another formation must be added to our present geological formations, the atmospheric or meteoric formation, to which must be referred all those singular geological anomalies which puzzle so much the systematic writers, when they find extraneous stones, soils, metals and other substances mixed or superincumbent over late or newer formations. It may perhaps in time be found necessary to ascribe to meteoric formation, those extensive substrata and upper strata of sand and gravel, which cannot properly be deemed alluvial nor volcanic. When our rocks were formed under water by deposition, many of their principles must have originated in the briny ocean; but some may have been derived from the atmosphere.[5]

Believe me, dear sir, respectfully yours, &c.

C. S. Rafinesque.

Transylvania University, 1*st October*, 1820.

Prepared for publication in the *Western Minerva* (1821), but not actually distributed until 1949, when proof sheets of that suppressed magazine were issued in photo-offset.

1. "Thoughts on Atmospheric Dust," *American Journal of Science*, 1 (1819), 397–400. The date "1809" is a misprint.
2. In a footnote to Asa Gray's obituary article on Rafinesque (*American Journal of Science*, 40 [1841], 221–241) Silliman explained that he closed the Journal to Rafinesque's contributions in 1819.
3. Julien Joseph Virey (1776–1847), Eugène Louis Melchior Patrin (1742–1815), and Jean André Deluc (1727–1817).
4. The South Seas explorer Jean François Galaup, Comte de Lapérouse (1741–1788).
5. Compare this final paragraph with the conclusion of Cosmonist VII.

23

Sea Serpents

At first glance, a "Dissertation on ... Sea Serpents" appears to be yet another example of Rafinesque's credulity, that flaw which, in the words of his memorialist S. S. Haldeman, "led him to believe the exaggerated accounts of the vulgar; and to write essays and found 'species,' upon grounds which should be beneath the notice of any naturalist." Perhaps so, but such was the attraction of the frightful monster reported as appearing in Gloucester Bay on August 10, 1817, that the Boston Linnaean Society appointed a committee to investigate the phenomenon. Harvard's professor of natural history, William Dandridge Peck, eventually wrote about the "enormous animal of the serpentine order" in the *Memoirs* of the Academy of Arts and Sciences, and his colleague Jacob Bigelow summed up the collective wisdom about the beast for America's leading professional journal, the *American Journal of Science*, edited at Yale by Benjamin Silliman.

To be sure, Rafinesque, writing from New York City, had not seen the monster himself, but neither had Peck or Bigelow, while the "research" of the Linnaean Society principally consisted in taking affidavits from seamen and others who claimed they *had* seen it. Edward Everett collected clippings about the Gloucester Sea Serpent for a German correspondent from about 300 newspapers before his effort was superceded by the fifty-page pamphlet the Linnaean Society published as its *Report of a Committee of the Linnaean Society of New England Relative to a Large Marine Animal Supposed to be a Serpent* (Boston, 1817).

Rafinesque's "Dissertation" offers a workmanlike summing-up of details about authentic sea snakes known in his time, concluding that the largest of these was only three to ten feet long. His article was sufficiently well thought of that in Britain Alexander Tilloch reprinted an abridged version of it in 1819 in his *Philosophical Magazine*, leaving out much of the summary of previously described sea snakes but including the description of the Gloucester reptile Rafinesque had gathered from "the various and contradictory accounts given of this monster by witnesses," and the name he gave it: *Megophias monstrosus* (the species epithet being slightly modified from Rafinesque's original).

Rafinesque could not have seen the Linnaean Society's pamphlet, because the Massachusetts savants also had named the "nondescript"; they called it *Scoliophis atlanticus*, a name based on their partial dissection of the monster's dead offspring found on the beach. They realized they were naming the Great Serpent by analogy, and they were aware that the anatomy of the baby "approaches nearly to the *Coluber constrictor*." Nevertheless, based on the eyewitness reports duly sworn before the Essex County Justice of the Peace and their partial dissection of the juvenile specimen, Boston's scientists, "considering this serpent nondescript, and as distinct from other genera of serpents in the flexuous structure of its spine, ... deemed it necessary to constitute a new genus."

Rafinesque's prediction proved true; Europeans did scoff at the "crude speculations" of the Americans. In Paris, for instance, the zoologist Henri de Blainville pointed out that there was no necessary connection between the "juvenile" and the "adult," and he was scornful of the notion that so large a creature would come ashore to lay its eggs. Even though the Linnaean Society eventually had to admit that the dissected juvenile was in fact a common blacksnake, sea serpent fever continued to run high in Massachusetts for another two years. By 1819 it became the subject of a burlesque: *The Sea Serpent; or, Gloucester Hoax*, by the Charleston playwright William Crafts. In 1826, when accounts of another American sea serpent reached England, the paleontologist Gideon Mantell dismissed it as just another "hoax—in other words, a Yankee lie."—*Editor*

* * *

Dissertation on Water Snakes, Sea Snakes and Sea Serpents. By C. S. Rafinesque, Esquire.

WHENEVER a singular phenomenon, or an extraordinary natural occurrence, happens to be observed in the U. S.: whether spots in the Sun, huge fossil bones or sea serpents, a crowd of superficial writers hasten to offer us, instead of facts, their own ideas and conjectures on the subject, which prove, sometimes, more or less ingenious; but often wild, incorrect, or ridiculous. They are generally so much taken up by their own fancy, that they forget entirely to consult former writers of eminence on the same subjects, should they even happen to know of their existence; what idea are we to entertain of their attempts to explain those subjects, without availing themselves of the valuable writings of Herschel or La Place, Cuvier or Pinkerton,[1] &c.? in whose works they had been previously and often completely illustrated. Let us listen to a group of children attempting to reason and argue on the rising of the sun, an eclipse of the moon, on the economy of the bees, or on the structure of a whale, without asking any previous questions to their parents, and we shall find a great similarity between their thoughts and those of many of our speculative writers. They often contribute to render contemptible the subject of their inquiries, at least towards the vulgar, while it would otherwise become at all times deeply interesting; and should their crude speculations ever reach Europe, they will certainly afford very unfavourable specimens of our knowledge and attainments in sciences. These reflections have naturally suggested themselves to my mind on the present occasion.

The ancients gave the name of Water-Snakes and Sea-Snakes to many fishes of the Eel tribe, which bear an apparent likeness with land snakes, although they differ materially on examination, by having fins and gills, and neither lungs nor scales.

Many land snakes are in the habit of going into the water, in pursuit of their food or to escape their enemies, and they have often been called Water Snakes when found it that element.

Real Water and Sea Snakes had been noticed at a very early period by navigators, in the Atlantic Ocean, and the Indian Seas; but as they had not been described, eminent naturalists had doubted their existence, believing that eels or similar fishes had been mistaken for snakes.

Russel[l][2] was perhaps the first writer who established their existence beyond a doubt, by describing and figuring many of them in his splendid work on the snakes of the Coast of Coromandel. Schneider[3] established for them his genus *Hydrus*, which wrong name has been with much propriety changed into *Hydrophis*. They have since been described in all the works on Erpetology, by [George] Shaw, [Pierre André] Latreille, [François Marie] Daudin, &c. and those last writers have divided them into four genera, *Enhydris*, *Platurus*, *Pelamis*, and *Hydrophis*; which form a peculiar tribe or natural family in the order of snakes, to which I have given the name of *Platuria* (Platurians, Flat tails of Water Snakes): they are completely distinguished from the land snakes, by having a compressed tail, which serves them as an oar and rudder, enabling them to swim with great swiftness, and from the fishes of the eel tribe, by having neither gills nor fins. They breathe through lungs, at remote periods, whence they generally live near the surface of the water, like the animals of the whale tribe. They prey on fishes and sea animals, and some of them have venomous fangs. Many are known to come on land as [do] turtles, to deposit their eggs.

About fourteen species of Water Snakes have been described by the above authors; ten more are noticed in the travels of Peron[4] to Australia or New-Holland, one of which was ten feet long; and lately several monstrous species have been seen near our shores. Many others appear to have been perceived by former travellers, and very probably a great variety are known to sailors. The knowledge of these animals is merely emerging into notice, and may yet be greatly improved. I shall not pretend to assert that they are as numerous as land snakes, but it is very likely that one hundred species at least of this tribe exist in the waters of the ocean, lakes and rivers. Intelligent travellers, seamen and fishermen, will gradually make us acquainted with them: meantime, I shall endeavour to give a concise account of those we know, which may facilitate their future observations, and I shall arrange my labour in a sinoptical order, concluding by some remarks on the Sea Serpents, which are merely Sea Snakes, of a very large size.

FAMILY PLATURIA.

Water Snakes, with *compressed or depressed tail, and a scaly body*. (No fins and no gills.)

I. Genus. ENHYDRIS Latreille, &c. (*Hydrus* Schneider. *Coluber* Pallas[,] Daudin.) Body with transverse scaly plates underneath, mouth with sharp teeth but no fangs, tail compressed, with two rows of scaly plates underneath, and often one or two nails at the end.

1. Sp. *Enhydris caspia* Latr. Caspian Enhydris. Back cinereous olivaceous, with 4 rows of round black spots, 180 abdominal plates, 70 pair[s] of caudal plates. Found by Pallas in the Caspian Sea, the Wolga [*sic*], &c. 3 feet long.

2. Sp. *Enhydris piscator* Latr. Fishing Enhydris. Yellowish brown, with many small round black spots, in oblique rows and black line, 152 abdominal plates, and 24 pairs of caudal plates. Found by Russel[l] in the swamps of India, 3 feet long.

3. Sp. *Enhydris palustris* Latr. Swamp Enhydris. Yellow brown, with rhomboidal brown spots, edged with black, tail whitish underneath, 140 abdominal plates, 49 pairs of caudal plates. Found by Russel[l] in the swamps of India, 2 or 3 feet long.

4. Sp. *Enhydris cerulea* Latr. Blue Enhydris. Body blue, belly and tail yellow, with a blue line in the middle, 159 abdominal plates, 52 pairs of caudal plates. Found by Russel[l] in the rivers of India, 2 feet long.

5. Sp. *Enhydris rhyncops* Latr. Beaked Enhydris. Head partly black, with a bill shaped snout, body dark gray, throat and belly yellowish, 144 abdominal plates, 59 pairs of caudal plates. Found in the East Indies by Russel[l], length four feet and half, perhaps a peculiar genus.

II. Genus. NATRIX Raf. (*Enhydris* Latr. Daud.) It differs from the foregoing, by having a broad head, (perhaps with fangs) a narrow neck, the abdomen carinated, &c.

1. Sp. *Natrix dorsalis* Raf. (*Enhydris dorsalis* Latr. Daud.) Dorsal Natrix. Dirty white, with a black sinuated dorsal stripe, 43 pairs of caudal plates. A very small species, about 1 foot long.

III. Genus. PLATURUS Latr. Daud. (*Hydrus* Schneider.) Differing from Enhydris, by having fangs, and the tail with two scales at the top.

1. Sp. *Platurus fasciatus* Latr. (*Hydrus colubrinus* Schn.) Zoned Plature. Cinereous above, with broad brown zones, tail acute. Length 2 feet, from South America and the East Indies: many species are probably blended here.

2. Sp. *Platurus taureali* Raf. Tail obtuse.

IV. Genus. HYDROPHIS Latr. Daud. (*Hydrus* Schneider.) Body cylindrical, with equal scales in parallel rows, mouth with fangs, tail compressed, scales as on the body.

1. Sp. *Hydrophis chittul* Latr. Chittul Hydrophis. White, with many zones of a light blue, tail obtuse, 306 scales in each row of the body, 48 in the caudal rows. Found in India by Russel[l], length 3 feet, very poisonous as well as the following; their bite kills in a few minutes.

2. Sp. *Hydrophis cyanura* Raf. (*H. hoglin* Latr.) Hoglin Hydrophis. Blue above, yellow underneath, 308 scales in each row in body; tail entirely blue, with 43 scales in each row. Also found in the East Indies by Russel[l], length two feet and [a] half.

V. Genus. PELAMIS Daud. (*Hydrophis* Latr. *Hydrus* Schneider.) Differing from *Hydrophis*, by having no fangs, and therefore being harmless.

1. Sp. *Pelamis bicolor* Daud. (*Hydrophis platura* Latr.) Bicolor Pelamis. Black above, white underneath, tail rounded at the end. Found by Forster[5] in the Pacific Ocean.

2. Sp. *Pelamis schneideri* Raf. (*Pelamis bicolor* Var. Daud.) Schneiderian Pelamis. From the East Indies.

3. Sp. *Pelamis fasciatus* Daud. (*Hydrophis lancicauda* Latr.) Zoned Pelamis. Sallow, with transverse brown zones, 200 scales in each row of the body; tail, lanceolate acute, with 50 scales in each row. Described by Vosmaer[6] and Russel[l], from the Indian Archipelago, &c.

4. Sp. *Pelamis marginatus* Raf. (*Hydrophis Shootur* Latr.) Shootur Pelamis. Blue, scales slightly edged with yellow, many narrow transverse yellow stripes on the back: very faint posteriorly, 332 scales in the rows of the body; tail lanceolate, with 40 scales in each row. Found by Russel[l] in the swamps of India, perhaps an Hydrophis.

5. Sp. *Pelamis fuscatus* Raf. Brown Pelamis. Entirely of an olivaceous brown, scales very small, tail obtuse. I have observed it in the Mediterranean, near the shores of Sicily, where it is called *Serpedemari* (Sea Snake,) along with many real fishes: length 2 feet.

VI. Genus. OPHINECTES Raf. Differing from *Pelamis* by having a compressed body and a carinated or angular abdomen.—I arrange in this new genus, all the Sea Snakes, mentioned in Peron's Travels; they were all found on the western and southern shores of Australia or New Holland; such as may have fangs ought to belong to the genus *Natrix*, and those with cylindrical bodies to the genus *Pelamis*.

1. Sp. *Ophinectes cinereus*, Raf. Cinereus Ophinectes. Entirely gray or ash colour.
2. Sp. *Ophinectes viridis*, Raf. Green Ophinectes. Entirely green.
3. Sp. *Ophinectes luteus*, Raf. Yellow O. Entirely yellow.
4. Sp. *Ophinectes cerulescens*, Raf. Bluish O. Entirely of a bluish colour.
5. Sp. *Ophinectes versicolor*, Raf. Versicolor O. Varied with many transverse zones, blue, white, red, green, and black. Many species are probably meant here.
6. Sp. *Ophinectes maculatus*, Raf. Spotted O. Covered with many irregular large spots.—Many species.
7. Sp. *Ophinectes punctatus*, Raf. Dotted O. Covered with numberless small dots.—Many species.
8. Sp. *Ophinectes erythrocephalus*, Raf. Red-head O. Head of a beautiful red, body———.
9. Sp. *Ophinectes dorsalis*, Raf. Backed O. Dark green with large spots of yellow and light green on the back.—Length 3 or 4 feet; near Dewitt's land.
10. Sp. *Ophinectes major*, Raf. Large Ophinectes. Green spotted with red and brown.—Length from 3 to 10 feet; also from the shores of Dewitt's land.

This last species appears to be the largest real Sea Snake which has fallen under the personal observation of naturalists as yet. But larger species still have been noticed at different periods. If I had the time and opportunity of perusing all the accounts of travellers and historians, I could probably bring many into notice; but this tedious labour must be postponed, and I must warn those who may be inclined to inquire into the subject, not to be deceived by the imperfect and exaggerated accounts of ancient or unknown writers. Whenever they mention neither the scales nor tail of their Sea Serpents, or when they assert they had no scales, or had gills or fins, you must in all those instances be certain that they are real fishes rather than Serpents. There might, however, be found some Sea Snakes without scales, since there are such land snakes; and there are fishes with scales and yet without fins; but there are no fishes without gills, and no snakes or serpents with gills! in that important character the classical distinction consists.

Nearly all the writers whom I can remember, have been unacquainted with that obvious distinction, and they have, in imitation of the ancient Greek and Roman writers, given the name of Sea Snakes to the large eels or fishes they happened to observe; this I apprehend is the case with Pontop[p]idan, in his Natural History of Norway; with Mongitore, in his remarkable objects of Sicily; with Leguat, in his travels to Rodriguez-Island, &c.[7] Their observations, and the facts they record, are notwithstanding equally valuable, since they relate to monstrous unknown fishes, which seldom fall under the observation of men. The individuals of huge species are not numerous in nature, either on land or in water; and it is probable they often become extinct for want of food or reproduction.

Among the four different animals which have lately been observed by Americans, and named Sea-Serpents, only one (the Massachusetts Serpent) appears to be such: another is evidently a fish, and two are doubtful. I shall offer a few remarks on each.

1. *The Massachusetts Sea Serpent.* From the various and contradictory accounts given of this monster by witnesses, the following description may be collected.— It is about 100 feet long; the body is round and nearly two feet in diameter, of a dark brown, and covered with long scales in transverse rows; its head is scaly, brown mixed with white, of the size of a horse's and nearly the shape of a dog's; the mouth is large, with teeth like a shark; its tail is compressed, obtuse, and shaped like an oar. This animal came in August last into the bay of Massachusetts, in pursuit of shoals of fishes, herrings, squids, &c. on which it feeds. Its motions are very quick: it was seen by a great many, but all attempts to catch it have failed, although $5000 has been offered for its spoils. It is evidently a real Sea-Snake, belonging probably to the genus *Pelamis*, and I propose to call it *Pelamis megophias*, which means great sea-snake Pelamis. It might however be a peculiar genus, which the long equal scales seem to indicate, and which a closer examination might have decided: in that case the name of *Megophias monstruosus* might have been appropriated to it.

Figure 17. Gloucester sea serpent, from Bernard Heuvelmans' *Le Grand Serpent-de-Mer* (1965), a book dedicated to Rafinesque.

2. *Capt. Brown's Sea Serpent.* This fish was observed by Capt. Brown in a voyage from America to St. Petersburg, in July 1818, near 60 N. latitude and 8 W. longitude, or north of Ireland. In swimming, the head, neck, and fore part of the body stood upright like a mast; it was surrounded by porpoises and fishes. It was smooth, without scales, and had 8 gills under the neck; which decidedly evinces that it is not a snake, but a new genus of fish! belonging to the eighth order *Tremapnea*, 28th family *Ophictia*, and third sub-family *Catremia*, along with the genera *Sphagebranchus* and *Symbranchus* of Bloch, which differ by having only one or two round gills under the neck. I shall call this new genus OCTIPOS (meaning 8 gills beneath), whose characters will be—body round, without scales (or fins,) head depressed, mouth transverse, large, 8 transverse gills under the neck.—And its specific name and definition will be *Octipos bicolor*. Dark brown above, muddy white beneath, head obtuse.—Capt. B. adds, that the head was two feet long, the mouth 15 inches, and the eyes over the jaws similar to the horse's—the whole length might he 50 feet.

3. *The Scarlet Sea Serpent.* This was observed in the Atlantic Ocean by the captain and crew of an American vessel, from New-York, while reposing and coiled up, near the surface of the water, in the summer of 1816. It is very likely that it was a fish, and perhaps might belong to the same genus with the foregoing; I shall refer it thereto, with doubt, and name it *Octipos? coccineus*.—Entirely of a bright crimson, head acute. Nothing further descriptive was added in the Gazettes where the account was given, except that its length was supposed to be about 40 feet.

4. *Lake Erie Serpent.* It appears that our large lakes have huge serpents or fishes, as well as the sea. On the 3d July, 1817, one was seen in Lake Erie, 3 miles from land, by the crew of a schooner, which was 35 or 40 feet long, and one foot in diameter; its colour was a dark mahogany, nearly black. This account is very imperfect, and does not even notice if it had scales; therefore it must remain doubtful whether it was a snake or a fish. I am inclined to believe it was a fish, until otherwise convinced; it might be a gigantic species of eel, or a species of the above genus *Ocitpos*. Until seen again, and better described, it may be recorded under the name of *Anguilla gigas*, or gigantic eel.

ADDITIONS.

1. The *Pelamis megophias*, or Great Sea-Snake, appears to have left the shores of Massachusetts, and to have baffled the attempts to catch it, probably because those attempts were conducted with very little judgement. But a smaller snake, or fish, 9 feet long, and a strange shark, have been taken, of which the papers give no description; let us hope that they will be described by the naturalists of Boston.

2. It appears that another large species of Water-Snake is noticed by D. Felix Azara,[8] in his travels in South America (Paris, 1809, 4 vol. 8vo.), under the name of *Curiyu*, which may belong to the genus *Pelamis*, although this worthy traveller has omitted to describe its tail and scales. It may be called and characterized as follows:

Pelamis curis. (*Curiyu*. Azara, trav. Vol. I. p. 226.) Spotted and variegated, of black and yellowish white.

It measures over 10 feet, and is of the size of the leg; it lives in the lakes and rivers of Paraguay, north of the 31st degree of latitude. It goes sometimes on land (and among shrubs), but moves heavily thereon; it has a dreadful aspect, but does not bite; it lives on fishes, young otters, apereas, and copibaras.

3. The Water-Snake of Lake Erie has been seen again, and described to be of a copper colour, with bright eyes, and sixty feet long. It is added, that at a short distance balls had no effect on him: but it is omitted to mention whether it was owing to having hard scales (in which case it might be a real snake of the genus *Enhydris* or *Pelamis*) or to the indexterity of the marksman.

4. Mr. W[illiam]. Lee has brought to notice[9] another Sea Snake, seen by him many years ago, near Cape Breton and Newfoundland, which was over 200 feet long, with the back of a dark green; it stood on the water in flexuous hillocks, and went through it with impetuous noise. This appears to be the largest on record, and might well be called *Pelamis monstruosus*; but if there are other species of equal size, it must he called then *Pelamis chloronotis*, or green-back Pelamis.

5. Dr. Samuel [Latham] Mitchill has exhibited to the Lyceum of Natural History, at the sitting of the 15th September, the specimen of a species of Sea-Snake from his museum, sent him some years ago from Guadeloupe, by Mr. Ricord de Mariana, which appears to be another species, belonging to the genus *Enhydris*, to which the name of *Enhydris annularis* may be given: we shall add its definition and description.

Enhydris annularis. Ringed Enhydris—whitish, ringed with black, rings broader on the back, which is cinereous and rather angular in the middle; tail broad, short, obtuse, with 70 pairs of scales underneath; more than 200 pairs of abdominal scales.

This animal is about 18 inches long, covered with smooth and roundish scales above, the head is depressed, obtuse, small, covered with similar scales, and nearly black; the lips are white; a white half ring sets on the nape of the neck, and extends on each side over the eyes; a black line connects the eyes with the nostrils; an oblong white band lays below the head, longitudinally; the nostrils are round, the mouth is small and with a few small teeth; the body is cylindrical, but the back is slightly carinated towards its centre, and of an ash colour; the black rings are narrow underneath. The tail is only two inches long, very compressed; the extremity is broader, obtuse, tipped with white, and has a slight lateral angle on each side, or a protruding longitudinal nerve; a similar appearance is perceptible on the upper and lower edges, which appear to be thickened; the whole tail is covered with large scales of a transverse and broad shape.

This snake is found in the West Indies, in the sea, particularly on the shores of the island of Guadaloupe.

6. A fabulous account of a great Water-Snake, that, according to the Indian tradition, dwelt in ancient times in a lake near Philadelphia, may be seen in Dr. [Benjamin Smith] Barton's Medical and Physical Journal, Vol. 2, p. 168. As other Indian traditions, relating to the mammoth, the megalon[y]x, &c. it may be partly founded on truth.

7. The great Sea-Snake has been seen again towards the middle of September, in the bay of Massachusetts, and three yellow collars observed on its neck, which has led some to believe it might he another individual and species; but this circumstance might have been overlooked before: it is not stated whether it had streaks of a lighter hue on the body, as the first was represented to have by some witnesses. It is therefore likely that the two characters of "streaks of a lighter hue on the body, and three yellow collars on the neck," may be added to its description. The collars are described as about 2 inches broad, and 1 foot apart.

8. Dr. Mitchill informs me that General Hawkins[10] has written a memoir on the Sea-Serpents of Massachusetts, which he has sent, with a drawing, to Sir Joseph Banks; it is a paper of some length, and much interest, as it relates facts and all the circumstances attending the appearance and natural history of those huge animals, taken upon oaths of eye witnesses. He attempts to prove, with much probability, that several individuals have been seen, and two at least, if not three species; one with three collars, another without any, and a smaller one.

Published in *The American Monthly Magazine and Critical Review*, 1 (October 1817), 431–435.

1. Astronomer Frederick William Herschel or astronomer Pierre Simon Laplace for such phenomena as "spots in the sun," paleontologist George Cuvier for "huge fossil bones," but Pinkerton? Perhaps what Rafinesque had in mind was John Pinkerton's six volumes titled *A General Collection of the Best and Most Interesting Voyages and Travels, in All Parts of the World* that had been reprinted in Philadelphia a few years earlier, 1810–1812, or Pinkerton's two-volume *Modern Geography* also published in Philadelphia in 1804.

2. Patrick Russell, *An Account of Indian Serpents, Collected on the Coast of Coromandel* (London, 1796).

3. Johann Gottlob Schneider, *Historiae amphibiorum naturalis et literariae* (2 vols.; Jena, 1799–1801), which, despite its title, also deals with "crocodilos, scincos, chamaesauras, boas, pseudoboas, elapes, angues, amphisbaenas et caecilias" in Volume II. However, Rafinesque could as well have learned that Schneider established the genus *Hydrus* from the later herpetologists he names.

4. François Péron, *Voyage de découvertes aux terres australes* (Paris, 1812–1815).

5. Johann Georg Adam Forster, *A Voyage Round the World, in His Britannic Majesty's Sloop, Resolution, Commanded by Capt. James Cook, during the Years 1772, 3, 4, and 5* (2 vols.; London, 1777).

6. Arnout Vosmaer, *Description d'un receuil exquis d'animaux rares: consistent en quadrupedes, oiseaux et serpents, des indes orientales, et occidentales* (Amsterdam, 1804).

7. Erik Pontoppidan, *The Natural History of Norway* (2 vols. trans.; London, 1755); Antonino Mongitore, *La Sicilia ricercata nelle cose più memorabili* (Palermo, 1742–43); François Leguat, *Voyages et aventures de François Leguat et de ses compagnons en deux îles désertes* (London, 1720).

8. Félix de Azara, *Voyages dans l'Amérique Méridionale* (4 vols.; Paris, 1809).

9. Reported also in *The American Monthly Magazine and Critical Review*, 1 (October 1817), 443.

10. Either Rafinesque heard the name wrong or he was forgetful. It was David Humphreys (1751–1818), who, in the War of 1812, had the odd title of captain-general of Veteran Volunteers for, in the Revolution, he had been Washington's aide-de-camp with the rank of lieutenant-colonel. His letter was not acknowledged by Banks. He put this together with other letters to Banks and published them as a pamphlet in 1817. See Chandos Michael Brown, "A Natural History of the Gloucester Sea Serpent," *American Quarterly*, 42 (September 1990), 402–36. Brown's article also is the source of most of the information in the headnote here.

24

Evolution

Those who wish to enhance Rafinesque's reputation, as a matter of course call attention to his understanding of biological variation. For one as engrossed as he was in establishing new species, it would have been surprising had he not been sensitive to small differences among several specimens of the same plant or animal. He early concluded that the categories called species and genera are man-made generalizations that, unlike individuals, have no physical existence. The question is, does this discernment make Rafinesque an "evolutionist"?

Many have thought so. After citing Rafinesque's best-known statement on the subject, Bernard Jaffe declared (*Men of Science in America*, p. 126) that "No other American scientist glimpsed this great truth of evolution as early as Rafinesque. None struggled so early for its acceptance. It may be said with a great deal of justice that, in a real sense, he was a forerunner of Charles Darwin." Rafinesque's bibliographer, T. J. Fitzpatrick, also would have us believe ("A Sketch of His Life," p. 43) that "Rafinesque was no ordinary man. He had fairly well defined opinions of the theory of evolution, thus antedating Darwin." And we must remember that Charles Darwin himself, in the historical sketch he prefaced to the third edition of *The Origin of Species*, named Rafinesque among the thirty-four authors he quoted as believing in the "modification of species." Of course, he also cited two other Americans—W. C. Wells and S. S. Haldeman—neither of whom has ever been accused of being a "forerunner" of Darwin.

In a footnote, Darwin cited Aristotle's *Physicae Auscultationes*, where, he said, we "see the principle of natural selection shadowed forth." It was the principal of natural selection, never attributed to Rafinesque or the other two Americans by Darwin or anyone else, that made *The Origin of Species by Means of Natural Selection* (1859) an epochal book and distinguished it from the evolutionary speculations of Lamarck, Geoffroy Saint-Hilaire, and, for that matter, Charles Darwin's own grandfather. While recognizing that variation, through time, leads to the development of what we identify as new species, Rafinesque made no attempt to explain this process by natural selection, though he did consider hybridity a possible mechanism and, without calling it that, he had what appears to be some perception of mutation.

However unwarranted it is to claim that Rafinesque out-Darwined Darwin, his reflections on the concepts of species and genera and his observations on biological variation were provocative at the time and are of interest yet. They have been brought together here in chronological order.—*Editor*

* * *

Nature has been able to create only Individuals or at the very most Species, all the other Denominations are only ideal notions invented by our imagination, to facilitate our knowledge of objects; but they are no less important and necessary, since without them we would never have had more than confused and uncertain ideas of Individuals and Species. [p. 13]

Species are the first Groups [of Individuals] and the most important, they are constituted by the uniting of all the individuals alike to each other in all essential points, and differing by several specific and constant characters from other congeneric individuals; their essence consists furthermore in reproducing themselves constantly according to the same original type, from whence one can deduce that they form the only true groups [which are] natural or owned to by nature; consequently these are they which the Somiologist has particularly in view in his studies, and which he endeavours to know and fix invariably. [p. 14]

The Genus is the original Type of Bodies and after the species the most essential group: sometimes it contains only one; but most often several species collectively alike in certain important and constant characters, which are called generic and the species possessing them are named congeneric Species. [p. 16]

Principes Fondamentaux de Somiologie (Palermo, 1814). Cain translation.

* * *

Extract of a letter to Dr. J. Torrey of New York dated 1st Dec. 1832

...I shall soon come out with my avowed principles about G[enera]. and Sp[ecies]. partly announced 1814 in my principles of Somiology, and which my experience and researches ever since have confirmed. The truth is that *Species and perhaps Genera also, are forming in organized beings* by gradual deviations of shapes, forms and organs, taking place in the lapse of time. There is a tendency to deviations and mutations through plants and animals by gradual steps at remote irregular periods. This is a part of the great universal law of PERPETUAL MUTABILITY in every thing.

Thus it is needless to dispute and differ about new G.[,] Sp. and varieties. Every variety is a deviation which becomes a Sp. as soon as it is permanent by reproduction. Deviations in essential organs may thus gradually become N. G. Yet every deviation in form ought to have a peculiar name, it is better to have only a generic and specific name for it than 4 [names, as] when deemed a variety. It is not impossible to ascertain the primitive Sp. that have produced all the actual; many means exist to ascertain it: history, locality, abundance, &c. This view of the subject will settle botany and zoology in a new way and greatly simplify those sciences. The races, breeds or varieties of men, monkeys, dogs, roses, apples, wheat ... and almost every other genus, may be reduced to one or a few primitive Sp. yet admit of several actual Sp. [N]ames may and will multiply as they do in geography and history by time and changes, but they will be reducible to a better classification by a kind of genealogical order or tables.

My last work on Botany if I live and after publishing all my N. Sp. will be on this, and the redaction of our Flora from 8000 to 1200 or 1500 primitive Sp. with genealogical tables of the gradual deviations having formed our actual Sp. If I cannot perform this give me credit for it, and do it yourself upon the plan that I trace.

C. S. R.

[pp. 11–12]

Some Principles of Natural Classification

See my principles of Somiology or Botany and Zoology 1814 [i.e. *Principes Fondamentaux de Somiologie*].

1. Individuals alone really exist among organized bodies, who vary slowly by reproduction, deviations, changes of soil, diseases, &c. So as to offer new typical forms, by gradual steps and evolutions.

2. We call a *Species*, the aggregate group of Individuals that offer the same forms.

3. When these forms are changed in or out of sight, in or out of gardens, we deem these new groups as they really are, *New Species*.

4. This we commonly do in Botany, but not always in Zoology: [Noel Joseph de] Necker wished to call these changes *Proles* or Breeds as in Zoology; but Botanists would not change their terms.

5. All the Sp. that have similar characters in the fructification or essential organs, form groups called *Genus* and *Genera*.

6. Every Sp. that offers peculiar characters, different from those of the G. to which it is referred, must form a New Genus.

7. Whether N. G. are forming yet as well as N. Sp. is a fact less obvious, because more rare and slow.

8. A variety is a slight deviation of a Sp. in form or peculiarities not essential, produced by Seeds or even Shoots of different years; it is the first step to form a N. Sp.

9. A Sp. may be new also by having escaped the attention of former Botanists, having been blended by inattention, overlooked, or not seen nor met with by not exploring a particular place and season. Such N. Sp. abound most in the least explored Countries.

10. Natural Orders or families consist of one or several Genera that have peculiar important features and characters.

11. A single G. may form a family, as a single Sp. may form a G. as it happens in man. Nay a single individual may form a Sp. if it can reproduce itself.

12. To multiply Sp.[,] G. and Families is as needful in the Science of Botany,

as in every other. Since they are made first by Nature! Like houses, Towns and Counties, yearly increase in Geography by the hand of man, while families, tribes, and Nations increased in History by the natural evolutions or revolutions of mankind.

13. Those who hate or omit N. Sp.[,] N. G.[,] and N. O. are like those who would prevent the building[,] or overlook when built[,] new Towns and Cities, [and] despise or neglect new Tribes and Nations, in writing Geography or History.

14. Towns and Nations rise and fall. So do Plants, Sp. and G. They may be ruined, destroyed or confined to a few individuals by the lapse of time. The fossil remains of Plants offer the buried ruins of former Sp. and Genera.

15. All our actual Sp. of Roses, Grapes, Oaks, Plumbs, Apples, Currants, Asters, Azaleas, Heaths, &c. have thus been formed. Nay it is so probably with every genuine Genus of many Species. [pp. 13–15]

Herbarium Rafinesquianum (Philadelphia, 1833).

* * *

Linn[a]eus elegantly says that minerals grow; Plants grow and live; Animals grow, live and feel. But later naturalists divide the bodies of our globe in two empires, Organized, and Inorganic. Each has a triple series of living forms.

These triple series are divided into a multitude of groups, called Classes, Orders, Tribes or Families, Genera and Species: Each formed by a cluster of individuals which in the Organic Beings reproduce and perpetuate themselves. But all these groups are factitious more or less, made by us by an inductive method of generalization: while in Inorganic Bodies we proceed by the contrary mode of analysis to seek the elementary principles, once thought only four or five, now increased to fifty or more.

Yet in this scientific process of grouping individuals we endeavour to follow the steps of Nature in their previous decompositions by gradual slow changes in reproduced individuals. We know not how many living forms existed at first, or were created on earth at the earliest period; but by the fossil relics of many, we ascertain that they were fewer and often different. Whatever was their original number and types, it is probable that these primitive individuals have produced all the actual various species, of which we have already ascertained nearly eighty thousand of animals, with a hundred and twenty thousand of plants. The proofs of this fact are found in the varieties and monstruosities, still proceeding under our eyes, or that have for ages past. Every species was once a variety, and every variety is the embryo of a new species. [p. 222]

Man forms a genus with a single species; as well as dogs and all domestic animals, as liable as he to vary. A genus in organized beings is the collection of all the varieties or species that differ essentially from others. A species is the collection of all the individuals acquiring distinct forms and colors, and all the deviations that can breed together. They are abstract terms of our own; Nature only acknowledges individuals, and vary them constantly; so as to produce new species now and then, particularly among plants. Genera vary also, but so slowly, as not to be easily perceived. It is probable that new genera are also forming, and that all our generic and specific form of animals and plants have been produced by successive deviations from the original types discovered among the fossils of the former earth. In birds and insects, the colors alone distinguish most of the species; yet they are known to be most variable characters. The zoologists and botanists begin to pay some attention to the relative importance and value of organs and characters; but they are still divided on the subject of species and varieties: whereby they fall into singular contradictions, and call varieties in men, dogs, grapes, cherries, what they call species in monkeys, birds, insects, roses, grasses, oaks. *Every species is a variety, and every variety is a species!* the only difference is in their age! and there is no actual limit between them: no more than between a hill and a mountain, a lake and a pond, a river and a creek! which are distinguished by mere relative size without demarcation. Therefore it is only a nominal question, and they became quite indifferent terms. Thus, if Bory [de Saint-Vincent] has made 15 species of men! let it be so; they are varieties, however, and all men form one procreating genus. Some naturalists neglect *varieties* altogether, and thus it is right to call them species in order to fix thereon the attention:

> Just like a tree, with many branches; most
> Of genera produce the various kinds
> Or species; varieties at first, like buds
> Unfolding, and becoming species, when
> By age, they may acquire the proper forms. [pp. 228–29]

The World, or Instability. A Poem in Twenty Parts, with Notes and Illustrations (Philadelphia, 1836).

* * *

Many botanists mistake real botanical species for varieties or vice versa. In fact[,] all species might have been varieties once, and many varieties are gradually becoming species by assuming constant and peculiar characters. This is an interesting feature of botanical philosophy, which I shall properly explain and prove hereafter. [I, 6]

I admit like Adanson, Necker and Linn[a]eus himself that plants do vary gradually and constantly, although often very slowly, both in the specific and generic characters. I refer to these authors for examples so often met in gardens; but I have chiefly attended to this in the woods and fields where [it is] quite spontaneous.

The process is by the seedling being somewhat different from the parents, and thus evincing a deviation of typical mould, that may be, or may not be, propagated again. If it is, this soon assumes a permanence, becoming a permanent variety if the deviation is slight, such as mere color of flowers, size of stem, leaves, &c.; but becoming a New Species! if at last several deviations are permanently combined. [I, 15–16]

The specific deviations which I could mention are numberless, this work [*New Flora of North America*] will be full of them, as all new species are in fact such permanent deviations of growth, unless they are widely different from all former species. The oldest type of the species may probably be found in the most common with most numerous individuals, while those called rare or with few individuals as yet must be the newest in order of time.

Generic deviations are more rarely observed or noticed, because less evident and very slowly produced, or not so easily propagated.... [I, 16]

Of all the European Floras, that in which generic and specific deviations are most explicitly stated and best detailed is the Flora Lithuanica [Inchoata. 2 vols.; Vilnius, 1781–82] of [Jean Emmanuel] Gilibert, where many … anomalies may be noticed.

Therefore the result will be that our species and genera are not quite permanent as supposed, but are gradually producing deviations of forms; when these are floral they are of course generic, and may after centuries form New Genera.

Hybridity also multiplies species: this process is much more common in plants than animals; but it is often difficult to detect in wild plants the real parents. Unless they are both found growing near the hybrid, and it has the traces of both forms, we may as safely ascribe it to a natural deviation of frame [form?], called Peloric if floral and generic. [I, 17–18]

SPECIES include all the individuals perfectly alike in all their parts—*Varieties* are slight casual deviations—*Proles* are permanent varieties, such as are called *Breeds* among animals.

GENERA are the groups of species that have similar floral characters and sometimes a similar habit. Whenever a species has different floral forms it must be a peculiar genus.

Such are the natural distinctions of those two groups, any others must be unnatural and improper! All species may have been varieties once, except the original types or ancestors of the genus, and all actual varieties may be incipient species. [I, 18]

289. PHRYMA MEDIA Raf. subpubescent, leaves petiolate equally serrate, lower cordate, upper ovate acute, last pair sessile, bracts subulate shorter than calix—in Kentucky, annual and estival like all the species. Our botanists admit of only one, but it offers so many deviations that I have collected 3 incipient sp. which I add with the real type *Phr. leptostachya*

290. PHRYMA PUBESCENS Raf.…

291 . PHRYMA PARVIFOLIA Raf.…

292. PHRYMA LEPTOSTACHYA L. &c. smooth, leaves difforme lower petiolate ovate base acute deeply serrate, upper sessile ovate often entire, bracts equal to calix—the most common kind, from New York to Carolina, 1 or 2 feet high. If all the above are mere varieties of this, they afford a fine illustration of incipient species forming under our eyes in our woods. [II, 37–38]

New Flora of North America (Philadelphia, 1836). [Part I was issued in December of that year, Part II during the second half of 1837.]

* * *

1. Vegetation produces only individuals! whose permanence is limited by their life. Our Species, Genera, Families, and Orders are well known to be mere abstract terms of successive groups, formed by a Synthetic operation of our mind[s], in order to study more conveniently such collective groups of Individuals. Their permanence in continual succession of forms can only be temporary: since their permutation of forms takes place spontaneously in their natal soils, as well as our gardens where it is increased by art; while new varieties and species were often met by me at long intervals in wild places well explored before, grown from seeds of akin species. See my remarks and facts collected in my new Flora. [The most significant of these are listed above. **Ed.**]

2. Plants vary gradually, in features, aspect, size, color, &c. by a natural spontaneous deviation from seedlings. This may happen quicker in annuals, less quick in perennials, slower still in trees, except when the tendency has already become active. These deviations may gradually form distinct varieties, next Breeds, at last becoming separate Species, when they assume a striking difference, and peculiar specific characters of a more permanent nature. The disparities in the descriptions and figures of old and modern botanists amply verify this.

3. Even perennials may vary slightly in annual shoots from the same root, and trees in different branches or annual growth. When a tendency to deviation by monstrosity, hybridity or variety is taken by an individual, the seeds produced will unfold them when growing, particularly if removed from the native place into gardens and new soils.

4. Pelorian Genera, or Generic Deviations in flowers and seeds, happen slower or more seldom; being often unnoticed, or the produced seed is not always fertile. When it is, the offspring may become the type of a New or distinct Genus. Many such perish before they reproduce the deviation by fertile seeds; but a few survive and are the types of akin Genera.

5. The periods of these deviations are doubtful, much fluctuating and various in length or existence. But we may assume as an average 30 to 100 years for the deviating or splitting range of specific deviation, and 500 to 1000 years for the Generic deviation; altho' their real permanence is much longer. Specific and generic Lives have not yet been calculated.

6. Therefore many of our actual or newly described Genera and Species, may be of recent origin, and all may have once sprung at the last rinovation or cataclysm of this Globe, from a lesser number of original types, perhaps found in the fossil plants of our Earth, which are far from being all known as yet, and whose seeds were preserved in mountains, earth, mud or water till the catastrophe was over.

7. It is even possible to ascertain the relative ages and affinities of actual species and Genera, sometimes their very parents or connections in the Genus or the tribe. Those we call hybrids are not always such, they may arise from other deviations; but artificial hybrids are evidently such. All these deviations are still less permanent.

8. As a general rule the real Genera (not the false ones of blending Botanists) of single or few species are the newest in order of time, and the most prolific the oldest in the Series. The same for tribes perhaps. False Genera like *Erica*, *Carex*, *Aster*, *Allium*, *Lichen*, *Euphorbia*, *Mimosa*, *Geranium* &c. comprizing a crowd of generic distinctions, are as many collections of related Genera, springing from very early sources or types of forms. Extensive natural Genera prolific in Sp[ecies] like *Rosa*, *Iris*, *Quercus*, *Salix*, *Oxalis*, *Malva*, *Vitis*, *Lactuca* &c. had also a very old or primitive source. Species prolific in individuals and varieties are always the oldest, and rare Species probably the newest of all, unless they are fragments of extinct groups. [I, 12–14]

INDIVIDUALS alone have a separate physical existence, all the other clusters are useful botanical groups of ideal abstractions, based on physical characters, by successive proportions of affinities; as political institutions collect men in successive clusters of families, clans, ranks or castes, communities, tribes, and States.

Therefore, Individuals are the main object and first aim of Botanical knowledge; the study of their clusters becomes the aim of systematic Botany: nomenclature and classification, which may be compared to a kind of Statistical Science, under a philosophical method, based on accurate principles

SPECIES are the collections of individuals perfectly alike in all their parts. *Varieties* are slight casual deviations. Breeds or *Proles* are permanent Varieties. Therefore Species are natural altho' variable.

GENERA are the collective groups of Species, that agree in the Characters of the fructification. No Species belongs to a Genus unless it agrees with all the others therein included. Sub-Genera are lesser groups or sections with some slight deviations chiefly in the habit, seldom in the floral organs. Therefore proper Genera are also natural.

NATURAL FAMILIES are groups of Genera having some striking characters in common, chiefly flora and organic.

NATURAL ORDERS are groups of families united by one or several important characters, chiefly floral and organic.

NATURAL CLASSES are groups of Orders, possessing some very peculiar floral characters, and common organization. [I, 38–39]

Nature in the spontaneous evolution of vegetation, baffles all our petty incongruities by making new Species out of varieties, and new Genera out of floral deviations! the process is not always so quick as to be perceived in a few years; but is very obvious to botanical observers who happen to study plants during 40 or 50 years. [I, 42]

Flora Telluriana (Philadelphia, 1836) [Part I actually was issued during the first quarter of 1837.]

* * *

As to varieties, *most of our species are such*, being natural deviations by seedlings, assuming peculiar forms, in the woods and wilds, as it is done constantly in our fields and gardens by the cultivated trees and plants. Those best known afford most of our noticed varieties or specific deviations; but it is only our ignorance or neglect that prevents us from ascertaining in others all consimilar varieties. One of the great aim[s] of accurate Botany is now to fix the typical and prototype species of each Genus; our subgenera are mostly such, when not based on floral disparities. When [they are] thus based they become real Genera, whose specific deviations should be traced. [pp. 156–57]

Sylva Telluriana (Philadelphia, 1838).

IX.

Metaphysics

25

The Psyche Papers

At the conclusion of his autobiography, among the numerous professions he claimed to have practiced Rafinesque listed that of "Philosopher." His earliest philosophical essays were unknown during his lifetime, because intended for the *Western Minerva*, a magazine suppressed before it could be distributed. There he printed an essay on the "Principles of Political Wisdom" ostensibly "collected from the works of Pythagoras and his disciples" and "translated from the Greek, by Benjamin Franklin." This was followed by an even longer essay on "Ethics, or Moral Philosophy," again, it was said, translated from the Greek of Pythagoras by the redoubtable Franklin. And that in turn was followed by an essay on "Metaphysics," signed by Leibnitz [*sic*]. Perhaps the dateline "Lexington, October 1820" on the last was intended to assure literal-minded readers that Franklin and Leibnitz were pen names.

In the "Leibnitz" essay on metaphysics, Rafinesque distinguished between ESSENCE ("necessary, eternal, efficient, primitive, essential," etc.) and SUBSTANCE ("willed, created, secondary, particular," etc.), then further defined these distinctions by saying that the former consists of monads ("elements of the spiritual substances"), the latter of atoms ("elements of the material substances"). This basic hypothesis from Gottfried Wilhelm Leibniz underlies the five essays which follow, and it was a concept much too subtle for James Fishback, the obtuse Christian fundamentalist who had provoked the controversy Rafinesque was addressing here.

It is understandable that little attention has been paid to Rafinesque's philosophical musings. The *Western Minerva* essays were unavailable until 1949. The Psyche essays, printed in five different numbers of Lexington's most cerebral newspaper, the *Western Monitor*, were entirely unknown until 2001, when they were first listed in *Mantissa*, the supplement to the revised Fitzpatrick bibliography (1982). Even then, they would be hard to come by for most readers. When the *Western Monitor* was filmed these issues were unknown. Unique copies of these particular numbers of the paper exist at the Filson Historical Society in Louisville.—*Editor*

* * *

Psyche—No. I.

Attend, reflect, compare and know your souls,
Eccentric man! immortal guides, received
From God, to lead you through this transient life.

A controversy has lately arisen[1] upon the nature of the Soul, which might perhaps have been allowed to sleep; but since the questions relating to the immateriality and consequent immortality of the human soul have been revived, and since unfortunately the opinions against these fundamental doctrines of correct morality, real philosophy and true religion have found yet some advocates and believers, it is become needful for the friends of truth to direct the judgment of those who have not yet investigated this abstruse subject, and to correct the sophisms of those who have mistaken its import and tendency.

The majority of mankind believe by an intuitive feeling of self knowledge that the human soul is immaterial and immortal; this is in itself no slender argument in favor of the opinion; but there is a minority who are so far misled, as to call this belief a delusion. To this part of mankind, I address myself by an appeal to their understanding and an invitation to study again the subject with me, and to consider it in the new aspect which I shall endeavour to present.

I do not claim any originality of thought on the subject, nor have any pretension to a display of new ideas; my arguments will mostly be borrowed from those enlightened writers who have, both in ancient and modern times, scrutinized with the greatest perspicuity and ability the various opinions sprung into the human mind or soul, upon its own nature and destiny.

I mean to present the mere results, and the most important only, of their numberless enquiries, divested of the tiresome incumbrance of verbosity—I shall begin by offering a kind of geometrical demonstration of the subject, to which I invite the attention of my impartial readers. It has often been asserted that this subject was not susceptible of mathematical demonstration; but this assertion will perhaps be allowed to have been erroneous by those who will be able to attend to the following chains of arguments.

1. Reason and judgment are two faculties of the mind or the soul, arising from the power to compare dissimilar perceptions.

2. Perceptions are received through the senses; but the senses have merely the power to know and compare their own sensations; while they are entirely unable to know and compare the perceptions of each other.

3. The sight cannot compare its own perceptions and sensations with those of the hearing, nor vice-versa, and so on with every other sense.

4. All the various sensations have therefore a point where they meet, and this point must be a simple indivisible substance.

5. This central point which compares, reasons and judges must therefore be a soul, a substance without parts and immaterial.

6. Such a simple indivisible being, must be incorruptible, unalterable, indestructible, and therefore immortal.

DEMONSTRATION.

A substance cannot compare two sensations, without having and proving or holding those two sensations, at the same time.

Two sensations may be supposed to become united only in two kinds of substance, a real indivisible substance or a false substance composed of parts which are themselves as many substances.

If it is allowed that the two sensations meet in a substance without parts, this substance will be simple, without extension, and immaterial, the immaterial soul.

If it is contended that they meet in a substance composed of other substances or parts; they must meet in one of these parts or in two among them, in the part A for instance or the parts B and C.

If they both meet in the part A, that part must have no other parts and be a simple indivisible substance, an immaterial soul.

Since if it should have parts, and the two sensations should meet in two different parts of the part A, say D or E or in the previous parts B and C, this would imply that the sensation is in the part D or the part B, while the sensation is in the part C or E.

If a sensation is in one part or substance and the other in another part and substance, it follows that they are not united in the same substance, and that a same substance does not hold them at the same time.

And if a substance does not hold at the same time the two sensations, it cannot compare them.

Whence it follows, as plainly as two and two make four, that the soul being a substance that compares, is not a substance extended and with parts.

The soul therefore is a simple, unextended, and indivisible substance, an immaterial and immortal being.

I invite those who contend that this cannot be proved philosophically, much less mathematically, to reflect on the evidence of these arguments, which had not yet been unfolded by the supporters of the late controversy, and will I trust carry conviction to every impartial mind.

CONSTANTINE.

* * *

Psyche—No. II.

Within the mind a latent feeling lies
Of consciousness that we exist and live;
An ardent wish for immortality
Unfolds itself, reveals the truth, and bids
Us hope that we shall live beyond the grave.

The numberless proofs of the spirituality & immortality of the soul, may be distinguished into mathematical, mental, physical, moral, and religious proofs. Having already stated one of the most evident demonstrations of the first order, I shall proceed to explain the evidence derived from mental consciousness.

Consciousness is that internal sentiment of the mind, through which it receives the conviction of its own existence, independently of the impressions received through the senses.

Our knowledge is derived therefore from two different sources; this opinion is now generally admitted by the most eminent philosophers, and it has the advantage of uniting the two sects of materialists and spiritualists, by proving that they were both in the right; but were led astray by taking only one view of the subject. The materialists believed that all knowledge came through the senses, and the spiritualists from mere internal sentiments; but by allowing that we derive knowledge from consciousness as well as sensations, we may hope to reach the truth, which always lies between extremes.

The study of human consciousness, of the sentiments which it evolves, and of that instinct which is its principal and primitive support, begins to draw the attention of all philosophers and naturalists. Instinct is however as yet less known than sentiment; it will be the subject of the next number.

Sentimental consciousness has been so well and so often analyzed and surveyed, that I cannot present any interesting arguments on that point of the question, but I will select two of the most striking.

First Argument.

We exist, we live and we may know it independently of the senses.

If a man was deprived of all his senses, without dying, he would still be conscious of his existence, of his life, and their succession.

This has happened to several persons who have fallen into a state of insensibility and apparent death, been deprived of all sensations; but have recovered their senses on being carried to their graves, and have remembered that state.

It happens with many animals deprived of all senses except an obtuse feeling; they must be aware of their existence even when their sense of feeling becomes latent or not acted upon by any external object.

It happens with men when they are immersed in deep thoughts, mental reflection, and do not make use of their sensorial organs.

It happens in sleep, when all the organs slumber, while we are conscious of our existence and dreams.

If we are or may be conscious of our existence independently of the senses, there must be something else in us besides sensations. It is consciousness.

If we derive knowledge from consciousness, this power cannot belong to matter. Stones and inorganized, inanimate bodies can not have any such consciousness nor any knowledge of internal or external objects.

If consciousness belongs only to a spiritual substance, our consciousness must be inherent in our soul, and our soul must be immaterial.

Second Argument.

All men have the constant desire to live, to be immortal, and a perpetual aversion for death and annihilation.

The exceptions to this general desire and aversion are so few, that they do not invalidate this sentiment, and they ought to be considered as moral diseases of the mind, arising from a morbid state of the intellect.

The inducements to lengthen our lives, and continue to exist hereafter are numerous, but our principal motive is in the natural instinct or internal sentiment producing a permanent aversion to pain, destruction, and death.

From this sentiment arise the hope and belief of a future existence, without term, and an intimate suggestion tells us that this hope and belief will be realized.

Mankind having reached the knowledge of God through several means, have received from this internal hope a prop to their belief, and have been inspired with religious sentiments, more or less eliminated by true piety and wisdom, or defaced by superstition and bigotry.

Since our mental consciousness has an internal hope, belief and evidence of its perpetuity, it must belong to an immortal being, or soul.

If the soul is immortal, it must be a spirit or immaterial substance, without parts, since every material aggregation has a definite existence.

CONSTANTINE.

* * *

Psyche—No. III.

Let truth be heard, and doubtful minds be told
How Instinct guides, untaught by senses, when
They are unable quite to feel and lead,
A helping hand receives from distinct source,
By sentiments innate reveals and proves
That we have souls with previous knowledge gifted.

All the branches of knowledge are so intimately connected, that we ought not to disdain borrowing from all sources in order to elucidate, explain, and prove any subject. Metaphysicians ought not therefore to neglect the study of the material beings upon which they frame their systems, and they ought to explore the wide field of natural science when they wish to know completely mankind and their fellow beings. It is often by the omission of this needful qualification that they advance various idle or erroneous theories respecting the faculties of the mind.

The correct knowledge of the innate instinct of man and animals, requires an accurate acquaintance with physiology, the functions of life, the theory of sentiments and passions &c. which cannot be properly acquired, without an attentive study of the animal creation, wherefore but few philosophers have been able to explain the real nature of the spontaneous phenomena of instinct. Hoping to be enabled to throw some light on this obscure subject, which has often engaged my attention, I shall proceed to state some results of my enquiries.

THEORY OF INSTINCT.

The mind has two active principles through which it receives perceptions, and evolves thoughts. These two principles are intellect and instinct.

Intellect has been called the head of the mind; its peculiar perceptions are called sensations, and its thoughts are ideas; they are received through the external senses & are therefore connected with matter.

Instinct has been called the heart of the mind; its peculiar perceptions may be called emotions, and its thoughts are sentiments, which are received through the internal senses, and are more or less unconnected with matter.

There are five internal senses, which are

I. Consciousness,
II. Intuitiveness,
III. Preservation,
IV. Happiness,
V. Sympathy.

Emotions (as well as sensations) may be distinguished into agreeable, indifferent, and painful, arising from the three inherent impressions or prompters of life, pleasure, indolence, and pain. These are generally spontaneous.

Sentiments (as well as ideas) are simple or complex. United to ideas they become notions.

Simple sentiments derive immediately from spontaneous emotions. The complex sentiments arise from emotions elaborated by the intellect, and being compared, abstracted or composed give rise to the numerous series of affections, wants, desires, inclinations, propensities, passions, actions, habits, &c.

The least complex are the reflective sentiments, which arise from emotions and reflections, such are the hope, energy, timidity, fear, despair, joy, uneasiness, grief, resentment, &c.

There are some men who are deprived of one or more internal senses: in the same manner that there are blind and deaf men. The born blind may sometimes deny light and colors; the man deprived of intuitiveness or sympathy may likewise deny intuition and innate perceptions. Let them believe those who have these internal eyes and ears.

It is a paradox to admit innate ideas, but innate sentiments are felt by all those who are not deprived of their internal senses. The innate ideas of the ancients were our innate sentiments under an erroneous appellation.

INTUITIVE SENSE.

The internal knowledge derived from instinct is called intuition. Intuitiveness differs from it, because it applies to that internal sense which teaches spontaneously without the help of external senses, experience, reflection, or education.

This sense is less unfolded in man than in many animals & becomes so soon blended with the results of external knowledge, that it is sometimes mistaken for it.

But it shows itself in the new born child, who sucks the breast of his mother, by an untaught pneumatical process.

In the mother who throws herself into the water, without being able to swim, if she sees her child drowning.

In those involuntary impulses, motions and frights which cannot be ascribed to the sense of preservation.

In the untaught and sudden impulsions of imagination and genius, which are deemed inspirations of the creative mind.

In those spontaneous surmises, hopes, presentiments, and convictions of the mind, otherwise unaccountable.

While the animal creation offers striking and numberless proofs of the most convincing and positive nature, relating to the existence of the internal sense of intuitiveness.

Many thousand kinds of insects, such as spiders, bees, flies, silk-worms, beetles &c. who have never seen their parents, nor have been taught, know by intuition the science and arts of geometry, architecture, weaving, hunting, digging, &c.

The birds build their nests in a regular plan without being taught: they foresee the seasons, & search climates which they have never seen; the new-born chickens distinguish seeds from sand, &c.

The quadrupeds, weaned from their parents, will unfold their faculties without experience: the beavers will build, the foxes will hunt and burrow, the squirrels will collect food for a winter, unknown to them, &c.

All this results from intuitiveness, that part of innate instinct, which is natural, independent of the will; learned without instruction, and perfect in its kind.

Since there are innate emotions, sentiments, notions and internal senses, the instinct from which they arise, is independent of the external senses and matter; it belongs therefore to a separate independent and spiritual existence, a soul.

CONSTANTINE.

* * *

Psyche—No. IV.

To search for many proofs, in aid of truth,
Must be the constant aim of wisdom's friends.

The moral and religious proofs of the spirituality and immortality of the soul are so well known, that I hardly need appeal to them. It is the particular duty of moral and religious teachers to inculcate them, and impress them on the minds of their hearers and readers.

The foundation of all our moral duties rests upon this belief and certitude. If the soul were not spiritual and immortal there would be no rule of conduct, and no ultimate aim to attain through this life. Whoever denies the moral proofs of this doctrine, denies the distinction between good and evil, virtue and vice: the consequence of which is obvious. Therefore the soul is immortal, since there is such a moral distinction, and it cannot be immortal without being spiritual.

All the religious have more or less admitted and inculcated the same doctrine; but there are some sects, which deny either the spirituality or the immortality of the soul or both. These sects have existed or exist yet, among the Christians, Jews and Mahometans, who all believe in revelation: while there are whole nations to which revelation is unknown and yet admit this doctrine.

If some religious sectarians deriving their belief from revealed books, could doubt a doctrine which appears to be so plainly taught in them, it would be improper to suppose that all our knowledge on the subject, is derived from them, as has been sometimes, and even lately contended. This opinion is similar to the pernicious doctrine, and false theory which supposes that moral distinctions depend on political enactment.

It is dangerous to endeavour to lessen the important proofs, which may be collected in support of their needful belief. It would be a very imprudent step to convince men that they ought to be atheists and materialists, before they can become christians. This was not the conduct of the Apostles; they appealed to the previous knowledge of men, and their acquired notions on religious subjects, when they went to preach the gospel among the heathen.

Some men are easily convinced or brought to this and other beliefs by religious proofs, faith and testimony; but there are some who require additional proofs or an accumulation of convincing proofs, before they admit a certitude.

Unbelief arises from ignorance or inaptitude to conceive and perceive, while belief is a partial conviction with few negative perceptions; and doubt is the state of mind fluctuating between belief and unbelief, of admitting by turns motives of affirmation and denial.

It is desirable that all men should have not merely a belief of the important doctrines of immortality; but the knowledge, certitude, and conviction of it; but this desirable state of the mind can only be produced by an intimate acquaintance with all the kinds of proof, or at least as many as are attainable and fit to be understood individually.

Whoever denies a great majority of such proofs, may be convinced in his own mind by a small number; but he ought to consider that all men are not alike, nor convinced by the same kind of proofs; and ought not to deny to others the use of such proofs as are best suited to their state of mind or actual aptitude.

It would be wrong and unjust for a man who wishes to live on meat alone, to deny the use of bread and vegetables to those who like them, require a mixed food, or cannot procure animal food.

I have been asked how I became convinced of this doctrine. The reply is easy: by reflecting upon all the various metaphysical, mathematical, physical, mental, logical, moral, and religious proofs, which I have been able to collect; and I could perhaps contend that my belief must be stronger and safer, than that of him who should trust to a single kind of proof, or be unable to understand any, but that kind.

I began these essays with the view of noticing merely some of the best and well known arguments on the subject; but I have been led to evolve some new ideas suggested by the course of my former reflections and late meditations. I hope therefore to be no longer accused of following common paths, and I shall conclude by presenting in my next number the outlines of my analysis of the mind, a subject in which there are yet discoveries to be made, as in every other branch of science, new distinctions to be drawn, better definitions and names to be given, convincing demonstrations to be framed, sound theories to be established, and correct opinions to be evolved, instead of mere assertions and hypotheses.

To do this, would require more time, and mental labour than I can at present bestow, therefore a mere sketch of the results of my mental observations and meditations well be given, and no attempt will be made to solve the difficulties, problems, and mysteries of psythology [*sic*].[2] Every science has such impediments, which can only be removed gradually and by able hands. A few can perhaps never be removed; since we cannot square the circle in geometry, nor find the relative proportion of all the lines, it would be presumptuous to suppose that similar difficulties are not met with in metaphysics, whence the endless disputes about free will, and fatality, the union of substances, the relative connections of causes and effects &c.

But our knowledge of the existence, functions, faculties, operations and destiny of our soul, is not problematical: we know that there is something within us that feels, thinks, acts, compares, judges, and reasons; but matter is unfit and unable to do this,—our soul is therefore immaterial or spiritual and immortal.

CONSTANTINE.

* * *

Psyche—No. V.

Celestial Truth! best holy aim of pure
And lofty minds, why ever hide thyself?
Beyond the reach of many panting souls,
That are compell'd to seek for thee thro' paths
Of errors, vain delusions oft renewed:
Yet all these paths may lead to thee at last.

My meditations on mental philosophy have convinced me that this sublime science is very far from having reached a state of perspicuous certainty: they have been so much extended by the perusal of the excellent work of the late Dr. Thomas Brown,[3] and a recapitulation of his omissions and errors, that the theme has become beyond the scope and limits of a newspaper essay. I must therefore postpone my remarks and reflections to a more convenient time and vehicle. This number will now close the series of these essays. But among the variety of mental discoveries which I presume to have made, I wish to lay claim upon a few, of perhaps a most important nature, of which I am very sorry to be compelled to delay the full explanation and illustration. Meantime, I hope that the mere hasty mention of their import will be sufficient to convince of their importance.

I conceive that all the various series of ideas or mutable states of the mind arise or are evolved from TWENTY SENSES or real definite immediate and active prompters of ideas: these senses may be classed into four orders, each order having five real senses besides three MONITORS or INDEFINITE SENSES, and producing a peculiar kind of Ideas, while every sense produces two peculiar kinds of situations, by its presence or absence. The following tabular view of these senses will explain my theory. The whole mind has besides four general attending powers, which are *Mutation*, *Perception*, *Consciousness*, and *Influence*.

I. ORDER.—EXTERNAL SENSES.

1. Sense. SIGHT. Presence *Vision*. Absence *Blindness*.
2. S. HEARING. Pres. *Audibility*. Abs. *Deafness*.
3. S. SMELL. Pres. *Smelling*. Abs. *Inodority*.
4. S. TASTE. Pres. *Tasting*. Abs. *Unsavidity* [*sic*].[4]
5. S. TOUCH. Pres. *Touching*. Abs. *Unfeelingness*.

The ideas produced by these senses are called SENSATIONS; their Monitors are *Pleasure*, *Pain* and *Intus-Sensation*.

II. ORDER.—CORDIAL SENSES.

1. S. IDENTITY. Pres. *Personality*. Abs. *Idiotism*.
2. S. PRESERVATION. Pres. *Vitality*. Abs. *Destruction*.

3. S. Sagacity. Pres. *Intuition*. Abs. *Stupidity*.
4. S. Comfort. Pres. *Happiness*. Abs. *Uneasiness*.
5. S. Sympathy. Pres. *Affection*. Abs. *Antipathy*.

The ideas produced by these senses are called Emotions; their Monitors are *Good, Evil,* and *Sensibility*.

III. ORDER.—INTELLECTUAL SENSES.

1. S. Conception. Pres. *Conceiving*. Abs. *Dullness*.
2. S. Suggestion. Pres. *Association*. Abs. *Separation*.
3. S. Relation. Pres. *Judgment*. Abs. *Folly*.
4. S. Memory. Pres. *Remembrance*. Abs. *Forgetfulness*.
5. S. Imagination. Pres. *Fancy*. Abs. *Vacuity*.

The ideas produced by these Senses are called Energies; their Monitors are *Attention, Will,* and *Susceptibility*.

IV. ORDER.—MORAL SENSES.

1. S. Rectitude. Pres. *Right*. Abs. *Wrong*.
2. S. Utility. Pres. *Usefulness*. Abs. *Uselessness*.
3. S. Honor. Pres. *Honesty*. Abs. *Dishonesty*.
4. S. Virtue. Pres. *Wisdom*. Abs. *Vice*.
5. S. Equity. Pres. *Justice*. Abs. *Injustice*.

The ideas evolved by these Senses are called Sentiments; their Monitors are *Praise, Blame,* and *Conscience*.

It is gratifying to perceive that the science which I have been led to illustrate, has lately been cultivated in Lexington by three gentlemen of great learning, and although it happens unfortunately that they all entertain a different opinion on the subject, some useful results may yet evolve by the conflict of opinions.

To the pious divine[5] who has occasioned these Essays, by the promulgation of some new opinions on the source of our metaphysical knowledge, I tender the tribute of the admiration due to his self-taught genius, the respect claimed by his great talents, and the veneration deserved by his pure motives. Since we think alike upon the great results of our enquiries, I trust that he will not misconceive the present discussion, which was merely intended to elicit truth, and if he cannot be convinced that there are several sources of metaphysical knowledge, he must at least confess that no harm to the holy cause of virtue and religion, can accrue from affording or endeavouring to afford several proofs in support of our common doctrine.

The accomplished scholar,[6] who teaches with success the science of mental philosophy; follows entirely the plan of the Scotch school, and writers; seldom noticing their mistakes, omissions, misnomers, and eccentricities; his course will do very well for his immediate purpose, but it is not likely to advance the science itself beyond its present limits.

Meantime a learned Physician,[7] now delivering a course of popular lectures on the branches of this science lately called by the peculiar appellations of Phrenology and Craniology, is endeavouring to convince us that the main points of their new doctrines are correct and true. His zeal is laudable, and his exertions praise worthy; but I must confess that I consider it a very difficult undertaking to prove even those points of this new hypothesis, until the supposed organs or intellectual tools of the brain, which are said to aid the mind in performing its functions, are specifically traced as easily as the well known organs of the external senses, and until it is shown that every new intellectual capacity or study of the mind, meets with an adequate corresponding organ unfolded on purpose! It is more simple and rational to suppose that our mental or intellectual and moral faculties, powers or functions are operations evolved or suggested spontaneously by the active perceptions and mutations of our intellect.

CONSTANTINE.

The five Psyche essays were published in Lexington's *Western Monitor* newspaper on these dates: No. I, February 5, 1822; No. II, February 12, 1822; No. III, February 26, 1822; No. IV, March 5, 1822; No. V, April 2, 1822. William Gibbes Hunt was editor of the weekly paper at this time.

1. The controversy was fomented by James Fishback (1776–1845), formerly a trustee of Transylvania University, where he also had been a professor of the medical faculty for a short time. In 1813 he published in Lexington a book titled *The Philosophy of the Human Mind, in Respect to Religion; or, a Demonstration, from the Necessity of Things, that Religion Entered the World by Revelation.* Now, just before Rafinesque wrote, Fishback had tested the curriculum of the university with his revelation hypothesis and found it sorely wanting. After denouncing his former colleagues from the pulpit, he published his strictures in a pamphlet issued under the title *The Substance of a Discourse, in Two Parts; Delivered in the Meeting-House of the First Baptist Church in Lexington February 3, 1822; to the Class of the Medical School of Transylvania University* (Lexington, 1822).

He also responded immediately to Rafinesque's newspaper articles. Commenting on Psyche—No. I under the pseudonym Philo-Constantine and printed in columns parallel with those of Psyche—No. II, he wrote that

> The only question of controversy with which I am acquainted on these subjects in the western country, is whether the independent and separate existence of the soul from matter, and its immortality, can be known without revelation, that is, without supernatural instruction from God communicated in the first instance immediately, and perpetuated by oral tradition or by writing.

It hardly needs to be said that Fishback's answer to the question was a resounding No. He went on to declare "that the true knowledge of" what Rafinesque had called "'these fundamental doctrines of correct morality, real philosophy, and true religion' do not belong to the circle of natural science, but to revealed religion alone." And he continued the same argument in a second article printed parallel with the columns of Psyche—No. III.

2. Probably a misprint, for elsewhere Rafinesque used *psychology* in a similar context. However, he denominated "new" sciences with such abandon that one cannot be sure.

3. The Scottish philosopher/psychologist Thomas Brown (1778–1820). The four volumes of his posthumously published *Lectures on the Philosophy of the Human Mind* (Edinburgh, 1820) are still in the Transylvania library.

4. This neologism was coined by Rafinesque from the *saveur* of his mother tongue.

5. Rafinesque probably meant it ironically, but whether or not a genius James Fishback was indeed "self-taught" in many respects. First a physician, he shifted to the practice of law in 1803, and in 1819 he was ordained a Baptist minister. Having found *true* religion, he was unrelenting in his attacks on the "Godless" professors of the university. Dean of the medical faculty, Charles Caldwell, answered him and other such critics in a pamphlet titled *Defense of the Medical Profession against the Charge of Irreligion and Infidelity* (Lexington, 1824) and challenged Fishback to a public debate, which the latter declined. Like other sectarians, when he discovered himself at odds even with his co-religionists he caused a split among the Lexington Baptists, and finally abandoned even that denomination in favor of the Campbellite Disciples of Christ, where he continued to preach until his death. With other fundamentalists, he succeeded in hounding Horace Holley, Transylvania's president, out of Lexington.

6. Horace Holley (1781–1827), a Unitarian minister from Boston (where he had been on the Board of Overseers of Harvard), made Transylvania into a real university. He was Professor of the Philosophy of Mind as well as the university's president, and he also filled in from time to time both in the medical and law faculties when there were vacancies. Never an enthusiast for Rafinesque or for natural history, Holley is unlikely to have been pleased by the condescending dismissal of his course here.

7. Charles Caldwell (1772–1853), Dean of the medical faculty who had been brought from Philadelphia by Holley to reform and energize that faculty, which he ably did. The previous year in Europe, where he had gone to buy books for the university, Caldwell learned about phrenology and now was lecturing on the subject—which he claimed to have introduced in America.

During the same month that Psyche—No. V was published, April 1822, Rafinesque addressed the public "On the Human Mind" as the inaugural lecture for a series on "The Natural History of Mankind." It appears that not enough auditors were willing to pay his $10 fee to make the series practical, but later that year, he added four pages to his "Human Mind" lecture and made it into a critical "Lecture on Phrenology" that was announced in the December 12, 1822, *Kentucky Gazette*. Both manuscripts are extant at the American Philosophical Society and have not been published.

Index of Genera

Animals

Acipenser 173
Agama 152
Anisoctus 69
Aphis 149
Ascalabetes 152

Catostomus 172
Chlidonias 141
Clupea 172, 173
Coluber 174, 233
Crocodilus 152

Didicla 151
Ditaxopus 138

Emyda 151
Endolobus 69
Enhydris 234, 235, 239
Eurycea 143

Formica 149–50

Hirudo 201
Hirundo 139
Hydrophis 234
Hydrus 234, 235

Isoctomesa 137
Isotelus 137

Larus 141
Lepidemy 151

Manis 69
Meanthes 142
Megalonyx 32
Megasaurus 152
Megophias 232, 237
Meloe 201
Monoclida 151
Motacilla 172

Natrix 235
Necturus 142

Octipos 238
Ophinectes 236

Pelamis 234, 235, 237, 238, 239
Perca 173
Platurus 234, 235
Podiceps 141
Polyprion 173

Rimamphus 139

Salamandra 142, 143
Salmo 172
Scincus 152
Scoliophis 233
Sirena 142
Sparus 173
Sphagebranchus 238
Sterna 141
Sutorites 69
Symbranchus 238

Talpa 69, 215, 217
Tessarops 69
Testudo 174
Todarus 69
Trionyx 151
Trituris 142
Tupinambis 151

Urotropis 142, 143

Plants

Acer 174
Acerates 213
Acerotis 213
Aconitum 183
Acorus 122, 214
Aegochloa 187
Æsculus 174
Aetyson 184
Agrostis 212
Aguilegia 174
Aira 212
Allionia 213
Allium 212, 247
Alnus 172, 173
Ammi 128, 212
Ammyrsine 215
Anapodophyllum 214
Andromeda 174, 212, 213
Andropogon 212, 213
Anemone 147, 174
Anthemis 145, 189
Antirrhinum 175
Arabis 173, 174
Arbutus 201
Aristida 212
Aronia 175
Arundinaria 215
Arundo 215
Asarum 181
Asclepias 123, 128, 201, 212, 213, 216
Asimina 188
Aspidium 173
Asplenium 173
Aster 247
Atropa 216
Aulaxanthus 213
Azalea 174, 186, 188

Baptisia 180, 212, 216
Bartonia 215
Bartsia 175
Bejaria 211
Bellis 187
Betula 173
Boerhavia 211
Brosimum 128
Brunnichia 211
Buchnera 182, 183

Cacalia 180
Cactus 187
Caenotis 189
Callicarpa 211

Calochortus 187
Caltha 147, 175, 186
Calycanthus 174
Camellia 174
Capraria 180
Capsella 174
Cardamine 174
Carex 174, 175, 247
Cassia 212
Catalpa 215
Catalpium 180, 215
Cauloma 183
Ceanothus 216
Centaurea 215
Centaurella 215
Centunculus 188
Cerastium 174, 175
Cercis 174
Ceropegia 213
Chaerophyllum 174
Chamaerops 211
Chenopodium 180, 214
Chloris 213
Chrysantheum 182
Cladorhiza 182
Cladrastis 156–57, 184, 187
Clarkia 187
Clethra 187, 216
Collinsia 187
Collinsonia 212, 216
Collomia 187
Comarum 186
Commelina 184
Comptonia 175, 179
Conostylis 213
Convallaria 172, 175
Convolvulus 214, 216
Coptis 186
Cornus 173, 201
Crocus 173
Cubelium 184
Cupressus 187
Cynanchum 213
Cyperus 212, 213

Daucus 189
Delphinium 183
Dentaria 174
Diapensia 216
Dichromena 212
Diervilla 186
Dilatris 213
Dionea 211
Dirca 184
Dodecatheon 183, 188
Draba 173, 175
Drosera 212
Dryas 148

Eclipta 183
Elephantopus 180, 183
Eleusine 212
Elliottia 216
Elytraria 211
Endopogon 216
Enemion 146–147
Epidendrum 211
Epigaea 174
Erianthus 212
Erica 247
Erigeron 174
Eriogonum 211
Eryngium 212
Eupatorium 145, 146, 180, 181, 184, 201
Euphorbia 127, 175, 247
Eustachya 69
Eutoca 187

Festuca 175
Floerkea 175
Fothergilla 187
Fragaria 175
Fucus 201
Fumaria 174

Galanthus 173
Garrya 187
Gaultheria 181
Gentiana 182, 183, 188
Geranium 247
Gerardia 181, 183
Geum 148
Gillenia 180
Glechoma 175
Gnaphalium 174
Gonolobus 214
Gordonia 211
Gratiola 212, 216

Halesia 187
Hamiltonia 188
Hedeoma 180
Helichroa 183, 188
Heliotropium 180, 184
Helonias 175
Hepatica 174
Heritiera 213
Herpestis 216
Hesperis 172, 180
Heuchera 186
Hicoria 186
Hottonia 212, 216
Houstonia 69, 174, 180, 183, 184, 212, 216
Hudsonia 180, 187
Hyacinthus 172, 174
Hydrangea 186
Hydrastis 174
Hydrocotyle 212, 216
Hydrolea 212
Hydrophyllum 175
Hypericum 183, 184, 189, 216
Hypogon 216
Hypopythis 214

Ilex 187
Ipomoea 216
Iris 173, 181, 214, 247
Isanthus 181
Jeffersonia 180
Juncus 212
Juniperus 173, 181

Kalmia 173, 181, 186, 188
Krigia 174
Kuhnia 182

Lachnanthes 213
Lactuca 186, 247
Lamium 173
Lantana 187
Larus 173, 174, 175
Lasthenia 187
Lechea 181
Leiophyllum 215
Lepachys 183
Leptocaulis 183
Lepuropetalon 213
Leucospora 187
Lewisia 187
Lichen 247
Ligusticum 122
Limnetis 215
Lindernia 212
Linnea 186
Liriodendron 174, 201
Lithospermum 175
Lobadium 184
Lobelia 127, 183, 187, 214
Lophiola 213
Ludwigia 187, 212, 213
Lupinus 187
Luzula 174
Lycopus 212
Lyonella 213
Lyonia 213, 217
Lysimachia 69, 212

Macbridea 213
Magnolia 174, 182, 187, 188, 201
Majanthus 215
Malva 247
Mariscus 212
Marshallia 187
Mayaca 215
Mayzea 190
Mespilus 174
Micranthemum 212
Miegia 123, 145, 180, 187, 215
Mimosa 247
Mimulus 187
Monocera 213
Monotropa 214
Monotropis 214, 215
Myosotis 175
Myrica 187

Narcissus 173, 174
Nelumbium 180, 190
Nemopanthes 186
Nepeta 122
Nicandra 216

Obolaria 174, 175
Onosma 215
Onosmodium 215
Ophiorhiza 212
Orchis 174, 181
Origanum 189
Orimaria 183
Osmodium 215
Oxalis 175, 183, 184, 247
Oxypolis 183

Pachysandra 183
Paeonia 147
Panax 174
Pancratium 180
Panicum 145, 212
Parnassia 186, 188
Paspalum 212, 213
Passiflora 187
Pavia 181, 187, 188
Pentostemon 187
Petalepis 214
Petalostemon 182
Phalaris 213
Phlox 174, 181, 212, 213
Phryma 246
Phytolacca 189, 201
Pinckneya 187
Pinus 173, 181
Pistia 211
Planera 187
Plantago 173, 175
Platanus 180
Poa 175, 212, 213
Podophyllum 174
Podostigma 213
Polanisia 180
Polemonium 175
Polygala 181
Polygonatum 215
Polygonum 212, 215
Polymnia 181
Polypodium 173
Polytenia 183
Pontederia 212
Populus 173, 174, 180
Porcelia 180
Potentilla 174
Pothos 173
Prenanthes 182
Primula 173
Prunus 180, 190
Pyrola 201
Pyxidanthera 213, 216

Quercus 175, 247

Ranunculus 174, 175
Rhamnus 216
Rhexia 187, 212
Rhododendron 173, 186, 188
Rhus 183, 201
Rhynchospora 212
Ribes 174, 186, 187
Rosa 153, 186, 247
Rubus 175, 186
Rudbeckia 181
Ruellia 181
Rumex 212

Sabbattia 212
Salix 173, 174, 247
Salvia 183, 212, 214, 216
Samolus 69
Sanguinaria 174
Sarothra 216
Saxifraga 173, 174, 175
Schubertia 187
Schweinitzia 214, 215
Scirpus 175, 212, 213
Scleranthus 175
Scutellaria 73
Sedum 174, 184
Senecio 175
Sigillaria 215
Silene 175, 182
Silphium 182
Sisyrinchium 183
Sium 212
Solanum 180
Solidago 181, 184, 186
Sophora 155
Spartina 215
Spartium 215
Spathyema 173, 174
Spermacoce 183
Spigelia 214
Staphylea 175
Statice 216
Stellaria 174
Stillingia 211
Stylisma 216
Stylosanthes 181
Stylypus 146–48
Syena 215
Symphoricarpos 213
Synandra 180
Syringa 174

Taraxacum 174
Thalia 211
Thalictrum 147
Therolepta 183
Thlaspi 174
Tiardium 184
Tillandsia 211, 212
Tofielda 212
Trachynotia 215
Tradescantia 183, 184, 188
Triatherus 213
Trichostema 182
Trifolium 175
Trillium 175, 186, 212, 213
Tulipa 172

Uniola 212
Utricularia 212
Uvularia 174

Vaccinium 174, 175, 181, 188
Veratrum 122
Verbascum 189
Vernasolis 183
Vernonia 145, 180
Veronica 173, 174
Viburnum 69, 173, 175, 187
Villarsia 212
Viola 174, 175, 212, 216
Virgilia 154–55, 156, 184
Viscum 69
Vitis 186, 247

Xerophylum 187
Xyris 212

Yucca 187

Zamia 211
Zizania 123

General Index

Abeel, David 42, 51
academic organization 104
Academy of Natural Sciences 6, 33, 67, 196, 197, 207, 218
Adanson, Michel 245
Adelung, Johann Christoph 57, 58
Adena site 17, 20
aerolites 160
aerosols 225–31
Afong Moy 40, 41, 51
Agathopolis, city of 87
agricultural schools 105, 106, 113
Aikin, Arthur 200, 208
Aitken, John 225
Akerly, Samuel 113, 201, 204
Albany, N.Y. 24
Alembert, Jean le Rond d' 166, 170
Alexander, Charles 9
Allegheny College 54
American Academy of Arts and Sciences 175, 195
American Antiquarian Society 23, 71
American Antiquities and Discoveries in the West 24, 25
American Florist 9
American Journal of Science 6, 144, 151, 171, 232
American Manual of the Grape Vines 9, 114, 127
American Manual of the Mulberry Trees 10, 114, 127
American Medical and Philosophical Register 199, 208
American Mineralogical Journal 207, 208
American Monthly Magazine and Critical Review 4–5, 6, 7, 8
American Nations 10
American Philosophical Society 39, 117, 195, 206, 219, 224, 262
Amiot, Joseph Marie 13
Analectic Magazine 199
Analyse de la Nature 13, 154, 224
Ancient Annals of Kentucky 23, 59
Andover Theological Seminary 131
Annales Générales des Sciences Physiques 7, 72, 154
Annals of Nature 140, 142, 199, 208
Annals of Philosophy 5
Annals of the Lyceum of Natural History 199
anthropology 201, 205
ants 148–50
Aratus 166, 169
archaeological sites, Kentucky 17–23, 33, 35
Archimedes 165, 168
Archives of Useful Knowledge 199, 208
Arcturus (star) 167
Aristotle 215, 241
Ashe, Thomas 32
astronomers 203, 204
Athenaeum of Boston 198
Athens of the West 73, 129
Atkinson, Samuel C. 9
Atlantic Journal and Friend of Knowledge 9, 11, 13, 24, 29, 97, 114, 138
atmospheric dust 143–44, 225–31
atoms 251
Atwater, Caleb 71, 73
Audubon, John James 11, 139, 140
Augusta College 54
Aurora (land company) 86–93
Australia 234, 236
Autikon Botanikon 1
Ayres, Samuel 79, 80
Azara, Félix de 238, 240

Bacon, Francis 70
Bailly, Jean Sylvain 166, 169
Baldwin, Loammi 203
Baldwin, William 5, 212
Bancroft, Edward 202
Banks, Sir Joseph 194, 206, 240
barrens (defined) 181
Barrow, Sir John 44, 47, 52, 200, 208
Barton, Benjamin Smith 172, 175, 193, 194, 195, 198, 199, 200, 201, 204, 206, 208, 239
Barton, William P.C. 5, 172, 195, 197, 204
Bartram, John 194, 197, 206
Bartram, William 150, 204, 211
Bayer, Johann 166, 169
Beck, Lewis Caleb 193
Belknap, Jeremy 194, 203, 206
Bentley, William 203
Bernardin de Saint-Pierre, Jacques Henri 200, 208
Big Bone Lick 32–36
Bigelow, Jacob 171, 172, 175, 197, 201, 204, 210, 217, 232
Binney, Amos 193
Binney, William G. 1
Biot, Jean Baptiste 166, 169
Birge, Benjamin 70, 72
Blainville, Henri de 233
Bledsoe, Jesse 136
Bloomington, Ill. 10, 90
Bogert, John G. 197, 204
Bonaparte, Charles Lucian 140
Bonpland, Aimé J.A. 190, 208
Boot, Francis 204
Bory de Saint-Vincent, Baron Jean B.G.M. 7, 67, 72, 245
Bosc, Louis Augustin Guillaume 150, 194, 204, 206
botanical gardens 113, 197
Botanical Nomenclature, International Code of 1
botanical regions: of Kentucky 145–46, 179–84; of North America 186–87
botanists 204, 205
botany, branches of 125–27
botany, utility of 121–24
bound feet, Chinese women's 40
Bowditch, Nathaniel 203
Brackenridge, Henry M. 203
Bradbury, John 69, 204
Bradley, Stephen Row 203
Brahe, Tyco 166, 169
Brendel, Frederick 158

Brickell, John 201, 204, 212, 217
Briggs, Edmund Lloyd 77, 79
Brongniart, Alexandre 138
Brown, Chandos Michael 240
Brown, Charles Brockden 224
Brown, Thomas 259, 261
Browne, Peter A. 113
Bruce, Archibald 197, 199, 203, 207, 208
Buckingham, James Silk 63, 64
Buffon, Georges L.L., Comte de 138, 140
Bulletin Nr. 3 of the Historical and Natural Sciences 114
Bulletin Nr. 7 of the Historical and Natural Sciences 108
Bulletin Universal des Sciences 7
Burckhardt, John Lewis 63, 64

Cabinet of Sciences of Philadelphia 196
Cain, Arthur 12, 154, 242
Caldwell, Charles 13, 260, 261, 262
Call, Richard Ellsworth 1, 11, 39, 59, 135
Candolle, Augustin-Pyramus de 90, 146, 185, 186, 189, 190, 210, 217
Canton, Ky. 18, 23
Caratteri di ... Nuove Specie di Animali e Piante 2, 13
Carver, Jonathan 194, 206
The Casket 9
Caspian Sea 26, 235
Cassini, Gian Domenico 166, 169
Castiglione, Luigi 194, 206
Catesby, Mark 138, 140, 194, 206, 211, 217
Catwabas 23
Celestial Wonders and Philosophy 10, 163
Central University of Illinois 10, 163
Cepheus (constellation, person, star) 165; (person) 168
Chambers, Ephraim 208
Chapman, Charlotte 13
Chapman, Isaac 201
Chapman, Nathaniel 201
Chaptal, Jean Antoine Claude 200, 208
Charleston, S.C. 196, 209
Charlevoix, Pierre François Xavier de 194, 206
chemistry defined 194
chemists 203, 204
Cherokees 23
Chinese nations 40–41
Chinese religions 41–43
Chloris Aetnensis 185
Chrétien, Gilles-Louis 14
Cincinnati 33, 54, 70, 73, 196
Cincinnati Almanac 158
Cincinnati Literary Gazette 9, 73, 157
Clairault, Alexis Claude 166, 169
Clark, William 202
classes, definition of 248
classification (plant), Linnaean 214–15
Clay, Henry 113
Clayton, John 194, 206
Cleaveland, Parker 200, 203, 208
Clements, James 204, 207
clergymen: contributions of to natural history 202
Clifford, John D. 3, 6, 8, 17, 18, 23, 33, 69, 70, 72
Clinton, DeWitt 70, 109, 113, 196, 200, 203, 208, 229
Clymer, George 207
coal, formation of 218
Colden, Cadwallader 194, 206
College of Physicians (Philadelphia) 77
college tuition 100–101
Collin, Nicholas 202
Collins, Zaccheus 6, 197, 203, 204
Columbian Chemical Society 195
Columbian Institute 196
concours 101
Confucius 27, 42, 166, 170
Connecticut Academy of Arts and Sciences 195
Conrad, Solomon W. 204
consciousness 253–54
Considerant, Victor Prosper 92, 93
Constantinople 129
Cooper, Thomas 203, 208
Cooper, William 33, 36, 202
Copernicus, Nicolas 166, 169
Coromandel Coast 234
Correa da Serra, José Francisco 198, 204, 207
cosmony 135, 136, 138, 194, 219, 224
Court de Gebelin, Antoine 73
Covington, Ky. 33
Coxe, John Redman 208
Coxe, William 197, 207
Crafts, William 233
Crockett, G.F.H. 143
Cumberland River 18, 23
Cutbush, Edward 196, 207
Cutbush, James 195, 203
Cutler, Manasseh 194, 202, 204, 206
Cuvier, Baron Georges 152, 220, 224, 233, 240

Darby, William 203
Darlington, William 5, 193
Darwin, Charles 241–42
Darwin, Erasmus 4, 241
Daudin, François Marie 234
Davidge, John 201
Davis, John 202
Davy, Sir Humphry 168, 200, 208
Day, Jeremiah 203
Decatur, Stephen 202
DeKay, James E. 138, 193
Delafield, Joseph 142
Delille, Jacques 166, 169
Deluc, Jean André 229, 231
Dencke, Christian F.H. 202
Derham, William 166, 169
Descartes, René 165, 168
Dewitt, Simeon 203
Discourse on ... the Rev. Horace Holley 13
Divitial Invention (and Institution) 83, 84, 86, 90, 114
Dobson, Judah 11
d'Olivet, Antoine Fabre 55, 57, 59, 64
Drake, Daniel 71, 72, 200, 203, 207, 208
Drayton, John 200, 208
Du Halde, Jean-Baptiste 47, 48, 52
Dumont de Courset, Georges L.M. 155
Dunbar, William 203
Du Ponceau, Pierre Étienne 39, 46, 47, 50
Du Pont, Eleuthère Irénée 203
Dupont de Nemours, Pierre Samuel 148
Du Pratz, Antoine Simon Le Page 194, 206
Dwight, Sereno E. 193
Dwight, Timothy 202
dyes 202

Eastburn, John & Co. 198, 207
Eaton, Amos 28, 186, 190, 208, 217, 218
Eclectic Repertory 199, 208
ecological succession 135, 145
Eddy, Caspar Wistar 204, 207
Eddy, John 197
Edinburgh castle 229
education systems compared 98–99
Edwards, George 138, 140
Eleutherium of Knowledge 10, 82, 89, 97, 106, 110
Ellicott, Andrew 200, 203, 208
Elliott, Stephen 5, 14, 185, 196, 197, 200, 204, 208
Ellis, Sir Henry 47, 52
Emporium of Arts & Sciences 199, 208
Enslen, Aloysius 211, 217
Enumeration and Account of

Some Remarkable Natural Objects 138
Esquimaux 27
Etna, Mt. 225
Euclid 166, 168
Euler, Leonhard 165, 168
European contributions to American natural history 200
Everett, Edward 232
Ewan, Joseph 2

Fairholme, George 31
Falopi (Sicilian artist) 14
families, definition of 243, 248
Farrar, John 203
Featherstonhaugh, George William 193
Férussac, Baron André 7
Ficklin, Joseph 77, 79
Filson Historical Society 179, 251
First Scientific Circular 106, 108
Fishback, James 251, 260, 261
Fitzpatrick, T.J. 1, 11, 39, 159, 218, 241
Flamsteed, John 166, 169
Flora Philadelphica Prodromus 172
Flora Telluriana 1, 13, 248
Florula Ludoviciana 2, 5, 210, 216, 217
Fontenelle, Bernard le Bouyer de 166, 170
Forrest, Thomas 113
Forster, Johann Georg Adam 235, 240
Forster, Johann Reinhold 190
Fothergill, John 202
Fourier, Charles Marie 92, 93
Foxe, John 54
Francis, John W. 201, 208
Franklin, Benjamin 32, 70, 168, 194, 206
Fraser, John 194, 206, 211, 217
Fraser, Simon 194, 206
French and Italian lexical influence 12
French language 7, 210
Fulton, Robert 203

Galileo Galilei 70, 166, 169
Gambold, Anna Rosina 203
Gannet, Caleb 203
Garden, Alexander 194, 206, 211
Gardner's Magazine 7
Gassendi, Pierre 168
Genius and Spirit of the Hebrew Bible 10, 59, 64
genus, definition of 214, 242, 244, 246, 248
geological surveys, practical use of 222–23
geologists 203, 204
geology, benefits of 222–23
Gibbs, George 197, 203, 204
Gilliams, Jacob 142
Glen, James 201
Goforth, William 32
Golovnin, Vasilii Mikhailovich 47, 52
Good Book, and Amenities of Nature 11, 13, 127, 138, 224
Gorham, John 203
government officials, contributions of to natural history 202
Graham, Christopher Columbus 141, 142
grammar, Chinese 50
Gray, Asa 158, 159, 193
Gray, John 83, 86
Greek temple, at Segesta 227
Greek War for Independence 129
Green, Ashel 201
Green, Jacob 113, 142
Griffen, Augustus Rupert 201
Griscom, John 203
Guignes, Joseph de 42, 47, 51, 52
Gützlaff, Karl Friedrich August 41, 46, 47, 51

Habersham, Joseph 212, 217
Haines, Reuben 203
Haldeman, S.S. 232, 241
Hall, Basil 47, 50, 52
Halley, Edmond 165, 168
Hamilton, William 175, 197, 203
Hare, Robert 199, 203, 207
Harris, Thaddeus Mason 203
Harrison, William Henry 113
Harvard, natural science at 198
Haworth, Adrian 146
Hayden, Horace 203
Hebrew alphabet 60–63
Hebrew language, antiquity of 56–57
Hebrew pronunciation 55–56, 58, 63
Heckewelder, John 27, 28
Herbarium Rafinesquianum 244
herbariums 197
Herbelot de Molainville, Barthélemy d' 73
Herbemont, Nicholas 203, 212, 217
Hermes 166, 170
Herschel, Caroline 169
Herschel, Sir Frederick William 164, 169, 233, 240
Herschel, Sir John Frederick William 169
Heuvelmans, Bernard 237
Hevelius, Johannes 168
Hipparchus 166, 168
Holley, Horace 3, 7, 8, 129, 260, 261
Holley, Orville L. 8
Holthuis, L.B. 2
Hooker, William Jackson 146, 193
Horsfield, Thomas 201
Hosack, David 193, 197, 198, 201, 208
Hourcastremé, Pierre 166, 169
Huber, Pierre 148
Humboldt, Friedrich Heinrich Alexander, Baron von 166, 170, 185, 190, 200, 208
Humphreys, David 203, 240
Hunt, William G. 8, 261
Hurons 23
Husar, Rudolf B. 225
Hutchins, Thomas 203
Hutton, James 220, 224
Huygens, Christiaan 168
hybridity 241, 246

Ichthyologia Ohiensis 1, 2, 8, 68
ideograms, Chinese 48–50
Improvements of Universities 10
Incombustible Architecture 113, 114
index fossils, concept of 218, 222
India 234, 235, 236
instinct 255–57
inventions, Chinese 43–44

Jackson, James 212, 217
Jaffe, Bernard 241
James, Edwin 27
Jefferson, Thomas 32, 34, 35, 67, 70, 113, 171, 194, 203, 206
Jillson, Willard Rouse 34, 36
Jones, Sir William 42
Jordan, David Starr 1
Jouett, Matthew 13
Journal de Physique, de Chemie et d'Histoire Naturelle 7, 140
Journal of the Academy of Natural Sciences 199
Journal of the Royal Institution 7

Kabala 59
Kaempfer, Engelbert 47, 52
Kalm, Pehr 194, 206
Kentucky Gazette 8, 135, 262
Kentucky Reporter 8, 73, 131
Kentucky, state of 70
Kepler, Johann 166, 169
Ker, Henry Bellenden 203
Kin, Matthias 211, 217
Kircher, Athanasius 50, 53
Kissam, B.P. 207
Knevels, D'Jurco 204, 207
Knickerbocker magazine 11, 39
knowledge, branches of 117–19

Lacaille, Nicolas Louis de 166, 169

Lacépède, Comte de Bernard G. E. de 194, 206
La Condamine, Charles Marie de 166, 170
Lake Erie 27
Lalande, Joseph Jérôme Lefrançais de 166, 167, 168
Lamarck, J.B.A.P.M. de 211, 217, 241
Lambert, Johann Heinrich 166, 169
Lambert, William 203
La Métherie, Jean Claude de 220, 224
languages of China 46–51
Lao-Tsze 43
Lapérouse, Jean François Galaup, Comte de 230, 231
Laplace, Pierre Simon, Marquis de 164, 168, 233, 240
Latin language 210
Latreille, Pierre André 148, 194, 206, 234
Latrobe, Benjamin 203
lawyers, contributions of to natural history 202
Leavy, William A. 80
LeConte, John Eatton 203, 204, 207, 212, 217
Lee, William 239
Leguat, François 237, 240
Leibniz, Gottfried Wilhelm 165, 168, 251
Lemonnier, Pierre Charles 166, 169
Lenni Lenape (Delawares) 23, 36
Lesueur, Charles Alexandre 68, 150, 204
Lewis, Meriwether 202, 203
Lewis, R. Barry 17
Lewis and Clark 152, 200, 208
Lexington Female Academy 129
Lexington, Ky. 67, 73–74, 129
libraries 198
Library of Congress 114
Liechtenstein, Prince of 124
life, definition of 120–21
Life of Travels 2, 11, 81, 128, 175
lightning, varieties of 159–61
Lining, John 201
Linnaeus, Carl 72, 155, 194, 206, 244, 245
Linnean Society of Bordeaux 7
Linnean Society of New England 195, 207, 232
Linnean Society of Philadelphia 195, 207
Literary and Philosophical Society of Charleston 196, 209
Literary and Philosophical Society of New-York 195–96
Livingston, Robert R. 197, 202
Lizard (constellation) 165
Longueuil, Charles LeMoyne de 36
Lowell, John 198, 207
Lyceum of Natural History 4, 5, 196, 207, 217, 239
Lyon, John 211, 213, 217

Macbride, James 201, 212, 214, 217
Mackenzie, Alexander 194, 206
Maclure, William 5, 14, 81, 82, 203, 207
Mairan, Jean-Jacques de 165, 168
Mantell, Gideon 233
Mantissa (Fitzpatrick supplement) 251
Maraldi, Giacomo Filippo 166, 169
Marshall, Humphry 194, 206
Maskelyne, Nevil 168
Massachusetts Society for ... Agriculture 195
Masson, Francis 194, 206
mastodons, Great Spirit destroys 36
materia medica illustrated 201
Maupertuis, Pierre Louis Moreau de 166, 169
McCulloh, James Haines 27, 28, 113
McMahon, Bernard 197, 207
Meade, William 201
Mease, James 113, 198, 200, 201, 203, 208
Medical and Physical Journal 199, 239
Medical Flora 9, 127, 128, 146
Medical Repository 4, 190, 199, 202, 213, 217
Melish, John 203
Melsheimer, Frederick Valentine 202, 204
merchants, contributions of to natural history 202
Merrill, Elmer D. 1, 39
Methodist Episcopal Church 54
Meton of Athens 168
Miamis 23
Michaux, André 194, 200, 204, 206, 208, 211, 217
Michaux, François André 154, 200, 204, 208
Milky Way described 163–67
Miller, Edward 199
Milton, John 4
mineral springs analyzed 201
mineralogists 203–204
Mississippian site 17, 21
Mitchell, John 194, 206
Mitchell, John S. 201
Mitchill, Samuel Latham 4, 12, 68, 70, 113, 152, 193, 194, 196, 197, 199, 200, 201, 203, 204, 206, 207, 208, 239
Molina, Ignazio 200, 208
monads 251
Monboddo, James Burnett, Lord 27
Mongitore, Antonino 237, 240
Monthly American Journal of Geology and the Natural Sciences 36
Morehead, Charles Slaughter 77, 79
Morrison, Robert 47, 50, 52
Morse, Jedidiah 200, 203, 208
Moses 26
Mott, Samuel G. 204
Muhlenberg, Frederick Augustus 197
Muhlenberg, Gotthilf Henry 175, 179, 194, 197, 200, 203, 204, 206, 208, 212, 217
Munsell, Joel 24
museums 196–97
mutability 242
mutation 242

Natchez, Miss. 77
nations, Chinese 41
natural history defined 194
natural history surveys, types of 110–12
natural selection 241
Necker, Noël Joseph de 245
New Flora of North America 1, 13, 127, 246
New Harmony Gazette 81, 86
New Harmony, Ind. 218
New-York Farmer 107
New-York Historical Society 196, 197, 207
New York natural history survey 108, 109
newspapers in Lexington 77
Newton, Sir Isaac 165, 168
Noah 25, 26, 29
North American Review 5
Notes on the State of Virginia 36
Nuttall, Thomas 153, 204, 211, 217

Oemler, Augustus Gottlieb 212, 217
Ohio, state of 70
Olbers, Heinrich Wilhelm Matthäus 166, 169
opium trade 131
Ord, George 139, 204
orders, definition of 243, 248
Origin of Species 241
Original Theory or New Hypothesis of the Universe 10, 162
Ortolani, Giuseppe Emmanuele 4
Osages 27
Ottoman Empire 129, 130

Owen, Robert 81, 82, 83, 86, 91, 218

Palisot de Beauvois, Ambroise M.F.J. 194, 206
Parkinson, James 137
Partridge, Alden 203
Pascalis, Felix 199, 201
Patrin, Eugène Louis Melchior 220, 224, 229, 231
Patterson, Robert 203
Peale, Charles Willson 196, 204
Peale's Museum 4, 196
Peck, William Dandridge 203, 204, 232
Peleg 26
Penn, William 27, 86
Pennell, Francis W. 9
perceptions 252
periodicals, general 199
periodicals, scientific 199
Péron, François 234, 236, 240
Peters, Richard 196, 198, 202
Philadelphia Society for ... Agriculture 196, 207
philosophers 203, 204
Philosophical Magazine 7
phonology, Chinese 51
phrenology 105, 260–61, 262
physicians, contributions of to natural history 201–202
Physick, Philip Syng 201
physics defined 194
physiognotrace 14, 220
phytogeography 135, 179–90
phytology 121
Piazzi, Giuseppe 164, 166, 167, 169
Pickering, Charles 185, 186, 190
Pike, Zebulon 202
Pinckney, Charles Cotesworth 212, 217
Pinkerton, John 233, 240
Pittsburgh 71, 73
Plan of the Philadelphia Land Company 10, 81, 97, 106
Plato 91, 166, 170
Pleasures and Duties of Wealth 10, 82, 93
Pontoppidan, Erik 237, 240
Popular Science Monthly 159
Port Folio 199
Porter, David 202
Portico 199
Potter's American Monthly 14
Priest, Josiah 11, 24, 28
Priestley, Joseph 194, 203, 206
Prince, William 197, 207
Principes Fondamentaux de Somiologie 242
Pringle, James S. 182
professors, appointment of 101–102
professors, contributions of to natural history 203
Profiles of Rafinesque 79
proles (or breeds) 243
Proposals to Publish by Subscription 3, 193
Ptolemy 166, 170
Pursh, Frederick 5, 153, 200, 208, 210, 211, 217
Pythagoras 166, 169, 251
Pytheas 165, 168

Rafinesque, C.S.: publications of reprinted 1–14; quality of the writing of 12–13; scientific specialties of 203, 204, 207
Rafinesque's Kentucky Friends 79
Ramsay, David 203
Rappites 81
Rea, Jeremiah 14
Rees, Abraham 200, 208
religions, Chinese 42–43
Rémusat, Abel 42, 50, 51, 52
Rensselaer School 104
Rich, Obadiah 202, 204
Richard, Louis Claude 200, 208
Richmond, Charles W. 1
Rittenhouse, David 166, 169, 194, 206
Robin, C.C. 200, 204, 208
rocks, classification of 221
roses 153–54
Rouelle, John 201
Rush, Benjamin 70, 113, 124, 201, 203
Russell, Michael 27, 28
Russell, Patrick 234, 235, 236, 240
Ruter, Martin 54, 59

Sabbath 30
Safe Banking 10
Saint-Hilaire, Étienne Geoffroy 151, 241
St. James Way (Milky Way) 164
Saint-Simon, Claude-Henri, Comte de 81, 92, 93
salamanders 142–43
Salt, Henry 200, 208
Saturday Evening Post 9, 24
Say, Thomas 159, 197, 204
Schaeffer, Frederick G. 202, 207
Schneider, Johann Gottlob 234, 240
School of Arts and Literature 196
Schöpf, Johann David 150, 194, 206
Schroeter, Johann Hieronymus 166, 169
Schultes, Joseph August 146
Schweinitz, Lewis D., von 190, 212, 217
science education 223
sciences, classification of 224
sciences, origin of 219
Scudder, John 196
Scudder's Museum 196
sea birds 140–42
Seaman, Valentine 201
senses 259–60
Seybert, Adam 202, 204
Shaw, George 234
Shawnees 23
Short, Charles Wilkins 33, 113, 171, 179, 182, 193
Shulz, Benjamin 201
Silliman, Benjamin 6, 171, 229, 232
Sims, John 211, 217
Smith, Rev. Ethan 24
Smith, Samuel Stanhope 201, 202
Smith, William 203
snakes distinguished from eels 236–37
social life, Chinese and Japanese 44–45
Société de Géographie 7, 17
South America 235, 238
South Carolina Academy of Fine Arts 209
South Carolina Society for ... Agriculture 195
Southern Review 209
Southern states, plant distribution in 210–11
Spafford, Horatio Gates 203
Speakman, John 207
Specchio delle Scienze 3, 4
species, age of 247
species, definition of 242, 244, 246, 248
species, incipient ones 246
Stastica Generali di Sicilia 4
state libraries 113
state museums 112
Staunton, Sir George 47, 52
Steel, John Honeywood 201
Steinhauer, Daniel 202, 204
Stout, Charles 17
Stuckey, Ronald L. 182
student discipline 103–104
Sullivan, James 203
Sutton, David 141, 142
Swainson, William 146
swallows 138–40
Sylva Telluriana 1, 13, 248

Tarascon, Lewis 113
Texas, Republic of 54
Theodosius of Bithynia 168
Thunberg, Carl Peter 45, 47, 52
Tilloch, Alexander 7, 232
Torrey, John 6, 7, 90, 113, 190, 193, 197, 204, 207, 242
Townsend, P.S. 207

Transylvania University 6, 7, 13, 54, 59, 67, 73, 76, 79, 102, 117, 129, 136, 218, 261
Trigg County, Ky. 17, 18
trilobites 136, 138
Troost, Gerard 207
Tryon, George W., Jr. 1
turtles 150–51

Ulloa, Antonio de 166, 170
Underwood, Lucien 159
University of Wisconsin 108
Ussher, Archbishop James 26

Vaincher, P. (Sicilian engraver) 14
Van der Schott, Joseph 124
Van Vleck, Jacob 202
variation in plants 245, 247, 248
varieties, definition of 243, 248
Vaughan, Benjamin 203
Vieillot, Louis Jean Pierre 138, 139, 140, 194, 200, 206, 208
Virey, Julien Joseph 225, 229, 231
Volney, Constantin-François Chassebœuf, Comte de 159, 194, 200, 203, 206, 208, 221, 223, 224
Vosmaer, Arnout 236, 240

Walch, Johann Ernst Immanuel 138
Waln, Robert 42, 52
Walter, Thomas 194, 206, 211, 217
Ward, John 72
Warden, David Bailie 202
Warren, John Collins 201
Waterhouse, Benjamin 204
Waterhouse, John Fothergill 201, 204
Waterhouse, T.W., Jr. 201
Webber, Samuel 203
Wells, W.C. 241
Welsh 27
Werner, Abraham Gottlob 220, 224
West Point (military academy) 104
Western Minerva 3, 8, 9, 11, 67, 77, 251, 162
Western Monitor 8, 131, 251
Western Museum Society (Cincinnati) 70, 72
Western Review and Miscellaneous Magazine 8, 68, 70, 127, 158, 171
Western Spy & Literary Cadet 73
Wetherill, Charles 10, 81, 90, 97, 110
Whitlaw, Charles 199, 207
Willdenow, Carl Ludwig 185, 190, 194, 206
Williams, Roger 27, 86
Williams, Samuel 203
Williamson, Hugh 200, 203, 208
Wilson, Alexander 138, 139, 140, 200, 204, 208
Winthrop, John 194, 203, 206
Wistar, Caspar 201, 204
women professors 105
women's education 105
Wood, George 203
Wood, John 203
Woodlands estate 174, 197
Woodward, Augustus Brevoort 202
Woodward, William W. 203
World or Instability 4, 11
Wright, Thomas, of Durham 10, 162
Wyandots 27

Yale University 129
year without a summer 171
Yellowwood 136, 154–57
Ypsilantes, Alexander 129

Zeno 215
Zoilus & Co. 73
zoologists 204